PENGUIN C

PRINCIPLES OF GEOLOGY

CHARLES LYELL was born in November 1797 at Kinnordy House near Kirriemuir, Forfarshire, Scotland, and was the son of a botanist. His family moved to Hampshire, on the border of the New Forest, before he was one year old. Lyell was educated at Ringwood, Salisbury, Midhurst and Exeter College, Oxford, from which he graduated in classics in 1819. He entered Lincoln's Inn to study law, but a weakness in his eyes temporarily prevented him from continuing, a difficulty which increased in later years. He published his first scientific papers in 1825 and in the following year became a Fellow of the Royal Society. Lyell also began contributing to the *Quarterly Review* and gradually abandoned legal practice, having found his vocation in authorship. The first volume of his masterpiece *Principles of Geology* was published in 1830, and the second and third volumes appeared in 1832 and 1833 respectively. Lyell's vision of a world shaped by uniformly acting natural processes became central to public controversy about the place of human beings in nature. Its impact is evident in a wide range of Victorian literary, philosophical and scientific works, notably those of Charles Darwin, who became a close friend. In 1831 Lyell was appointed a professor of geology at King's College, London (although he soon resigned), and a year later he married Mary Horner, who assisted in his work and became a leading hostess in metropolitan intellectual society. He was elected President of the Geological Society of London in 1835–7 and 1849–51. A great admirer of the United States, Lyell lectured extensively there, and in 1845 and 1849 he published lively descriptions of his travels in North America. He continued to revise *Principles of Geology*, which appeared in its final two-volume form in 1867–8. Among his other books, all of which appeared in many editions, are *Elements of Geology* (1838) and *The Antiquity of Man* (1863). He was knighted in 1848 and became a baronet in 1864. His health deteriorated rapidly after the death of Lady Lyell in 1873, but he continued with research until his death in 1875. He is buried in the nave of Westminster Abbey.

JIM SECORD is a lecturer in the Department of History and Philosophy of Science at the University of Cambridge. Born in Madison, Wisconsin, he studied geology, literature and history at Pomona College and Princeton University, and has lived in England since 1980. He is the author of *Controversy in Victorian Geology* (1986), editor of Robert Chambers's *Vestiges of the Natural History of Creation* (1994) and co-editor (with N. Jardine and E. Spary) of *Cultures of Natural History* (1996).

Charles Lyell
PRINCIPLES OF GEOLOGY

Edited with an introduction by
JAMES A. SECORD

PENGUIN BOOKS

PENGUIN BOOKS

Published by the Penguin Group
Penguin Books Ltd, 80 Strand, London WC2R 0RL, England
Penguin Putnam Inc., 375 Hudson Street, New York, New York 10014, USA
Penguin Books Australia Ltd, 250 Camberwell Road, Camberwell, Victoria 3124, Australia
Penguin Books Canada Ltd, 10 Alcorn Avenue, Toronto, Ontario, Canada M4V 3B2
Penguin Books India (P) Ltd, 11 Community Centre, Panchsheel Park, New Delhi – 110 017, India
Penguin Books (NZ) Ltd, Cnr Rosedale and Airborne Roads, Albany, Auckland, New Zealand
Penguin Books (South Africa) (Pty) Ltd, 24 Sturdee Avenue, Rosebank 2196, South Africa

Penguin Books Ltd, Registered Offices: 80 Strand, London WC2R 0RL, England

www.penguin.com

First published 1830–33
This selected text published in Penguin Classics 1997

036

This edition copyright © James A. Secord, 1997
All rights reserved

The moral right of the editor has been asserted

Set in 9.5/12 pt PostScript Monotype Baskerville
Typeset by Rowland Phototypesetting Ltd, Bury St Edmunds, Suffolk
Printed and bound in Great Britain by Clays Ltd, Elcograf S.p.A.

ISBN 978-0-14-043528-3

www.greenpenguin.co.uk

Contents

Charles Lyell's *PRINCIPLES OF GEOLOGY*

VOLUME I (1830)

VOLUME II (1832)

* Numbers in italics refer to chapters included only as summaries or
in brief extracts.

VOLUME III (1833)

List of Illustrations*

---

* Lyell's numbering and descriptions of his illustrations are retained both in this list (which
derives from those provided by Lyell in each volume) and in captions. In the few instances
where Lyell does not provide captions in the text, they are added from the wording of his
list.

VOLUME II

VOLUME III

Introduction

'The great merit of the Principles,' Charles Darwin once said, 'was that it altered the whole tone of one's mind, & therefore that, when seeing a thing never seen by Lyell, one yet saw it partially through his eyes.'[1] Charles Lyell's *Principles of Geology* shaped Darwin's vision of nature as he circumnavigated the globe on the *Beagle* and as he later created his theory of evolution. Modern ecologists have looked to Lyell as an important interpreter of the economy of nature. He provided the starting point for the American conservationist George Perkins Marsh, whose *Nature and Man* of 1864 contested Lyell's claim that human impact on the landscape was minimal. The *Principles* continues to inform debates ranging from the death of the dinosaurs to the possibility of environmental catastrophe. As evolutionary biologist, geologist and historian Stephen Jay Gould has said, Lyell 'doth bestride my world of work like a colossus.'[2]

The *Principles*, as Darwin recognized, is about seeing. In it, theology, political economy and the philosophy of perception are united with natural history, anthropology, geography and travel. The title proclaimed heroic ambitions – no less than doing for the study of the earth what Isaac Newton had done for astronomy and natural philosophy in the *Principia Mathematica* (1687). Practitioners, Lyell argued, should carry out their investigations under the assumption that causes now visible around us (volcanoes, rivers, tidal currents, earthquakes, storms) are of the same kind that have acted in the past, and have done so with the same degree of intensity as in the present. In the *Principles*, uniformity was not a theory about the actual history of nature, but a policy for securing the philosophical foundations of geology: Lyell aimed to define, as he said in a letter, the 'principles of reasoning in the science'.[3]

When first published in three volumes from 1830 to 1833, the *Principles* attracted a wide range of readers in Europe, America and Australia, from explorers and engineers to poets and artists. Celebrated for its startling analogies and the elegance of its prose, it compares in significance to writings by Thomas Carlyle, Thomas De Quincey and Thomas Babington Macaulay. Alfred Lord Tennyson brought its themes of universal decay and the vastness of time into *The Princess* (1847) and *In Memoriam* (1850). In prison for blasphemy in 1842, the atheist agitator Charles Southwell asked for a copy along with his accordion and some cigars.[4]

The *Principles* can be and has been read in different ways and in different places. For any book to survive as a classic, it needs to escape its intended audience, to be used for purposes that the author never could have foreseen. But many readings of Lyell's work narrow its meaning. The *Principles* is often seen solely as a precursor to Darwin's *On the Origin of Species* (1859) – a view that would have surprised Darwin himself. Even more implausibly, the *Principles* is sometimes read as the work that created the modern earth sciences, established their first 'paradigm', or somehow 'discovered time'. Alternatively, historians who recognize that geology was a seriously researched subject in 1830 tend to ignore the widespread acknowledgement by Lyell's contemporaries of his role in establishing its philosophical credentials in the English-speaking world. The reception of the *Principles* is poorly understood, so that it is hard to see why the book had such a significant impact on literature and the arts. Only five or six reviews are generally thought to have appeared in Britain, when in fact there were over thirty.[5] Lyell wrote for a wide audience, and his book needs to be seen as part of public controversies about the reform of knowledge.

I

The *Principles* is often read as a Victorian work, but its origins are in the Romantic era, in the religious, political and literary history of the Regency. The first edition belongs on the shelf with De Quincey and Scott, not Dickens and Darwin. Its foundational project marks the book as part of political debates that culminated in the 1832 Reform Bill.

Lyell was born in 1797, during one of the most turbulent decades of European history, and grew up under the shadow of the French Revolution and fears of foreign invasion. The Revolution, with its ferocious anti-clericalism and religion of Reason, was widely traced to the works of Thomas Paine, Baron d'Holbach, Jean-Jacques Rousseau, Voltaire and other Enlightenment writers. Although Britain escaped revolution, cam-

paigns for press freedom, repeal of restrictive taxes and Catholic eman-
cipation were widespread. The reform of Parliament, condemned by
many in the middle and working classes as corrupt and unrepresentative,
became the key issue. In this setting, any argument for the constancy of
natural laws was bound to create controversy.

Lyell's father Charles was a wealthy gentleman and opponent of reform,
who had inherited a large estate in Scotland. He was a Dante scholar and
keen botanist. The young Charles Lyell, both on the family estate and at
his childhood home in Hampshire in southern England, grew up in a
cultured environment typical of the moderate Tory gentry, who aimed
to adapt the best traditions of church and state to changing circumstances.
When Lyell matriculated as a gentleman commoner at the age of seventeen
at Exeter College, Oxford, classics and theology dominated the curriculum, as might be expected at an institution devoted to the Establishment.
Lyell read classics and his writing received sufficient praise for him to
think of winning prizes and becoming a poet. His interest in natural
history developed through browsing in his father's library, where he read
Robert Bakewell's *Introduction to Geology* in 1816.

Geology was lively, popular and controversial, a product of the more
general transformation of the study of nature in the decades around 1800.
Its practices had been freshly confected from cosmological theorizing,
mineral surveying, natural history collecting, biblical exegesis and continental mining traditions. Like other natural history disciplines in the first
three decades of the century, geology engaged a network of practitioners
ranging from physicians and aristocrats to engineers and farmers. A focus
on strata, as exemplified in the publications of the Geological Society of
London, founded in 1807, gave these diverse constituencies a practical
programme of research and a common goal.

In being open to so many, however, geology was in danger of gaining
a reputation for philosophical promiscuity. Underlying disagreements
about wider issues could erupt into speculative excess or religious scepticism. For all the focus on strata, public debate about the meaning of
geology remained embedded in controversy about Creation, the Fall and
the Flood, as illustrated by Byron's notorious unperformed play *Cain*
(1821). Who was to interpret the meaning of a science whose findings
could so flagrantly be used to contradict the opening verses of the Bible?
How was a scientific view to be given of the history of life which did not
lead to soul-denying materialism and atheism?

Geology had been introduced at Oxford to arm undergraduates against the infidels, and like other natural sciences it was an extracurricular option which did not lead to a degree. Lyell attended the flamboyant lectures of the Rev. William Buckland, whose daring reconstructions of extinct monsters and lost worlds attracted an enthusiastic following. Buckland stressed ties to classical learning and agricultural utility; and because many feared that the new science undermined the truth of the Mosaic narrative in Genesis, he contended that geology evidenced divine design and a universal Deluge. The result was a romantic vision of the progress of life through countless ages, strange animals perfectly adapted to even stranger physical conditions, and culminating in the creation of the human race. The earliest articles Lyell wrote show how much he had learned from Buckland about reconstructing extinct animals, dating strata by fossils and charting the progressive history of life.

By the early 1820s Lyell had graduated and moved to London so that he could prepare to become a barrister. Poor eyesight and ambitions to shine in literary circles led him to shift his career to science, despite his father's worry that he was abandoning a secure profession. In centring his identity around geology – while hoping to make money from it – Lyell was doing something new. Most activities that might be called professional in science during the early nineteenth century were seen as low status, involving specimen-selling, instrument-making, curating collections and hack writing. Lyell hoped to raise authorship into a calling fit for gentlemen, much as the mathematician and natural philospher John Playfair had done through scientific reviewing and as Carlyle and Macaulay were doing through their celebrated essays. Conversely, gentility itself was to be redefined around notions of intellectual leadership. The major quarterlies, especially the *Edinburgh* and *Quarterly*, played a crucial part in defining this new role for the author.

Living in the metropolis Lyell associated with 'Lawyers, Geologists & other sinners'[6] and moved away from the moderate Toryism of his family. By the mid-1820s he had become an ardent liberal Whig, advocating electoral reform and disestablishment of the Anglican church. His experience of the ancient universities, which he thought were in a bad way, was instrumental in his change. To his mind only reform could salvage the established order from the horrors of 'mob-rule'. Although his privileged background shielded him from the rawer aspects of day-to-day political struggles, they could not be ignored. Lyell witnessed violent revolutionary demonstrations in Paris, while in Scotland an angry crowd stoned a carriage carrying his sisters, who were hated as Tory gentry.[7]

The most immediate conflicts which Lyell faced were within his family, for his father and brothers had no sympathy for reform. At a crucial by-election in Forfarshire in 1831, when the rest of the family canvassed for the Tories, Lyell confronted a difficult choice. He could not violate his beliefs by voting Tory; but neither could he slight his father by siding with the Whigs, at the height of the Reform Bill agitation and in an open election with less than a hundred voters. Lyell abstained.[8]

Similarly, in devoting himself to science Lyell believed that he had found a vocation that could be used to escape the strife of ordinary affairs. The study of nature was presented as politically and theologically neutral. 'As for public affairs,' he told his fiancée Mary Horner (herself from a famous family of reforming Whigs), 'I have long left off troubling myself about them, as knowing that one engaged in scientific pursuits has as little to do with them, in point of influencing their career, as with the government of hurricanes or earthly motion . . .'[9] For Lyell, science offered an indirect way of forwarding reform without betraying his father or his teachers. Since a gentleman's right to independent judgement was acknowledged, he remained welcome both at home and in fashionable Tory circles in the metropolis. The centre of his social life was the Athenaeum, a club for literary, scientific and artistic gentlemen formed in 1824 by the President of the Royal Society, Humphry Davy, and the Tory politician and reviewer John Wilson Croker. It grew out of meetings in the Albemarle Street premises of the publisher John Murray and encouraged the dominance of British intellectual life by clubbable gentlemen.

Lyell put his family connections and strategic place on the political map to good use, taking the battle for reform into what he now privately saw as the enemy camp. When he briefly assumed an academic chair as part of his campaign to become a new kind of professional man of science, it was at King's College, founded by Anglican Tories to counter the utilitarian University of London. Lyell reviewed for the Tory *Quarterly*, not the Whig *Edinburgh*, where reforming views on education and science had been trumpeted for decades. He brought out his book with Murray, a Tory publisher, not with the useful knowledge merchant Charles Knight. Lyell of course knew about a much wider spectrum of political and religious groups, from radical socialists to the millenarian followers of Edward Irving and Joanna Southcott. But the first edition of the *Principles* was not written for any of these; instead, it targeted a conservative and respectable readership, made up of gentlemen and ladies who feared that geology was anti-Bible and anti-Christian, and needed to be convinced that science had nothing to do with materialism. He hoped to reach an audience among the gentry, aristocracy and professional classes who might be familiar with his early reviews. Behind the scenes Lyell manoeuvred to get sympathetic reviewers for his book in the *Quarterly* and the religiously orthodox *British Critic*. 'It is just the time to strike,' he told his ally George Poulett Scrope, 'so rejoice that, sinner as you are, the Q. R. is open to you.'[10]

Murray, publisher of both the *Quarterly* and the *Principles*, became one of Lyell's greatest assets. Murray wished to counteract the Whig-dominated Society for the Diffusion of Useful Knowledge, on whose behalf Knight and others were flooding the market for cheap books with inexpensive 'libraries' of Entertaining Knowledge and Useful Knowledge. (Knight in term aimed to stamp out the illegal 'cheap trash' from radical working-class publishers.) The chief monument of Murray's efforts was the Family Library, fifty-three volumes of original non-fiction issued between 1829 and 1834. The aim of the Family Library, as Scott Bennett has argued, was counter-revolutionary; it embodied 'a remarkable effort to publish across class lines at a time when class divisions were newly felt to be threatening the fabric of national life'.[11]

At five shillings each the little Family Library volumes were much cheaper than the three octavo volumes of the *Principles*, each of which cost about four times that price. But Murray instantly recognized the affinities of Lyell's project with his own ambition to beat the useful knowledge publishers at their own game. 'There are very few authors, or ever have been,' Lyell reported him as saying, 'who could write profound science and make a book readable.'[12] Murray made sure that the third edition (the first reissuing of the whole work) was in a cheap format of four small volumes, priced at just six shillings apiece. The *Principles* could then compete effectively and push hack compilers out of the market. This strategy accorded with Lyell's own financial demands – he needed big sales to maintain his gentlemanly style of life – and with his lofty conception of the man of science as arbiter of truth. He had originally planned his book as a series of didactic 'conversations' of the kind popular at this time and 'instructing the millions' remained his lifelong mission.

Lyell thus hoped to encourage the creation of an enlightened clerisy of truth-seekers – like himself – who would diffuse knowledge to the 'vulgar'. This élitist, authoritarian vision of 'popular science', increasingly dominant from the 1830s, contrasted with older traditions in natural history, which had encouraged participation by a wide range of people in the making of knowledge. These paternalist practices were embodied in the vast correspondence networks of the botanist Joseph Banks, the mineralogist Robert Jameson and Lyell's own father.[13] Young Lyell, in contrast, tended to dismiss provincials, colonials, women and working people as mere sources of information, whose facts gained meaning by being drawn together and reasoned upon by the few. (Irate colleagues in America, tired of an itinerant bigwig plying them for hard-won findings, nicknamed him 'The Pump'.) Lyell saw no need for ordinary readers to master all the research and reasoning that had gone into the making of knowledge; rather, experts should 'communicate the results, and this we are bound to do'.[14]

The liberal Whig weekly *Spectator* agreed, stressing the moral value of the *Principles*:

The earth is an old reformer; her constitution has been subjected to innumerable changes; the signs of radical movements are to be detected everywhere, yet it is by no means easy to ascertain either the course or the causes of the revolutionary phenomena that so perpetually meet the eye of the inquirer. There are other investigations which more nearly affect our social happiness than the philosophy of geology, but perhaps there is none which in an indirect manner produce a more wholesome and beneficial effect upon the mind . . . After the perusal of Mr. Lyell's volume, we confess to emotions of humility, to aspirations of the mind, to an elevation of thought, altogether foreign from the ordinary temper of worldly and busy men . . . So disposed mentally, the heart overflows with charity and compassion; vanity shrivels into nothingness; wrongs are forgotten, errors forgiven, prejudices fade away; the present is taken at its real value; virtue is tried by an eternal standard. There are sermons in stones and tongues in brooks, but they want an interpreter: that interpreter is the enlightened geologist. Such a man is Mr. Lyell.[15]

Only if a science was gentlemanly, disinterested and apolitical could it take its place as a force for reform.

II

Making science 'philosophical' was a way of making it respectable. Lyell's ambitions need to be understood in the light of the contested character of knowledge at the time he wrote. Geologists had to demonstrate, through introductory books, tracts and sermons, that they could go beyond strata-hunting to treat sensitive issues safely. Published at a pivotal moment in the public history of the sciences in Britain, the *Principles* was the most thoroughgoing and sophisticated of these vindications. More than any of his associates, Lyell worried that an impressive edifice of practical research could be accused of resting on shaky foundations. In the specific sense of research conducted according to clearly articulated rules of reasoning (such as those in Newton's *Principia*), he believed that geology was not a science. His ambition was, single-handedly, to make it one.

What did it mean to attempt to make a subject into a science in 1830? The subtitle announced how Lyell planned to go about his task. Most geological works offered a connected narrative of progress; or they were historical in the older, descriptive sense of 'natural history'. In contrast the *Principles* was a treatise on natural philosophy, what the subtitle called 'an attempt to explain the former changes of the earth's surface, by reference to causes now in operation'. The key word here was 'causes'. Lyell, like Newton, depended on the notion of a *vera causa* or 'true cause', defined as a cause in operation which could be *observed*. The notion of a *vera causa* had been developed in the eighteenth century by Scottish common sense philosophers such as Thomas Reid, to whom Lyell was much indebted through the writings of John Playfair and Dugald Stewart.[16]

Three rules of reasoning underpinned Lyell's proposals for reforming the perceptions of his readers. First, geologists must assume that the basic *laws* of nature (such as gravity) had not changed over time. The sole exception was the origin of new species, where Lyell was non-committal, claiming that the human span of observation has been too short. Even here, though, he left open the possibility of an explanation through natural law; thus there might be a middle way between creation and evolution. Since, however, there was no observable basis for theorizing on this issue – no *vera causa* – Lyell's genteel silence on these alternatives appeared consistent with his philosophical position. Almost all geologists accepted the constancy of the laws of nature, an assumption that directly tied the *Principles* with the *Principia*.

The second proposed rule of reasoning involved the *kinds* of causes to be used in interpreting the earth's past. Here Lyell took a far more controversial stance, arguing that no causes other than those that we can now see acting should be employed in explanations. Of course, he recognized that different causes from those occurring now might be responsible for the past condition of the earth; almost all of his fellow practitioners thought that this was the case. But the *Principles* argued that such causes could be part of science only if they were observed by reliable witnesses. Recourse to such causes, without the backup of trustworthy testimony, encouraged investigators 'to cut, rather than patiently to untie, the Gordian knot' (p. 355). Lyell's scientific critics complained that this was unduly restrictive: why assume that human observation was the only measure of untold eons of earth history?

Most surprising of all, Lyell argued that interpretation must proceed on the basis that causes had not varied in *degree* either; in short, the intensity of observable earthquakes, volcanoes, floods and other geological agents should be the measure of their action in the past. Again, knowledgeable commentators such as the Rev. Adam Sedgwick of Cambridge pointed to twisted mountain strata and giant erratic boulders as good evidence for episodes of 'feverish spasmodic energy' against which modern causes paled into insignificance.[17] Quiet stasis was punctuated by violent catastrophe and sudden death. But for Lyell, recourse to such unobserved events had made geology a byword for unrestrained speculation. We can see, he claimed, forces adequate to produce the Andes and the Alps, especially if the cumulative effects of time were taken into account.

Lyell developed the implications of this policy in opposition to some of the best-supported generalizations of his fellow geologists. In early reviews he had agreed with their advocacy of progress – invertebrates giving way to fish, reptiles, mammals and then man. But in the *Principles* he switched sides to argue that this was an unphilosophical assumption which ignored the patchy nature of the fossil evidence. Where most geologists dreamed of filling in the lost page in the book of life, Lyell believed that they should work under the assumption that almost the entire volume had been destroyed. Gaps in the record, such as that between the Cretaceous and Tertiary, might represent vast ages, longer than the Tertiary itself. Similarly, the absence of the remains of mammals, birds and other 'high' forms from the very oldest rocks did not necessarily mean that such creatures had not lived then; their remains just might not have been preserved. (This suggestion was not unreasonable, as the discovery of a fossil opossum from the so-called 'Age of the Reptiles' had already demonstrated that mammals had existed long before geologists had thought possible.)

While denying any need to assume progress within the history of life, Lyell accepted that hotter temperatures had probably prevailed in the past. How was this to be explained while preserving the rigorous application of his rules of reasoning? He needed to show that climate changes could have been produced by the ordinary action of observable causes – not by a comet or a change in the earth's axis (a cause differing in kind) nor by a cooling earth (a cause differing in degree). His solution, given pride of place in his first volume, was one of the triumphs of the *Principles*. Deploying a geographical argument, Lyell showed how shifting relations between land and sea could produce the required changes. As he told the surgeon Gideon Mantell, he had 'a receipt for growing tree ferns at the pole, or if it suits me, pines at the equator; walruses under the line, and crocodiles in the arctic circle'.[18]

With such arguments Lyell rejected the possibility of constructing any continuous 'story of the earth' at all; in this specific sense the *Principles* was profoundly ahistorical. The sole defining *narrative* event, he claimed, was the advent of the human race. Denying any shape to earth history kept humans special because as moral beings they were separate from the rest of creation and outside the reign of natural law. Within the random, shifting balance of forces, the uplift of mountains and the erosion of coasts, the creation of humanity marked the beginning of one era and the end of another; for human origins were kept outside the physical flux that constituted earth history. Notably, human history provided the only sustained forward narratives in the *Principles*, and these were militantly Whiggish, developmental and progressive. Thus the 'historical sketch of the progress of geology' in the first volume and the history of Tertiary geology in the third celebrated the triumph of truth in Lyell's own system.

The earth, according to Lyell, exhibited no such signs of progress. He underlined the lack of direction by speculating on the future, when a warmer climate might once again prevail. 'Then might those genera of animals return,' he wrote, 'of which the memorials are preserved in the ancient rocks of our continents. The huge iguanodon might reappear in the woods, and the ichthyosaur in the sea, while the pterodactyle might flit again through umbrageous groves of tree-ferns' (p. 67). This is often taken to mean that Lyell advocated a cyclical theory of earth history. (A cartoon privately circulated by one of Lyell's opponents showed a future 'Professor Ichthyosaurus' lecturing to an audience of reptiles on the comparative anatomy of the extinct human race.)[19] However, the conditional 'might' in this famous passage shows that there was no reason why such changes *must* or *will* come about. For all his romantic evocation of the return of the pterodactyls in a cycle of a 'great year', Lyell's substantive claims were models of philosophical caution.

This point needs to be stressed, for many of the best modern commentators on the *Principles* – notably Reijer Hooykaas, Martin Rudwick and Stephen Jay Gould – have taken Lyell's public statements about the pattern of earth history out of context, reading them in terms of private letters and journals so that their function as thought-experiments about the past is obscured. Lyell's recommendations in the *Principles* about method are turned into doctrines or 'isms' and compared with those of his contemporaries: the Cambridge polymath William Whewell began this process in the 1830s by dubbing Lyell a 'uniformitarian' in opposition to his own 'catastrophism'. Recent historians have set up dichotomies between progression and non-progression, or between directionalism and steady-state, which are more nuanced but serve the same purpose. They make the *Principles* a cosmological book, which points towards the construction of a connected narrative history of the world. However, in his public statements during the 1830s Lyell no more advocated a steady-state, cyclical or non-progressionist cosmology than he did progression itself. Indeed, the *Principles* claimed that *any* kind of global narrative would prove impossible to reconstruct, as too much of the record had been lost. Lyell, despite what Gould has argued, was not the 'historian of time's cycle'.[20]

Lyell's public application of his principles was thus almost entirely regulative: that is, geologists should carry out their investigations *as though* visible causes are the same kinds as those that have acted in the past, and of the same degree of intensity. Uniformity of law, kind and degree had to be assumed, Lyell argued, to make geology scientific. His subtitle cautiously spoke of 'an *attempt*' to explain former changes; and the third and all subsequent editions softened this still further. In essence, he hoped to make the study of the pre-human past amenable to the same rules of inductive inference that continental scholars, notably the great historian of Rome, Barthold Georg Niebuhr, were using to put the study of human history on a new basis. The positive aim was to make geology into an inductive science grounded in the observation of causes. Lyellian uniformity was thus a method, not a doctrine about the shape of earth history.

All his readers needed to do, Lyell argued, was to reason on what they could see everywhere around them. The sheer length of the *Principles* – over 1400 pages in the original edition – was essential to this programme of perceptual reform. Text and pictures, through the cumulative effect of hundreds of examples, made readers into witnesses to the power of modern changes. Large parts of the *Principles* (over half of the original text) were devoted to assembling a comprehensive toolkit of observed forces for use in interpreting the past. Lyell's descriptions of rivers, tides, currents and delta-formation were so extensive that the civil engineer Sir John Rennie used the *Principles* as a professional reference work while constructing waterworks on the Thames.[21] Aqueous causes were followed by a thorough account of their igneous counterparts. An entire chapter patiently documented the devastation wrought by earthquakes in one year in a single region in Italy. Lyell turned in his second volume from the inorganic realm to the organic, discussing causes now in operation involving plants and animals. The main focus was on the relation between organisms and their environment, especially in relation to geographical distribution and impact of humans and other animals in altering the face of the globe. Long chapters were also devoted to the fossilization of organisms ranging from shipwrecked sailors to corals. This kind of detailed exposition occupied a central place in Lyell's strategy, for as Dugald Stewart had said, most people learned best from examples rather than general principles.[22]

Even those who disagreed with Lyell admired the empirical richness of his book, which mustered research from hundreds of sources in five modern European languages and two ancient ones. The prestigious metropolitan weekly *Athenaeum* praised it as 'one of the noblest accumulations of facts of modern times, interwoven with highly ingenious theories and truly philosophical speculations'. 'It is essentially,' the *Scotsman* newspaper noted, 'a book of facts.'[23]

Lyell drew extensively on French, German and Italian sources, for he believed that continental researchers were often far ahead of the British. Much of his history was taken (with scant acknowledgement) from Giovanni Battista Brocchi's work on fossil conchology; information about inorganic causes derived from Karl Ernst Adolf von Hoff's massive compilation; Augustin De Candolle was the main source for plant geography; and Gérard Paul Deshayes provided tables of shells and assisted in establishing the innovative percentage system for dating strata used in the *Principles*. Lyell's greatest and earliest debt was to a group of geologists he had met in Paris, notably Constant Prévost.[24] These men had shown, against the celebrated naturalist Georges Cuvier, that the strata of the Paris Basin could be explained without reference to catastrophe, through comparisons with modern lakes, rivers and seas.

Lyell's own fieldwork also impressed his readers, who were invited to picture an author climbing volcanoes, hammering chalk cliffs, measuring ancient temples and excavating fossils with his own hands. Lyell had three pieces of advice for aspiring geologists, and had followed them at every opportunity: travel, travel, travel.[25] Travel had been central to the making of the *Principles*, with tours to France, Italy and above all Sicily being critical to the shaping of Lyell's own vision. His readers in turn gained a sense of direct contact with nature. In all this Lyell's place in the vogue for 'Baconian' induction and empiricist theories of perception is clear.

Lyell's most obvious philosophical debt was to Scottish traditions of theorizing about the earth. John Playfair, mathematician and natural philosopher at the University of Edinburgh and one of the most influential commentators on science during the Regency, had published *Illustrations on the Huttonian Theory of the Earth* in 1802. This work had brought the Enlightenment natural philosopher James Hutton's system of the earth into the mainstream of nineteenth-century debate, recasting it in inductive, empiricist terms acceptable to the post-Enlightenment era. Hutton had argued for a cyclical system of strata consolidation, mountain uplift and decay, all produced by gradual forces operating under a beneficent Deity. As a natural philosopher, Playfair presented geology as a science which dealt with stable systems operating under unvarying laws. Lyell never seems to have read Hutton in the original – his quotations were all at second-hand – but he used Playfair's works extensively.[26]

The philosophy of the *Principles* is often summed up, as is that of Hutton and Playfair, as 'the present is the key to the past.' But this slogan is unspecific and indeed almost meaningless: causal keys can function in many ways. For example, they can license reasoning by simple analogy, which is not at all what Lyell had in mind. Rather he claimed that his key worked because present causes could be – or had been – witnessed in action by reliable observers. It was the visibility of modern causes that made them the *only* legitimate basis of explanation. In fact, Lyell thought that humans were not ideally placed to be good geologists: 'an amphibious being' would be better able to 'arrive at sound theoretical opinions' because so much of the earth is covered by water, while an underground gnome would be the best geological philosopher of all (p. 32). Lyell imagined such beings to make the point that any true system of the earth depended on seeing what was going on in the present.

To extend our inferences into the past, Lyell claimed, we could rely only on induction from the evidence of our eyes. Here, at a crucial point in his argument, he drew on Dugald Stewart's *Elements of the Philosophy of the Human Mind* (1792–1827). 'Any naturalist will be convinced,' Lyell said, by the observation that 'the same species have always retained the same instincts, and therefore that all the strata wherein any of *their* remains occur, must have been formed when the phenomena of inanimate matter were the same as they are in the actual condition of the earth.' Continuity of animal instincts thus implied continuity of 'the physical laws of the universe', including those governing climatic change (p. 98). It was this view of instinct, drawn from the most influential British philosopher of the early nineteenth century, that underpinned Lyell's induction about the constancy of natural laws.

The emphasis on *vera causa* reasoning, the philosophical heart of the *Principles*, appealed to views about scientific method that were widely canvassed during the decade before the Reform Bill. Lyell himself had little formal philosophical training and his references to philosophers are tactical and implicit. At Oxford he had been bored by formal logic and found Aristotle (whose writings still dominated the curriculum) 'an astonishingly stiff author'. But he would have known that Stewart and Reid were being discussed by authors across the religious and political spectrum. At Oxford, the enlightened Tory Edward Copleston and his followers (the High Church Noetics) applied common sense philosophy as part of their project to create a Christian version of political economy.[27] After moving to London, Lyell met the young astronomer John Herschel, whose *Preliminary Discourse* of 1831 praised the *vera causa* theory of climate in the *Principles*. Herschel welcomed the book as 'one of those productions which work a complete revolution in their subject by altering entirely the point of view in which it must henceforward be contemplated'.[28]

Few reviewers were as enthusiastic about *vera causa* explanations as Herschel was, but almost all agreed that Lyell had transformed the philosophical status of debate in geology. Whewell celebrated the creation

of a new science of 'geological dynamics': just as Galileo's work on motion
had led to Newton's explanation of the planetary motion, so would Lyell's
account of terrestrial forces provide the basis for reconstructing earth
history. Moreover, Lyell had proceeded 'not in an occasional, imperfect,
and unconnected manner, but by systematic, complete, and conclusive
methods'. Only in this way could geology 'be a Science, and not a
promiscuous assemblage of desultory essays'.[29] This was high praise –
even if Lyell was cast as the Galileo of geology and not its Newton.

Whewell's approval of the *Principles* was widely shared. At the British
Association for the Advancement of Science in 1832, the geologist William
Daniel Conybeare hailed the book (albeit before launching a critical
barrage) as 'in itself sufficiently important to mark almost a new aera in
the progress of our science'. The *Monthly Magazine* called it 'a masterly
performance' whose publication 'will form an epoch in the history of a
science.' The *Athenaeum* hailed 'the work of no ordinary mind . . . a rich
harvest of instruction and delight'. In an appropriate analogy, the weekly
Literary Gazette praised Lyell for 'removing much of the delta which has
obstructed the current of geological knowledge'. The *New Monthly Magazine*,
despite serious reservations based on Scripture, noted that 'Professor
Lyell's work is the nearest approach towards establishing geology as a
science, of any thing we have met with.'[30]

No one adopted Lyell's views wholesale, and the *Principles* has accordingly been presented in recent historical literature as an idiosyncratic
failure, a singularity, or even 'bad science'.[31] It would be misleading,
however, to judge the effect of the book – or the reception of any text –
by the number of converts or the comments of a few specialists. As with
most widely-read books, the effects were subtle and pervasive. Notably,
critics of all shades of opinion agreed that debate about the earth was
pursued on a new plane of sophistication as a result of the *Principles*.
Geology could claim to be a 'science' worthy of the pursuit of gentlemen,
according to new criteria being established at the time of the Reform Bill.
Lyell's intervention was a key part of that process.

III

The chief obstacle to a philosophical science of the earth, Lyell believed,
was posed by the clergy. Nowhere was reform more overdue than in the
politics of knowledge, he thought, for intellectual life in Britain was

dominated by the established Anglican church. Geology could not be given principles without redefining the relations between human observation and divine revelation, between religious and secular education, between Church and State. In arguing these points Lyell saw no inherent conflict between science and theology; rather, he wished to redefine their respective domains. What was to be the relation between a new subject, largely imported from suspect continental sources, and native traditions of text-based theological criticism about earth history?

The authority of Scripture among the learned, Lyell thought, had its foundations in popular culture (or as he condescendingly said, among the 'vulgar'). Sermons, children's books, chronological charts on schoolroom walls, metropolitan shows, engravings of visionary paintings: the Bible structured the narrative history of the world in everyday experience. Working people were usually taught to read from the first verses of Genesis in the King James translation and assumed the literal truth of the creation story.

Lyell, eager for the dissemination of what he saw as the truth, lamented to American friends that his own nation was 'more parson-ridden than any in Europe except Spain'.[32] Thousands of clergymen preached Genesis as a matter of course, while no more than a handful (and here he included Sedgwick and Buckland) professed scientific views of the Flood or the earth's antiquity. Of course Lyell also knew that exegetical theology had a far longer and more distinguished history than did geology, an upstart science associated with infidelity and revolutionary atheism. For most readers, the authority of Scripture continued to outweigh that of strata-maps and sections, so that biblically-oriented accounts of earth history predominated in publishers' lists right through the first half of the century. Sharon Turner's *Sacred History of the World* of 1832, 'firmly attached to the great Newtonian principle, of the Divine causation of all things', went into its eighth edition two years before Lyell's book did.[33] Books in the same tradition were written by Thomas Chalmers, Edward Hitchcock, Granville Penn, John Bird Sumner, Andrew Ure and Nicholas Wiseman – respected authors whose writings often sold more copies and were better known than those of Lyell and his friends.

No line could distinguish theological works from geological ones: Lyell's passionate desire to 'free the science from Moses' was by no means empty or non-controversial.[34] His teacher Buckland had made the most sophisticated dovetailing of scientific findings with the literary evidence of the Bible in the *Reliquiæ Diluvianæ* (1823). Buckland based his defence of the physical reality of a universal Deluge on impeccable reasoning from existing causes, with texts and theological exegesis accorded a subsidiary role. He observed the actual behaviour of live hyenas in the zoo to reconstruct cave environments and ancient forms of life, and like most practising men of science, employed a time scale far beyond the 6000-year span accorded by tradition. There was good evidence for the Flood, too, in the unsorted debris or 'diluvium' spread across Europe. Buckland's brilliant strategy had the support of many in the moderate Church establishment, including most of the clerical geologists from Oxford and Cambridge.[35]

The Geological Society witnessed lively debates about the Flood in the 1820s, clashes as vigorous as those taking place in Parliament over electoral reform. Together with the retired soldier Roderick Murchison and the political economist George Poulett Scrope, Lyell emerged as leader of the opposition to the diluvialist party. The *Principles*, as a manifesto for fundamental change in the organization of intellectual life, capped their campaign to sever all links of geology to a theology based on Scripture. Throughout his life Lyell continued to advocate an overhaul of the ancient universities, the centre of clerical authority; and he became a great admirer of the American republic. He saw the future of science in the hands of gentlemen of character and independence, such as Herschel, Murchison, Scrope and himself, who would draw on the best continental work without succumbing to 'unmanly' materialism or dreamy idealism.

The boundary Lyell drew around the science implicitly put Buckland, Conybeare and Sedgwick with those who had retarded progress. He did not, however, say this directly, but ended the polemical history which opened the *Principles* just before they had begun their work, tactfully leaving readers to draw their own conclusions. 'I conceived the idea five or six years ago,' he told his ally Scrope, 'that if ever the Mosaic geology could be set down without giving offence, it would be in an historical sketch.' Lyell had to tread softly. The *Principles* could appeal to its audience only by respecting 'the weakness of human nature' and 'the sentiments of our neighbours in the ordinary concerns of the world and its customs'.[36]

Take the issue of the age of the earth. At the Geological Society, all parties in the diluvial debates had spoken freely of millions of years, but Lyell believed that they had failed to comprehend the meaning of such a vastly expanded time-scale. The classic site for demonstrating the antiquity of the world was Etna, the largest volcano in Europe, and Lyell turned to it in the *Principles* to exhibit his fieldwork skills and the virtues of his new philosophy. His highly-wrought depictions of grand scenery created the imaginative depth of past time which was at the heart of his message. As he recalled:

In the course of my tour I had been frequently led to reflect on the precept of Descartes, 'that a philosopher should once in his life doubt every thing he had been taught;' but I still retained so much faith in my early geological creed as to feel the most lively surprise, on visiting . . . parts of the Val di Noto, at beholding a limestone of enormous thickness filled with recent shells . . . resting on marl in which shells of Mediterranean species were imbedded in a high state of preservation. All idea of attaching a high antiquity to a regularly stratified limestone, in which the casts and impressions of shells alone were discernible, vanished at once from my mind.[37]

This was a Cartesian moment, but one especially suited to the man of knowledge in action. The philosopher did not sit in a stove-heated room, but explored bandit-infested country with hammer in hand. Conversion was prompted not by abstract contemplation, but by limestones, shells and marls. And behind the reference to Descartes was a biblical one: like Saint Paul, Lyell changed his 'creed' on the road. The most controversial claims of the *Principles* concerned what John McPhee has called 'deep time';[38] other practising geologists, having abandoned the biblical time-scale, had failed to realize the interpretive power of the ages at their disposal. Through the force of Lyell's rhetoric, deep time was vouchsafed as a revelation from nature, realized through a secular pilgrimage.

In underlining the immensity of past time Lyell needed to distance himself from claims that the earth was eternal, for these had been common-place in Enlightenment diatribes against religion. The issue of eternalism arose most acutely in relation to the crystalline rocks often found at the heart of mountain chains. Following Hutton and Playfair, Lyell argued that these rocks – previously called 'Primary' – were not remnants of an original molten state of the planet, but could in fact be formed at any time. Those that had a stratified appearance, he suggested, should be termed 'metamorphic' to emphasize that they were altered strata of diverse ages. This meant (as he said, misquoting Hutton as usual) that 'I can find no traces of a beginning, no prospect of an end' (p. 16). To say that science should abandon the search for origins did not imply that there never was a beginning, or that there would never be an end – questions 'worthy a theologian'.[39] Lyell thereby reaffirmed that the *Principles* was not a cosmological treatise, but a manifesto for pursuing science.

Rude attacks upon cherished beliefs before the time was ripe, Lyell believed, were unjustified, unproductive and ungentlemanly. He contrasted his strategy with that of the freethinking Edinburgh physician George Hoggart Toulmin, whose *Eternity of the World* (1785) had been reissued by a radical publisher in the 1820s. In a letter Lyell compared Toulmin with English soldiers who had raided Burmese temples for war trophies:

> To insult their idols was an act of Christian intolerance, and, until we can convert them, should be penal. If a philosopher commits a similar act of intolerance by insulting the idols of an European mob (the popular prejudices of the day), the vengeance of the more intolerant herd of the ignorant will overtake him, and he may have less reason to complain of his punishment than of its undue severity.

As Lyell concluded, 'This is not courage or manliness in the cause of Truth, nor does it promote its progress.'[40]

The dangers of 'insulting the idols of a European mob' were illustrated just before the *Principles* reached the booksellers. One of Lyell's closest associates, the Rev. Henry Milman, introduced German-language theological scholarship into Britain in his *History of the Jews* (1830), which did for Old Testament studies more generally what the *Principles* aimed to accomplish for the early chapters of Genesis. Milman treated the Jews as an 'eastern tribe', used documentary evidence and minimized the miraculous. Denounced by critics from the evangelical Sharon Turner to the young John Henry Newman, Milman seemed to have forfeited all chance of clerical preferment – even though it was said that nothing he had written went beyond what could be found in the notes to an expensive Bible edited by one of his most learned opponents. Milman's work, however, had been published by Murray from the first at a cheap price in the Family Library. 'The crime,' as Lyell recognized, 'is to have put it forth in a *popular* book.'[41]

In contrast, the early editions of the *Principles* were expensive octavos with fine paper and a handsome typeface. These sold well, so that the 1500 copies of the first edition of volume one were soon exhausted. Using his 'scientific' rules of reasoning, Lyell had successfully recast for a genteel audience arguments associated with the popular realm, and especially with the radical materialist cosmologies that had sprouted up in the wake of the French Revolution. Only *after* the first two editions had been safely launched did Murray repackage the set for a wider market, in accordance with Lyell's own views on the 'communication' or popularization of knowledge by gentlemen.

Lyell's concern with respectability permeated the *Principles*. Divisive issues were introduced in measured prose, with quotations from Horace, Ovid, Pindar, Pliny, Virgil, Thucydides, Dante, Milton and Shakespeare. The popular travel author Captain Basil Hall praised 'the calm, dispassionate, gentlemanlike style in which he handled, not one, but every controversial subject'.[42] The Temple of Serapis, featured as the frontispiece to the first volume, reminded readers that classical structures could survive even dramatic changes. The Temple had been subjected successively to subsidence and elevation, all within the Christian era. But these upheavals, however sudden, had left the pillars standing. In human affairs as in earth history, stability was critically important.

Lyell thought that his step-by-step approach would insure that his book 'will be thought quite orthodox' and would '*only* offend the ultras.'[43] In this he proved correct. Wordsworth's friend Henry Crabb Robinson, a liberal Unitarian who heard Lyell lecture during his brief tenure at King's, praised his sensitivity:

> He decorously and boldly maintained the propriety of pursuing the study without any reference to the Scriptures; and dexterously obviated the objection to the doctrine of the eternity of the world being hostile to the idea of a God, by remarking that the idea of a world which carries in itself the seeds of its own destruction is not that of the work of an all-wise and powerful Being. And geology suggests as little the idea of an end as of a beginning to the world.[44]

The religious periodical press, which blossomed in the late 1820s and early 1830s, was equally enthusiastic. As the monthly *Metropolitan Magazine* told its readers, 'All those who wish to lift up their souls in adoration to the Great First Cause, should attentively study this work.'[45]

Even journals that supported the supreme authority of Scripture acknowledged the virtues of Lyell's book. The High Church *British Critic*, in its survey of the completed work, was almost alone among the quarterlies in confronting the *Principles* with Old Testament miracles, a six-thousand-year-old earth and a universal Deluge. However, the reviewer had rarely met with 'a more philosophical, patiently inductive chain of reasoning'.

The *New Monthly Magazine* thought Lyell 'has occasionally betrayed a wildness of speculation which sets him in needless hostility to the cosmogony of Moses, without advancing his claims as a philosopher.' Even so, the reviewer praised him for writing the best book on geology ever published. The *Presbyterian Review and Religious Journal*, a new quarterly representing Scottish evangelicalism, acknowledged that the book 'is indeed a work of no common interest', 'well executed' and 'very valuable'. At the same time, the reviewer drew very different conclusions from Lyell's history of 'guessing and erring' in geological theory: 'are they all wrong but the author?' The English Nonconformist *Eclectic Review* found the *Principles* – despite its unwarranted speculations – 'invaluable as a collection and arrangement of facts and geological phenomena'.[46]

Lyell achieved his greatest theological triumph within the Geological Society. The *Principles* precipitated a series of declarations from the chair, as former opponents abjured their use of the Flood as a middle way between Genesis and geology. As Sedgwick admitted, the 'double testimony' of the Bible and of physical traces of diluvial action had led to 'erroneous induction'. 'Having myself been a believer,' he told the Society, 'and, to the best of my power, a propagator of what I now regard as a philosophic heresy, and having more than once been quoted for opinions I do not now maintain, I think it right, as one of my last acts before I quit this Chair, thus publicly to read my recantation.'[47] The *Principles* thus proved decisive in boundary disputes between science and theology.

IV

If the Flood and the age of the earth ceased to be debated at the Geological Society, they occupied the centre stage of public concern for another three decades. (In many parts of the United States, they remain controversial to the present day.) Lyell's views on these matters were canvassed in Genesis commentaries, sermons, tracts and other theological writings. Such discussions remained urgent partly because atheists and deists, writing in the working-class penny press, used geology to give the stamp of authority to unbelief. But the infidels had hard work to make a book whose last sentence compared 'the finite powers of man and the attributes of an Infinite and Eternal Being' (p. 438) into an ally of atheism. The *Principles*, in offering new proofs of divine power and goodness, had distanced geology from scriptural exegesis, miracles and creation – not from all theology. At the same time, though, it cut away supports that science had traditionally provided for belief. In the *Principles* the rock of faith rested solely on God's maintenance of the economy of nature and on the status of humans as moral, rational beings.

In stressing the special character of the human species, Lyell drew on John Bird Sumner's *Records of Creation* (1816), which used the distinction between humans and animals as evidence for a divine creator. To think otherwise had been the mistake of the French materialists, most notoriously Jean-Baptiste Lamarck. In his *Philosophie zoologique* of 1809 and the introductory essay to his definitive seven-volume work on invertebrates, Lamarck had argued for the evolution of one species into another, or 'transmutation' as it was generally called, within the context of a broader philosophy of nature which emphasized the power of environmental circumstance. An attack on evolution became central to the *Principles* because of the threat Lamarck posed to the special status of humanity.

During visits to Paris, Lyell had been shocked to discover that evolution 'has met with some degree of favour from many naturalists' (p. 196), among them Etienne Serres, Geoffroy Saint Hilaire, Jean Baptiste Bory de Saint Vincent, Pierre Boitard and Jean Baptiste d'Omalius d'Halloy. In Britain, Robert Jameson's lively scientific monthly, the *Edinburgh New Philosophical Journal*, had published an anonymous article advocating evolution. Perhaps most threateningly, the declared Lamarckian Robert Edmond Grant was on the Geological Society Council and taught fossil zoology at the University of London.[48]

But Lyell had few worries about these views spreading among his intended readership, where evolution was typically dismissed as 'the most stupid and ridiculous' idea to have been hatched by 'the heated fancy of man'.[49] The second volume of the *Principles* began by noting that Lamarck's views were 'not generally received' (p. 184). Even Lyell had had to read a borrowed copy of the *Philosophie zoologique*, which was not translated into English until 1914. By refuting the French naturalist at length, Lyell made evolution accessible throughout the English-speaking world, beyond a narrow circle of naturalists, medical lecturers and political radicals.

The real question is not why Lyell should reject Lamarck, but why he did so at such length and so publicly. One answer becomes evident if the audience for the *Principles* is kept in mind. Lyell did not expect to be read by atheists or deists, but by conservative, genteel subscribers to the *Quarterly*. This audience needed to be shown that geology was safe. Lyell anticipated clerical attacks on his advocacy of the uniformity of nature, views which could lead to suspicions that the groundwork was being laid for a naturalistic explanation of species. However minor an issue evolution might be within science, annihilating Lamarck established Lyell's orthodox credentials. Given the prestige of French science and its importance in his attempted reform of geology, Lyell's fear that he might be condemned as a closet Lamarckian becomes understandable. If uniform causal laws applied to all other parts of the natural world, why not to humanity itself?

The need to combat evolution more directly had become clear to Lyell only at a very late stage in the composition of his first volume. One factor was certainly the impending publication, by Murray, of Humphry Davy's posthumous *Consolations in Travel*. Davy, a famous chemist, explicitly linked the 'absurd, vague, atheistical doctrine' of evolution with Huttonian views on uniformity.[50] Lyell must have been shown the relevant passages in proof and realized that views similar to his own were suspected of tending to materialism. By this point, too, Murray would have had in hand the manuscript of Thomas Hope's *Essay on the Origin and Prospects of Man*, which advocated organic evolution as part of a system of abstract metaphysics. Hope was famous as a novelist, connoisseur and champion of neoclassical design, and intended the *Essay* as his philosophical masterpiece. After his death in 1831 alarm about the work's heterodoxies became manifest. Carlyle called it the 'highest culminating-point of the Mechanical Spirit of this age; as it were, the *reductio ad absurdum* of that whole most melancholy doctrine'.[51] Printed in a tiny edition of 250 copies and withdrawn by Hope's executors immediately after publication, the *Essay* was the last pro-evolutionary book Murray published before Darwin's *Origin*.

Lyell countered the growing alarm about transmutation with last-minute additions to his first volume. He did not explicitly contest Davy's claim that uniform natural laws entailed evolution, but undercut the argument at a more fundamental level by denying the assumption of a progressive history of life (pp. 84–91). No progress, no evolution: the method of uniformity, far from leading to Lamarck, made the Frenchman's views impossible.

Lyell strategically refrained from mentioning the evolutionary spectre at this stage of his argument, despite abundant opportunities to do so. His increasing awareness of the dangers of being tarred with a Lamarckian brush, however, soon led to major changes in his second volume. His original plan, submitted to Murray in 1827, had devoted no more than four or five chapters to changes in the organic world with (at most) a brief dismissal of Lamarck. At a later stage – probably when he returned to his old drafts in November 1830 – these chapters were greatly expanded, with the attack on evolution extended well beyond its allotted space. Lyell cited evidence from animal and plant hybrids, the similarity of mummified animals from ancient Egypt to present-day forms, and the variability of domestic forms – all to demonstrate that there was no evidence for evolution operating as a *vera causa*.

Reactions to the first volume, especially by Sedgwick in a Geological Society presidential address, established the virtues of the new strategy. For Sedgwick, as for Davy, the thoroughgoing application of 'what we commonly understand by the laws of nature' to the problem of life would seem to demand 'the doctrines of spontaneous generation and transmutation of species, with all their train of monstrous consequences'.[52] As Lyell must have recognized, distancing himself from Lamarck was just as imperative as liberating science from Scripture.

Lyell also confronted career pressures on the issue of evolution early in 1831. As part of his campaign to establish himself in the metropolis, he was negotiating for a geological chair at King's College. One of the electors was Copleston, who by this time was Bishop of Llandaff and Dean of St Paul's. Copleston feared that geologists undermined beliefs concerning the creation of man and the reality of a universal Deluge. Lyell could satisfy him on the first point, but only in part on the latter (the Flood had not been universal, although it might have covered the inhabited globe).

On species, however, Lyell had been reassuring. 'I combat in several chapters of my second volume,' he wrote, 'the leading hypothesis which has been started to dispense with the direct intervention of the First Cause in the creation of species, & I certainly am not acquainted with any physical evidence by which a geologist could shake the opinion generally entertained of the creation of man within the period generally assigned.'[53] Sedgwick and Conybeare intervened on Lyell's behalf with the electors at King's, aware that noxious Lamarckian dogmas were dispatched in the new chapters. Cannily and cautiously, Lyell never said in public if a law-like origin for species was possible, preferring his views 'to be inferred'. In correspondence he tailored his message to the recipient. To liberal friends like Herschel, he advocated 'the intervention of intermediate causes'; to vigilant clerics like Copleston, he appeared to confirm 'the direct intervention of the First Cause'.[54] When cornered, Lyell could allay orthodox suspicions by arguing that origins were not part of science.

The anti-Lamarckian additions were sometimes recognized as tangential to the main argument of the *Principles*. 'Lyell is amusing,' the Edinburgh naturalist Robert Jameson wrote, 'but it is not what geologists want.'[55] The *Gentleman's Magazine* was disappointed too, after being 'compelled to wade through one half of the volume among *disjecta membra* of the vegetable and animal kingdoms, which might do, as speculations on natural history, or under any other title than that of "Principles of Geology" '. A handful of reviewers complained that Lamarck was an out-dated straw man and that Lyell would have been more candid had he chosen a more credible opponent. The *Literary Gazette*, which had even praised Hope's *Essay* for advocating evolution, pointed out that the real debate in 1832 centred on continental discoveries about monstrosities and embryology. Analyzing the sophisticated doctrine of the unity of type in the writings of Geoffroy Saint Hilaire, Friedrich Tiedemann, Etienne Serres and Bory de Saint Vincent would have prevented Lyell from making crude and unphilosophical jokes about savages and orang-utans. 'It is always much more easy to ridicule a theory,' the reviewer noted, 'than to prove its soundness.' Even the *Spectator*, for all its enthusiasm, thought that Lyell had been successful only in undermining Lamarck's evidence – not evolution itself. But most critics, including Whewell and Scrope in the *Quarterly*, felt that Lyell had successfully parried Lamarck. The middle-class weekly *Chambers's Edinburgh Journal* was one of many mass-circulation periodicals that quarried criticisms of evolution from the *Principles* over the next three decades.[56] The discussion of Lamarck in the *Principles* – with its elegant rhetoric, strategic silences and wealth of evidence – was frequently praised as a model of clear argument.

Yet there was another side to Lyell's response that remained unknown until his private notebooks on species were published in 1970. These extraordinary documents reveal an author appalled by the religious, political and ethical consequences of evolution. If evolution was true, Lyell believed, no divinely implanted reason, spirit or soul would set human beings apart; they would be nothing but an improved form of the apes that he watched, fascinated, at the newly opened London Zoo. Evolution was a dirty, disgusting doctrine, which raised fears of miscegenation and sexual corruption. Not only did evolution repel Lyell's highly refined aesthetic sense, it undermined his lofty conception of science as the search for laws governing a perfectly adapted divine creation. With humans no more than better beasts and religion exposed as a fable, the foundations of civil society would crumble, just as they had done in revolutionary France. As Lyell admitted to Darwin many years later, it was Lamarck's conclusion 'about man that fortified me thirty years ago against the great impression which his arguments at first made upon my mind'.[57] Historians have shown that these secret fears shaped not only the explicit attacks on evolution in Lyell's second volume, but his entire geological system.

Lyell read Lamarck in 1827, and in a letter to the surgeon Mantell compared the experience with the pleasures of light fiction. 'His theories delighted me more than any novel I ever read,' he told his friend, 'and much in the same way, for they address themselves to the imagination, at least of geologists who know the mighty inferences which would be deducible were they established by observations.'[58] This has sometimes been seen as a positive reaction, but in the light of contemporary scepticism about novel-reading it was about as damning a comment as a man of science could make. Lamarck addressed the imagination rather than the reason and relied on invention rather than observation. Lyell admired 'even his flights' but thought him wrong.

The shock of reading Lamarck led Lyell to take up some of his most controversial positions. It was for this reason that he abandoned Buckland's grand narrative of progress, for progress implied a regular gradation of organisms in time, with all that meant for the status of human reason and the soul. Instead, Lyell stressed the geographical causes of climate change and the piecemeal nature of the geological record. These, as we have seen, became key tenets of the *Principles*. Confronting Lamarck also led Lyell to put the human species outside the flux of natural causation, a position many thought was strangely inconsistent with his views on natural law. Even the humble molluscan fossils of Etna were pressed into his anti-evolutionary crusade. As he wrote from Paris in 1830 as revolutionary crowds once again thronged the streets, if species of shells remained unchanged after thousands of years, it 'must therefore have required a good time for Ourang-Outangs to become men on Lamarckian principles.'[59]

This profound distaste for bestial origins remained deeply private, a matter of personal faith inappropriate for airing in the public world of quarterly reviews, scientific meetings or London clubs. Lyell thus concealed his fears behind a public stance that all recognized as scientific. On such grounds, he could mount a credible case against the widespread assumption of a progressive history of life and reject Lamarckian evolution as an imaginative fantasy. Contemporary readers, even Lyell's closest friends, never learned of the pervasive significance of the status of the human species for his science.

V

In the context of the Regency the *Principles* was a Trojan horse. It had the classically balanced prose, learned references and gentlemanly form of publication to shift the terms of debate among those whom Lyell recognized as the holders of power. Opposing divine intervention in nature, combating the Flood and hammering home the consensus of geologists on the antiquity of the world, the *Principles* could have been seen as allied to Enlightenment freethought, and suffered a fate like that of Milman's *History of the Jews*. But the reaction was overwhelmingly positive. The cautious rhetoric, the clever demolition of Lamarck, and the sheer empirical exuberance of the *Principles* made it possible to revitalize a lawful system based on actual causes, to exorcize the demons of revolutionary deism which had dogged Huttonianism since the 1790s. Only someone with impeccable credentials as a gentleman, with all the independence of character and sense of propriety that being a gentleman entailed, could have used such an unlikely vehicle to advocate controversial views. The *Principles* had the imprint of conservative classicism, but hid within a secret army of reform.

For the next fifty years the *Principles* held its place as a standard work, one of the key titles on Murray's list. Henry Adams, Robert Chambers, Charles Dickens, George Eliot, Edward Fitzgerald, John Stuart Mill, John Ruskin and almost all the other leading authors of the Victorian period in Europe and North America read the work. Opinions on its literary merits varied: Thomas Carlyle dismissed Lyell as 'a twaddling circumfused *ill*-writing man', while Harriet Martineau praised his 'expanding liberality of opinion and freedom of speech'.[60] Numerous editions appeared in the United States, particularly after D. Appleton & Co. took over publication there from the 1850s. However, the *Principles* never made much of an impact in continental Europe, presumably because of the very different theological and political setting of the sciences – although French and German translations were eventually issued.

In terms of Lyell's ambitions to establish himself as a professional author the *Principles* did indeed 'prove an annuity', with the ten Murray editions after the first crucial to his income. Darwin, having an independent fortune, could afford to wonder why his friend kept on revising to keep the book up to date, but Lyell could not be so sanguine. Almost everything he and his wife Mary Lyell did during the rest of their lives was related to the *Principles*, including trips to Scandinavia, Iceland, Germany, France and America. Tours in the United States and Canada (where the *Principles* was even more widely read than in Britain) resulted in two major travel books. Lyell also published in periodicals ranging from the *Philosophical Transactions* and the *Philosophical Magazine* to *The Times* and the *Athenaeum*.

Among the next generation of naturalists, Lyell established frames of reference that took scientific writing in new directions. Darwin became his most enthusiastic advocate, and found the *Principles* immensely liberating from the moment of landing on his first tropical island during the *Beagle* voyage. The young naturalist used the book to develop his own causally-oriented style of interpretation, comparing the uplifted but scarcely disrupted rocks in the mid-Atlantic Cape Verde Islands with the Temple of Serapis. Later in the voyage he reinterpreted the origin of coral reefs, overturning the model that Lyell had advocated; but the style of reasoning was that of the master, who altered his next edition accordingly.[61] When Darwin began to consider the possibility of species evolution after his return home, he did so in private dialogue with the *Principles*.

Not only did Darwin build his approach to species around the book, but so too did the generation of naturalists who came of age in the 1840s, including Asa Gray, T. H. Huxley, Joseph Hooker, John Lubbock, Andrew Ramsay, Herbert Spencer and Alfred Russel Wallace. By 1863 a disgruntled friend could complain that young geologists were 'imbibing like pap' Lyell's 'inconceivable nonsense'.[62] Almost no one accepted all the specific doctrines in the *Principles*, but the questions Lyell asked set the agenda for the evolutionary debate after 1859. The triumph of evolution (ironically) secured his place in the pantheon. As Darwin said in the *Origin*, 'He who can read Sir Charles Lyell's grand work on the Principles of Geology, which the future historian will recognize as having produced a revolution in natural science, yet does not admit how incomprehensibly vast have been the past periods of time, may at once close this volume.'[63] At the end of the century the *Principles* was recast as a textbook for the science degrees that had become available in the new universities. Taken out of its original context in Regency reform and post-Revolutionary paranoia, it became the work of a hero of late Victorian scientific naturalism.

The transition to secular sainthood proved acutely painful for Lyell. Throughout the 1840s and 1850s he fought against the assumption of progress in the fossil record, driven by his secret obsession with Lamarck and the status of the human species. Although he could point to discoveries that buttressed his argument that all the major fossil groups might be found from the earliest strata onwards, this became a rearguard action and he convinced only a few – notably the young Huxley. In 1850 and 1851 Lyell delivered his final case against progression from the presidential chair of the Geological Society. The tenth edition of the *Principles*, written in the wake of the *Origin* and published in the 1860s, finally abandoned the idea that mammals might be found in the oldest fossil-bearing rocks. Chapters once devoted to attacking the idea of progress were turned around to support it. Moreover, progress was tied to evolution, and although Lyell remained more cautious than the Darwinians would have liked, this still represented a remarkable change of mind. He used his influence to gain natural selection a fair hearing, not least by communicating the famous joint paper of Darwin and Wallace to the Linnean Society in 1858. 'When I came to the conclusion that after all Lamarck was going to be shown to be right,' he confessed to Darwin, 'that we must "go the whole orang", I re-read his book, and remembering when it was written, I felt I had done him injustice.'[64]

The change of mind was all the more striking in that Lyell had staked his personal faith on the primacy of reason. Implanted by God, reason had provided the moral basis for setting the human species apart. Reason found its apotheosis in the pursuit of science; but the highest and best science – Darwin's theory of natural selection – had now suggested that humans were descended from the animals. As younger naturalists looked set to adopt evolution as the new banner of scientific truth, Lyell pondered the implications in his private notebooks:

November 1, 1858. If the geologist dwelling exclusively on one class of facts, which might be paralleled by the existing creation arriving at conclusions derogating from the elevated position previously assigned by him to Man, if he blends him inseparably with the inferior animals & considers him as belonging to the earth solely, & as doomed to pass away like them & have no farther any relation to the living world, he may feel dissatisfied with his labours & doubt whether he would not have been happier had he never entered upon them & whether he ought to impart the result to others.[65]

As Lyell later said, 'it cost me a struggle to renounce my old creed.' He could follow Darwin's reasoning, but his 'sentiments and imagination' revolted against removing man from the exalted position in which the seventeenth-century philosopher Pascal had placed him as 'the archangel ruined'.[66]

Lyell defined the role of geology in the 'crisis of faith' for many Victorians precisely because the *Principles* held traces of its author's private anguish over the consequences of science for the place of humanity in nature. He had always sensed the ambiguities involved in the quest for knowledge. As a public moralist, he tempered his self-appointed role as an advocate of truth with a keen sense of responsibility about what such a role should entail. Balancing reason with feeling, science with silence, his book became the most eloquent exploration of the significance of the pre-human past.

By the mid-century almost any invocation of mundane natural forces acting over time was identified with the *Principles*. Should man, Tennyson asked in *In Memoriam*,

> Who trusted God was love indeed
> And love Creation's final law –
> Tho' Nature, red in tooth and claw
> With ravine, shriek'd against his creed –
>
> Who loved, who suffer'd countless ills,
> Who battled for the True, the Just,
> Be blown about the desert dust,
> Or seal'd within the iron hills?

In the first of these stanzas, the laws of morality and religion appear irrelevant against the macabre panoramas of the past, the nightmare of 'Dragons of the prime,/ That tare each other in their slime' revealed by the geology of Buckland, Conybeare and Mantell. But the second stanza – with its bleak Lyellian landscape of wind, sand and rain – is ultimately even more terrifying. Lyell almost inadvertently moved the human observer into what Isobel Armstrong has called 'a world of non sequitur', a continual state of repositioning and flux in which nothing could be completely known or controlled.[67]

> 'So careful of the type?' but no.
> From scarped cliff and quarried stone
> She cries, 'A thousand types are gone:
> I care for nothing, all shall go.'[68]

The vast ages called up by science dwarfed all human activity, so that the vocation of the poet became meaningless and the fame of even the greatest writings of antiquity seemed ephemeral. Tennyson's associate Edward Fitzgerald wrote in a letter of 1847:

Yes, as I often think, it is not the poetical imagination, but bare Science that every day more and more unrolls a greater Epic than the Iliad – the history of the World, the infinitudes of Space and Time! I never take up a book of Geology or Astronomy but this strikes me ... So that, as Lyell says, the Geologist looking at Niagara forgets even the roar of its waters in the contemplation of the awful processes of time that it suggests. It is not only that this vision of Time must wither the Poet's hope of immortality – but it is in itself more wonderful than all the conceptions of Dante and Milton.

The man of science would outrun the poet, 'distance all his imaginations, dissolve the language in which they are uttered.'[69]

Towards the end of his life, Lyell's de-centred vision became so associated with Darwinism that his reluctance to apply his law-bound system to humanity was often forgotten. George Eliot, who drew on the *Principles* in her novels, wondered in a late notebook whether Lyellian geology really could serve as a key to 'the interpretation of man's past life on earth'. Was there, she wrote, 'something incalculable by us from the data of our present experience? Even within comparatively recent times & in kindred communities how many conceptions & fashions of life have existed to which our understanding & sympathy has no clue!'[70] Lyell, of course, always wanted the answer to be clear: the principles of geology do not apply, human history is different, reason sets us apart. But he never could be sure, and the questions that Eliot and Tennyson asked had haunted him throughout his life. In 1860, the year after Darwin's *Origin*, Eliot concluded *The Mill on the Floss* with a great flood that sweeps away the novel's two main characters. The narrator reflects on how nature 'repairs her ravages', so that new trees replace the old and torn hills are covered with spring grass. But below the surface are 'marks of past rending', so that nothing is ever the same. The sense of human loss maintained through memory here connects with Lyell's understanding that contingency, change and dislocation must frame our vision of the order of nature. As Eliot's narrator says, 'To the eyes that have dwelt on the past, there is no thorough repair.'

NOTES

1. Darwin to L. Horner, 29 Aug. [1844], in F. Burkhardt and S. Smith (eds), *Correspondence of Charles Darwin* (1987), vol. 3, p. 55.

2. Gould, *Time's Arrow, Time's Cycle* (1987), p. 179.

3. Lyell to R. I. Murchison, 17 Jan. 1829, in L. G. Wilson, *Charles Lyell: The Years to 1841: The Revolution in Geology* (1972), p. 256, where the text is quoted from the original manuscript.

4. Southwell, 'A View from Bristol Gaol', *Oracle of Reason* 1 (19 Feb. 1842), pp. 78–9.

5. See Bibliography of Reviews.

6. Lyell to W. Whewell, 20 Feb. 1831, in Wilson, *Charles Lyell*, p. 308.

7. A. Desmond, *Politics of Evolution* (1989), p. 328.

8. Wilson, *Charles Lyell*, pp. 320–22.

9. Lyell to M. Horner, 17 Nov. 1831, in K. M. Lyell (ed.), *Life Letters and Journals of Sir Charles Lyell, Bart.*, 2 vols (1881), vol. 1, pp. 352–3; henceforth referred to as *LLJ*.

10. Lyell to Scrope, 14 June 1830, *LLJ*, vol. 1, p. 271.

11. Bennett, 'John Murray's Family Library and the Cheapening of Books in Early Nineteenth Century Britain', *Studies in Bibliography* 29 (1976), p. 141.

12. Lyell to M. Horner, 17 Feb. 1832, in Wilson, *Charles Lyell*, p. 344.

13. D. Allen, *The Naturalist in Britain* (1976); A. Secord, 'Corresponding Interests: Artisans and Gentlemen in Nineteenth-century Natural History', *British Journal for the History of Science* 27 (1994), pp. 383–408.

14. Lyell, *A Second Visit to the United States of America*, 2 vols (1850), vol. 2, p. 317; R. H. Dott, 'Lyell in America', *Earth Sciences History* 15 (1996), p. 115.

15. *Spectator*, 14 Jan. 1832, p. 39.

16. R. Laudan, *From Mineralogy to Geology* (1987), pp. 201–21; *id.*, 'The Role of Methodology in Lyell's Science', *Studies in History and Philosophy of Science* 13 (1982), pp. 215–49.

17. Sedgwick, 'Address to the Geological Society', *Proceedings of the Geological Society of London* 1 (1831), p. 307.

18. Lyell to Mantell, 15 Feb. 1830, *LLJ*, vol. 1, p. 262.

19. M. Rudwick, 'Caricature as a Source for the History of Science: De la Beche's Anti-Lyellian Sketches of 1831', *Isis* (1975), vol. 66, pp. 534–60. On cycles, see especially D. Ospovat, 'Lyell's Theory of Climate', *Journal of the History of Biology* 10 (1977), pp. 317–39.

20. Gould, *Time's Arrow, Time's Cycle* (1987), pp. 99–179. I should emphasize that my claim here concerns only Lyell's strategy in the text of the *Principles*, not private comments in letters or in reviews by his associates.

21. *LLJ*, vol. 1, p. 377.

22. D. Stewart, *Elements of the Philosophy of the Human Mind*, ch. 3 (p. 30 in 1850 ed.).

23. *Athenaeum*, 6 Dec. 1834, p. 881; [G. Maclaren], *Scotsman*, 25 Sept. 1830, p. 1.

24. M. Rudwick, 'Charles Lyell's Dream of a Statistical Palaeontology', *Palaeontology* (1978), vol. 21, pp. 225–44. The best way to appreciate the remarkable diversity of Lyell's sources is through Rudwick's 'Bibliography of Lyell's Sources', in the University of Chicago Press facsimile reprint of the first edition of the *Principles* (3 vols, 1990–91), vol. 3, pp. 113–60.

25. Lyell to R. Murchison, 12 Jan. 1829, *LLJ*, vol. 1, p. 232.

26. D. Dean, *James Hutton and the History of Geology* (1992), p. 239.

27. Lyell to C. Lyell, Sr, 31 Oct. 1816, *LLJ*, vol. 1, pp. 38–9; P. Corsi, 'The Heritage of Dugald Stewart: Oxford Philosophy and the Method of Political Economy', *Nuncius* 2 (1987), pp. 89–144; A. M. C. Waterman, *Revolution, Economics and Religion: Christian Political Economy, 1798–1833* (1991).

28. Herschel to Lyell, 20 Feb. 1836, in W. F. Cannon, 'The Impact of Uniformitarianism', *Proceedings of the American Philosophical Society* 105 (1961), p. 305.

29. W. Whewell, *History of the Inductive Sciences*, 3 vols (1837, 3rd ed. 1857), vol. 3, p. 451.

30. *1832 British Association Report*, p. 406; *Monthly Magazine* n.s. 10 (Dec. 1830), p. 701; *Athenaeum*, 25 Sept. 1830, p. 595; *Literary Gazette*, 14 Aug. 1830, p. 526; *New Monthly Magazine*, 1 June 1832, p. 241.

31. See, for example, M. Bartholomew, 'The Singularity of Lyell', *History of Science* 17 (1979), pp. 276–93; M. T. Greene, *Geology in the Nineteenth Century* (1982); N. Rupke, *The Great Chain of History* (1983), p. 190.

32. Lyell to G. Ticknor, 1850, *LLJ*, vol. 2, pp. 168–9.

33. Sharon Turner, *The Sacred History of the World, Attempted to be Philosophically Considered, in a Series of Letters to a Son*, 3 vols (1848, 8th ed.), vol. 1, p. x.

34. Lyell to Scrope, 14 June 1830, *LLJ*, vol. 1, p. 268.

35. Rupke, *The Great Chain of History*.

36. Lyell to Scrope, 14 June 1830, *LLJ*, vol. 1, p. 271; Lyell to Mantell, 29 Dec. 1827, *LLJ*, vol. 1, p. 174.

37. *Principles*, vol. 3 (1833), pp. x–xi. See M. J. S. Rudwick, 'Lyell on Etna and the Antiquity of the Earth', in C. Schneer (ed.), *Toward a History of Geology* (1969), pp. 288–304.

38. McPhee, *Basin and Range* (1981).

39. Lyell to Scrope, 14 June 1830, *LLJ*, vol. 1, p. 269.

40. Lyell to Mantell, 29 Dec. 1827, *LLJ*, vol. 1, pp. 173–4.

41. Lyell to E. Lyell, 26 Feb. 1830, *LLJ*, vol. 1, p. 263.

42. Hall to L. Horner, 7 Sept. 1833, *LLJ*, vol. 2, p. 466.

43. Lyell to E. Lyell, 11 May 1830, *LLJ*, vol. 1, p. 267.

44. Quoted in D. Dean, ' "Through Science to Despair": Geology and the Victorians', in J. Paradis and T. Postlewait (eds), *Victorian Science and Victorian Values* (1985), p. 115.

45. *Metropolitan Magazine* 10 (May 1834): 'Literature', pp. 4–5.

46. *British Critic* 15 (1834), p. 363; *New Monthly Magazine*, 1 June 1832, p. 241; *Presbyterian Review and Religious Journal* 2 (1832), p. 339; *Eclectic Review* n.s. 6 (1831), p. 79.

47. *Proceedings of the Geological Society of London* 1 (1831), p. 313.

48. A. Desmond, *Politics of Evolution* (1989); J. A. Secord, 'Edinburgh Lamarckians: Robert Jameson and Robert E. Grant', *Journal of the History of Biology* 24 (1991), pp. 1–18.

49. *Monthly Review* (Mar. 1832), p. 353.

50. Davy, *Consolations in Travel* (1830), pp. 149–50.

51. Carlyle to M. Napier, 8 Oct. 1831, in C. R. Sanders and K. J. Fielding (eds), *The Collected Letters of Thomas and Jane Welsh Carlyle* (1977), vol. 6, p. 13.

52. *Proceedings of the Geological Society of London* 1 (1831), p. 305.

53. Lyell to Copleston, 28 Mar. 1831, in Wilson, *Charles Lyell*, p. 310; for more on the incident, see M. J. S. Rudwick, 'Charles Lyell, F.R.S. (1797–1875) and his London Lectures on Geology, 1832–33', *Notes and Records of the Royal Society of London* 29 (1975), pp. 231–63.

54. Lyell to Herschel, 1 June 1836, *LLJ*, vol. 1, p. 467; Lyell to Copleston, 28 Mar. 1831, in Wilson, *Charles Lyell*, p. 310.

55. Jameson to J. Murray, 18 Feb. 1832, in J. Browne, *Charles Darwin: Voyaging* (1995), p. 188.

56. *Gentleman's Magazine* 102 (1832), pt 1, p. 45; *Spectator*, 14 Jan. 1832, p. 39; *Literary Gazette*, 28 Jan. 1832, p. 49; 'Popular Information on Science. Transmutation of Species', *Chambers's Edinburgh Journal* 26 (Sept. 1835), pp. 273–4.

57. Lyell to Darwin, 15 Mar. 1863, *LLJ*, vol. 2, p. 365.

58. Lyell to Mantell, 2 Mar. 1827, *LLJ*, vol. 1, p. 163.

59. Lyell to C. Lyell, 19 Oct. 1830, *LLJ*, vol. 1, p. 308.

60. Carlyle to J. W. Carlyle, 11 July 1843, in C. de L. Ryals and K. J. Fielding (eds), *The Collected Letters of Thomas and Jane Welsh Carlyle* (1990), vol. 16, p. 260; H. Martineau, *Autobiography*, 2 vols (1877), vol. 1, 355.

61. D. R. Stoddart, 'Darwin, Lyell, and the Geological Significance of Coral Reefs', *British Journal for the History of Science* 9 (1976), pp. 199–218; J. A. Secord, 'The Discovery of a Vocation: Darwin's Early Geology', *British Journal for the History of Science* 24 (1991), pp. 133–57.

62. Murchison to H. Reeve, 29 June 1863, in L. Page, 'The Rivalry between Charles Lyell and Roderick Murchison', *British Journal for the History of Science* 9 (1976), p. 162.

63. J. Burrow (ed.), *Origin of Species* (1968), pp. 293.

64. Lyell to Darwin, 15 Mar. 1863, *LLJ*, vol. 2, p. 365.

65. L. G. Wilson (ed.), *Sir Charles Lyell's Scientific Journals on the Species Question* (1970), p. 196.

66. Lyell to T. S. Spedding, 19 May 1863, *LLJ*, vol. 2, p. 376; see M. Bartholomew, 'Lyell and Evolution', *British Journal for the History of Science* 6 (1973), pp. 261–303.

67. I. Armstrong, *Victorian Poetry: Poetry, Poetics and Politics* (1993), p. 262.

68. *Poems of Tennyson*, ed. C. Ricks, 3 vols (2d ed., 1987), vol. 2.

69. Fitzgerald to E. B. Cowell, [24 July 1847], in A. M. Terhune and A. B. Terhune (eds), *The Letters of Edward Fitzgerald* (1980), vol. 1, p. 566.

70. Quoted in Jonathan Smith, *Fact and Feeling: Baconian Science and the Nineteenth-century Literary Imagination* (1994), p. 150.

Further Reading

For those who wish to explore the *Principles* in more detail, a facsimile reprint (1990–91) of the first edition of the three-volume original is conveniently available from the University of Chicago Press. This includes an important introduction by Martin Rudwick explicating the strategy of the book, together with a comprehensive list of recent literature.

Stephen Jay Gould, *Time's Arrow, Time's Cycle* (Cambridge, MA: Harvard University Press, 1987) discusses the *Principles* in an engaging history of ideas about time. Relevant theological issues are dealt with by John Brooke in his *Science and Religion: Some Historical Perspectives* (Cambridge: Cambridge University Press, 1991). Rachel Laudan, in *From Mineralogy to Geology: The Foundations of a Science, 1650–1830* (Chicago: Chicago University Press, 1987) analyses the philosophical bases of Lyell's project.

The best analysis of the context in natural history and geology is Martin Rudwick's *The Meaning of Fossils: Episodes in the History of Palaeontology* (New York: Science History, 1976); see also several of the essays in N. Jardine, J. Secord and E. C. Spary (eds), *Cultures of Natural History* (Cambridge: Cambridge University Press, 1996).

A generous selection of Lyell's correspondence is published in Katherine M. Lyell (ed.), *Life Letters and Journals of Sir Charles Lyell, Bart.* (London: John Murray, 1881); this is supplemented by Leonard G. Wilson, *Charles Lyell: The Years to 1841: The Revolution in Geology* (New Haven: Yale University Press, 1972). A spirited account of Lyell's dilemmas in self-presentation is offered by Roy Porter in 'Charles Lyell: The Public and Private Faces of Science', *Janus* 69 (1982), pp. 29–50. Michael Bartholomew, 'Lyell and Evolution', *British Journal for the History of Science* 6 (1973), pp. 261–303 is an influential account of Lyell's tortuous musings on human origins.

Study of the reception of the *Principles* has been distorted by its significance for Charles Darwin. There is a sensitive account of the Lyell/Darwin relationship in Janet Browne, *Charles Darwin: Voyaging* (London:

Jonathan Cape, 1995). Several of the essays in the Lyell Centenary issue of the *British Journal for the History of Science* 9, pt 2 (1976), pp. 89–242 focus on debates sparked off by the *Principles*. Jonathan Smith, *Fact and Feeling: Baconian Science and the Nineteenth-century Literary Imagination* (Madison, WI: University of Wisconsin Press, 1994) has good chapters on Darwin and George Eliot as readers of the *Principles*. The reviews listed at the end of the present edition reveal the remarkable range of responses in contemporary periodicals.

A Note on this Edition

This selected text is from the first edition of Charles Lyell's *Principles of Geology: Being an Attempt to Explain the Former Changes of the Earth's Surface, by Reference to Causes now in Operation*, published in London by John Murray, 1830–33. When the first volume appeared in July 1830, the title page announced that a single further volume would complete the work, but additions (especially to the evolutionary chapters) led to a second volume being issued in January 1832 and a third in April 1833. The books sold well, so that the first two volumes were already in second editions by the time the third volume was published.

Widely recognized as an engaging, accessible and readable work, the sheer length of the *Principles* has challenged the stamina of all but the most persistent modern readers. This abridgement aims to provide an accurate text of the most significant parts of the book. While retaining only slightly more than one-third of the original, it has proved possible to include the full text of virtually all those chapters that have attracted discussion and debate. These include the core theoretical chapters from the first volume, the influential analyses of evolutionary theory and ecological issues from the second, and the novel techniques for reconstructing the past introduced in the third.

Throughout, the aim has been to retain complete chapters and continuous chains of argument, rather than scattered extracts. Thus the main text of all the chapters in this edition is complete and unabridged (with all illustrations) unless otherwise indicated. Those few places where a paragraph or a few pages within a chapter have been omitted are marked by three asterisks. Lyell's inconsistent use of quotation marks has been regularized, and species names in the first volume have been italicized in accordance with his own policy in the rest of the work; otherwise his original spelling and punctuation have been retained.

Much of the argument of the *Principles* depends upon the cumulative force of examples, and many chapters devoted to material of this kind have had to be omitted. The overall effect is to underline the theoretical, speculative aspects of the book. Wherever possible, however, at least one characteristic empirical example from each part of the work is retained in full, usually in an unabridged chapter. Thus Lyell's celebrated account of geology in the late eighteenth century stands in for his history of the science from antiquity; his description of Niagara Falls exemplifies the power of aqueous processes; the case of Etna shows how the Tertiary period is analysed. The contents of all missing chapters are summarized in bracketed, italicized editorial passages – so that the argument of the book as a whole can be kept in mind.

Lyell's chapter and figure numbers have been retained, to facilitate comparison with the three volumes of the first edition. The chapters in the first edition had no titles, and were headed instead by lengthy outlines of subjects. The titles used here are adapted from the third edition of 1834, where they appeared for the first time. That edition features different versions of the titles in its table of contents and in the main body of the work; I have chosen those that in my view best express the contents of each particular chapter.

To include as much of Lyell's main text as possible, most of his original scholarly apparatus has had to be omitted. This material includes prefaces, indices and tables of contents, as well as a sixty-page table of fossil shells. Lyell's glossary is included in an abridged form.

Like Coleridge, Lyell often incorporated long passages from other authors with minimal acknowledgement, and his citations are often erratic. Most of his extensive, incomplete and often inaccurate footnotes have been silently deleted from this edition. Those few that have been retained are marked by a single asterisk and still treated as footnotes. Readers interested in Lyell's sources should consult the complete text, and in particular Martin Rudwick's helpful bibliography of works cited by Lyell, appended to the third volume of the University of Chicago Press reprint.

For the present edition, it seemed more useful to add clear references (in the form of editorial end-notes) for all the passages marked by Lyell as quotations. Notably, most of his citations of poetry were unattributed (and in the case of Latin and Italian, untranslated), which has hitherto obscured the significance of his frequent references to classic works of literature.

PRINCIPLES

OF

GEOLOGY,

BEING

AN ATTEMPT TO EXPLAIN THE FORMER CHANGES
OF THE EARTH'S SURFACE,

BY REFERENCE TO CAUSES NOW IN OPERATION.

BY

CHARLES LYELL, Esq., F.R.S.

FOR. SEC. TO THE GEOL. SOC., &c.

IN TWO VOLUMES.

VOL. I.

LONDON:

JOHN MURRAY, ALBEMARLE-STREET.

MDCCCXXX.

Present state of the Temple of Serapis at Puzzuoli

VOLUME I

'Amid all the revolutions of the globe the economy of Nature has been uniform, and her laws are the only things that have resisted the general movement. The rivers and the rocks, the seas and the continents have been changed in all their parts; but the laws which direct those changes, and the rules to which they are subject, have remained invariably the same.'

Playfair, *Illustrations of the Huttonian Theory*, § 374[1]

CHAPTER I

Objects and Nature of Geology

Geology is the science which investigates the successive changes that have taken place in the organic and inorganic kingdoms of nature; it enquires into the causes of these changes, and the influence which they have exerted in modifying the surface and external structure of our planet.

By these researches into the state of the earth and its inhabitants at former periods, we acquire a more perfect knowledge of its *present* condition, and more comprehensive views concerning the laws *now* governing its animate and inanimate productions. When we study history, we obtain a more profound insight into human nature, by instituting a comparison between the present and former states of society. We trace the long series of events which have gradually led to the actual posture of affairs; and by connecting effects with their causes, we are enabled to classify and retain in the memory a multitude of complicated relations – the various peculiarities of national character – the different degrees of moral and intellectual refinement, and numerous other circumstances, which, without historical associations, would be uninteresting or imperfectly understood. As the present condition of nations is the result of many antecedent changes, some extremely remote and others recent, some gradual, others sudden and violent, so the state of the natural world is the result of a long succession of events, and if we would enlarge our experience of the present economy of nature, we must investigate the effects of her operations in former epochs.

We often discover with surprise, on looking back into the chronicles of nations, how the fortune of some battle has influenced the fate of millions of our contemporaries, when it has long been forgotten by the mass of the population. With this remote event we may find inseparably connected the geographical boundaries of a great state, the language now spoken by the inhabitants, their peculiar manners, laws, and religious opinions. But far more astonishing and unexpected are the connexions brought to light, when we carry back our researches into the history of nature. The form of a coast, the configuration of the interior of a country, the existence and extent of lakes, valleys, and mountains, can often be traced to the former prevalence of earthquakes and volcanoes, in regions which have long been undisturbed. To these remote convulsions the present fertility of some districts, the sterile character of others, the elevation of land above the sea, the climate, and various peculiarities, may be distinctly referred. On the other hand, many distinguishing features of the surface may often be ascribed to the operation at a remote era of slow and tranquil causes – to the gradual deposition of sediment in a lake or in the ocean, or to the prolific growth in the same of corals and testacea. To select another example, we find in certain localities subterranean deposits of coal, consisting of vegetable matter, formerly drifted into seas and lakes. These seas and lakes have since been filled up, the lands whereon the forests grew have disappeared or changed their form, the rivers and currents which floated the vegetable masses can no longer be traced, and the plants belonged to species which for ages have passed away from the surface of our planet. Yet the commercial prosperity, and numerical strength of a nation, may now be mainly dependant on the local distribution of fuel determined by that ancient state of things.

Geology is intimately related to almost all the physical sciences, as is history to the moral. An historian should, if possible, be at once profoundly acquainted with ethics, politics, jurisprudence, the military art, theology; in a word, with all branches of knowledge, whereby any insight into human affairs, or into the moral and intellectual nature of man, can be obtained. It would be no less desirable that a geologist should be well versed in chemistry, natural philosophy, mineralogy, zoology, comparative anatomy, botany; in short, in every science relating to organic and inorganic nature. With these accomplishments the historian and geologist would rarely fail to draw correct and philosophical conclusions from the various monuments transmitted to them of former occurrences. They would know to what combination of causes analogous effects were referrible, and they would often be enabled to supply by inference, information concerning many events unrecorded in the defective archives of former ages. But the brief duration of human life, and our limited powers, are so far from permitting us to aspire to such extensive acquisitions, that excellence even in one department is within the reach of few, and those individuals most effectually promote the general progress, who concentrate their thoughts on a limited portion of the field of inquiry. As it is necessary that the historian and the cultivators of moral or political science should reciprocally aid each other, so the geologist and those who study natural history or physics stand in equal need of mutual assistance. A comparative anatomist may derive some accession of knowledge from the bare inspection of the remains of an extinct quadruped, but the relic throws much greater light upon his own science, when he is informed to what relative era it belonged, what plants and animals were its contemporaries, in what degree of latitude it once existed, and other historical details. A fossil shell may interest a conchologist, though he be ignorant of the locality from which it came; but it will be of more value when he learns with what other species it was associated, whether they were marine or fresh-water, whether the strata containing them were at a certain elevation above the sea, and what relative position they held in regard to other groups of strata, with many other particulars determinable by an experienced geologist alone. On the other hand, the skill of the comparative anatomist and conchologist are often indispensable to those engaged in geological research, although it will rarely happen that the geologist will himself combine these different qualifications in his own person.

Some remains of former organic beings, like the ancient temple, statue, or picture, may have both their intrinsic and their historical value, while there are others which can never be expected to attract attention for their own sake. A painter, sculptor, or architect, would often neglect many curious relics of antiquity, as devoid of beauty and uninstructive with relation to their own art, however illustrative of the progress of refinement in some ancient nation. It has therefore been found desirable that the antiquary should unite his labours to those of the historian, and similar co-operation has become necessary in geology. The field of inquiry in living nature being inexhaustible, the zoologist and botanist can rarely be induced to sacrifice time in exploring the imperfect remains of lost species of animals and plants, while those still existing afford constant matter of novelty. They must entertain a desire of promoting *geology* by such investigations, and some knowledge of its objects must guide and direct their studies. According to the different opportunities, tastes, and talents of individuals, they may employ themselves in collecting particular kinds of minerals, rocks, or organic remains, and these, when well examined and explained, afford data to the geologist, as do coins, medals, and inscriptions to the historian.

It was long ere the distinct nature and legitimate objects of geology were fully recognized, and it was at first confounded with many other branches of inquiry, just as the limits of history, poetry, and mythology were ill-defined in the infancy of civilization. Werner appears to have regarded geology as little other than a subordinate department of mineralogy, and Desmarest included it under the head of Physical Geography. But the identification of its objects with those of Cosmogony has been the most common and serious source of confusion. The first who endeavoured to draw a clear line of demarcation between these distinct departments, was Hutton, who declared that geology was in no ways concerned 'with questions as to the origin of things.'[2] But his doctrine on this head was vehemently opposed at first, and although it has gradually gained ground, and will ultimately prevail, it is yet far from being established. We shall attempt in the sequel of this work to demonstrate that geology differs as widely from cosmogony, as speculations concerning the creation of man differ from history. But before we enter more at large on this controverted question, we shall endeavour to trace the progress of opinion on this topic, from the earliest ages, to the commencement of the present century.

CHAPTERS 2–4

Historical Sketch of the Progress of Geology

[*Beginning with the cosmologies of ancient India and Egypt, Chapters 2 and 3 analyse ideas about the earth up to the final decades of the eighteenth century. This is presented as a history of error and absurdities. Cosmology, Lyell argues, has been confounded with geology; Scripture has been used as evidence in science; the earth and its productions (such as fossils) have been attributed to dramatic and even supernatural causes. Through this polemical history, the opponents of uniformity are implicitly identified as inheritors of a tradition that has 'retarded' progress in science. Chapter 4, included here, turns to more recent events. It deals with the generation preceding Lyell's own, focusing on the work of Abraham Gottlob Werner in Saxony and James Hutton in Scotland:*]

The art of mining has long been taught in France, Germany, and Hungary, in scientific institutions established for that purpose, where mineralogy has always been a principal branch of instruction.*

Werner was named, in 1775, professor of that science in the 'School of Mines' at Freyberg in Saxony. He directed his attention not merely to the composition and external characters of minerals, but also to what he termed 'geognosy,' or the natural position of minerals in particular rocks, together with the grouping of those rocks, their geographical distribution, and various relations. The phenomena observed in the structure of the globe had hitherto served for little else than to furnish interesting topics for philosophical discussion; but when Werner pointed out their application to the practical purposes of mining, they were instantly regarded by a large

---

* Our miners have been left to themselves, almost without the assistance of scientific works in the English language, and without any 'school of mines,' to blunder their own way into a certain degree of practical skill. The inconvenience of this want of system in a country where so much capital is expended, and often wasted, in mining adventures, has been well exposed by an eminent practical miner. – See 'Prospectus of a School of Mines in Cornwall, by J. Taylor, 1825.'

9

class of men as an essential part of their professional education, and from that time the science was cultivated in Europe more ardently and systematically. Werner's mind was at once imaginative and richly stored with miscellaneous knowledge. He associated everything with his favourite science, and in his excursive lectures he pointed out all the economical uses of minerals, and their application to medicine; the influence of the mineral composition of rocks upon the soil, and of the soil upon the resources, wealth, and civilization of man. The vast sandy plains of Tartary and Africa he would say retained their inhabitants in the shape of wandering shepherds; the granitic mountains and the low calcareous and alluvial plains gave rise to different manners, degrees of wealth and intelligence. The history even of languages, and the migrations of tribes had, according to him, been determined by the direction of particular strata. The qualities of certain stones used in building would lead him to descant on the architecture of different ages and nations, and the physical geography of a country frequently invited him to treat of military tactics. The charm of his manners and his eloquence kindled enthusiasm in the minds of all his pupils, many of whom only intended at first to acquire a slight knowledge of mineralogy; but, when they had once heard him, they devoted themselves to it as the business of their lives. In a few years a small school of mines, before unheard of in Europe, was raised to the rank of a great university, and men already distinguished in science studied the German language, and came from the most distant countries to hear the great oracle of geology.

Werner had a great antipathy to the mechanical labour of writing, and he could never be persuaded to pen more than a few brief memoirs, and those containing no development of his general views. Although the natural modesty of his disposition was excessive, approaching even to timidity, he indulged in the most bold and sweeping generalizations, and he inspired all his scholars with a most implicit faith in his doctrines. Their admiration of his genius, and the feelings of gratitude and friendship which they all felt for him, were not undeserved; but the supreme authority usurped by him over the opinions of his contemporaries, was eventually prejudicial to the progress of the science, so much so, as greatly to counterbalance the advantages which it derived from his exertions. If it be true that delivery be the first, second, and third requisite in a popular orator, it is no less certain that to travel is of threefold importance to those who desire to originate just and comprehensive views concerning the structure of our globe, and Werner had never travelled to distant countries. He had merely explored a small portion of Germany, and conceived, and persuaded others to believe, that the whole surface of our planet, and all the mountain chains in the world, were made after the model of his own province. It was a ruling object of ambition in the minds of his pupils to confirm the generalizations of their great master, and to discover in the most distant parts of the globe his 'universal formations,' which he supposed had been each in succession simultaneously precipitated over the whole earth from a common menstruum, or 'chaotic fluid.' Unfortunately, the limited district examined by the Saxon professor was no type of the world, nor even of Europe; and, what was still more deplorable, when the ingenuity of his scholars had tortured the phenomena of distant countries, and even of another hemisphere, into conformity with his theoretical standard, it was discovered that 'the master' had misinterpreted many of the appearances in the immediate neighbourhood of Freyberg.

Thus, for example, within a day's journey of his school, the porphyry, called by him primitive, has been found not only to send forth veins or dikes through strata of the coal formation, but to overlie them in mass. The granite of the Hartz mountains, on the other hand, which he supposed to be the nucleus of the chain, is now well known to traverse and breach the other beds, penetrating even into the plain (as near Goslar); and nearer Freyberg, in the Erzgebirge, the mica slate does not mantle round the granite, as the professor supposed, but abuts abruptly against it. But it is still more remarkable, that in the Hartz mountains all his flötz rocks, which he represented as horizontal, are highly inclined, and often nearly vertical, as the chalk at Goslar, and the green sand near Blankenberg.

The principal merit of Werner's system of instruction consisted in steadily directing the attention of his scholars to the constant relations of certain mineral groups, and their regular order of superposition. But he had been anticipated, as we have shewn in the last chapter, in the discovery of this general law, by several geologists in Italy and elsewhere; and his leading divisions of the secondary strata were at the same time made the basis of an arrangement of the British strata by our countryman, William Smith, to whose work we shall return by-and-by. In regard to basalt and other igneous rocks, Werner's theory was original, but it was also extremely erroneous. The basalts of Saxony and Hesse, to which his observations were chiefly confined, consisted of tabular masses capping the hills, and not connected with the levels of existing valleys, like many in Auvergne and the Vivarais. These basalts, and all other rocks of the same family in other countries, were, according to him, chemical precipitates from water. He denied that they were the products of submarine volcanos, and even taught that, in the primeval ages of the world, there were no volcanos. His theory was opposed, in a two-fold sense, to the doctrine of uniformity in the course of nature; for not only did he introduce, without scruple, many imaginary causes supposed to have once effected great revolutions in the earth, and then to have become extinct, but new ones also were feigned to have come into play in modern times; and, above all, that most violent instrument of change, the agency of subterranean fire. So early as 1768, before Werner had commenced his mineralogical studies, Raspe had truly characterized the basalts of Hesse as of igneous origin. Arduino, as we have already seen, had pointed out numerous varieties of trap-rock in the Vicentin, as analogous to volcanic products, and as distinctly referrible to ancient submarine eruptions. Desmarest, as we stated, had, in company with Fortis, examined the Vicentin in 1766, and confirmed Arduino's views. In 1772, Banks, Solander, and Troil, compared the columnar basalt of Hecla with that of the Hebrides. Collini, in 1774, recognised the true nature of the igneous rocks on the Rhine, between Andernach and Bonn. In 1775, Guettard visited the Vivarais, and established the relation of basaltic currents to lavas. Lastly, in 1779, Faujas published his description of the volcanos of the Vivarais and Velay, and shewed how the streams of basalt had poured out from craters which still remain in a perfect state.

When sound opinions had for twenty years prevailed in Europe concerning the true nature of the ancient trap-rocks, Werner by his dictum caused a retrograde movement, and not only overturned the true theory, but substituted for it one of the most unphilosophical ever advanced in any science. The continued ascendancy of his dogmas on this subject was the more astonishing, because a variety of new and striking facts were daily accumulated in favour of the correct opinions first established. Desmarest, after a careful examination of Auvergne, pointed out first the most recent volcanos which had their craters still entire, and their streams of lava conforming to the level of the present river-courses. He then shewed that there were others of an intermediate epoch, whose craters were nearly effaced, and whose lavas were less intimately connected with the present valleys; and, lastly, that there were volcanic rocks still more ancient, without any discernible craters or scoriæ, and bearing the closest analogy to rocks in other parts of Europe, the igneous origin of which was denied by the school of Freyberg.

Desmarest's map of Auvergne was a work of uncommon merit. He first made a trigonometrical survey of the district, and delineated its physical geography with minute accuracy and admirable graphic power. He contrived, at the same time, to express, without the aid of colours, a vast quantity of geological detail, the different ages, and sometimes even the structure of the volcanic rocks, distinguishing them from the fresh-water and the granitic. They alone who have carefully studied Auvergne, and traced the different lava streams from their craters to their termination, – the various isolated basaltic cappings, – the relation of some lavas to the present valleys, – the absence of such relations in others, – can appreciate the extraordinary fidelity of this elaborate work. No other district of equal dimensions in Europe exhibits, perhaps, so beautiful and varied a series of phenomena; and, fortunately, Desmarest possessed at once the mathematical knowledge required for the construction of a map, skill in mineralogy, and a power of original generalization.

Dolomieu, another of Werner's contemporaries, had found prismatic basalts among the ancient lavas of Etna, and in 1784 had observed the alternations of submarine and calcareous strata in the Val di Noto in Sicily. In 1790, he also described similar phenomena in the Vicentin and in the Tyrol. Montlosier also published, in 1788, an elegant and spirited essay on the volcanos of Auvergne, combining accurate local observations with comprehensive views. In opposition to this mass of evidence, the scholars of Werner were prepared to support his opinions to their utmost extent, maintaining in the fulness of their faith that even obsidian was an aqueous precipitate. As they were blinded by their veneration for the great teacher, they were impatient of opposition, and soon imbibed the spirit of a faction; and their opponents, the Vulcanists, were not long in becoming contaminated with the same intemperate zeal. Ridicule and irony were weapons more frequently employed than argument by the rival sects, till at last the controversy was carried on with a degree of bitterness, almost unprecedented in questions of physical science. Desmarest alone, who had long before provided ample materials for refuting such a theory, kept aloof from the strife, and whenever a zealous Neptunist wished to draw the old man into an argument, he was satisfied with replying, 'Go and see.'[3]

It would be contrary to all analogy, in matters of graver import, that a war should rage with such fury on the continent, and that the inhabitants of our island should not mingle in the affray. Although in England the personal influence of Werner was wanting to stimulate men to the defence of the weaker side of the question, they contrived to find good reason for espousing the Wernerian errors with great enthusiasm. In order to explain the peculiar motives which led many to enter, even with party feeling, into this contest, we must present the reader with a sketch of the views unfolded by Hutton, a contemporary of the Saxon geologist. That naturalist had been educated as a physician, but, declining the practice of medicine, he resolved, when young, to remain content with the small independence inherited from his father, and thenceforth to give his undivided attention to scientific pursuits. He resided at Edinburgh, where he enjoyed the society of many men of high attainments, who loved him for the simplicity of his manners and the sincerity of his character. His application was unwearied, and he made frequent tours through different parts of England and Scotland, acquiring considerable skill as a mineralogist, and constantly arriving at grand and comprehensive views in geology. He communicated the results of his observations unreservedly, and with the fearless spirit of one who was conscious that love of truth was the sole stimulus of all his exertions. When at length he had matured his views, he published, in 1788, his 'Theory of the Earth,' and the same, afterwards more fully developed in a separate work, in 1795. This treatise was the first in which geology was declared to be in no way concerned about 'questions as to the origin of things;'[4] the first in which an attempt was made to dispense entirely with all hypothetical causes, and to explain the former changes of the earth's crust, by reference exclusively to natural agents. Hutton laboured to give fixed principles to geology, as Newton had succeeded in doing to astronomy; but in the former science too little progress had been made towards furnishing the necessary data to enable any philosopher, however great his genius, to realize so noble a project.

'The ruins of an older world,' said Hutton, 'are visible in the present structure of our planet, and the strata which now compose our continents have been once beneath the sea, and were formed out of the waste of pre-existing continents. The same forces are still destroying, by chemical decomposition or mechanical violence, even the hardest rocks, and transporting the materials to the sea, where they are spread out, and form strata analogous to those of more ancient date. Although loosely deposited along the bottom of the ocean, they become afterwards altered and consolidated by volcanic heat, and then heaved up, fractured and con-

torted.'[5] Although Hutton had never explored any region of active volcanos, he had convinced himself that basalt and many other trap-rocks
were of igneous origin, and that many of them had been injected in a
melted state through fissures in the older strata. The compactness of these
rocks, and their different aspect from that of ordinary lava, he attributed
to their having cooled down under the pressure of the sea, and in order
to remove the objections started against this theory, his friend Sir James
Hall instituted a most curious and instructive series of chemical experiments, illustrating the crystalline arrangement and texture assumed by
melted matter cooled down under high pressure. The absence of stratification in granite, and its analogy in mineral character to rocks which he
deemed of igneous origin, led Hutton to conclude that granite must also
have been formed from matter in fusion, and this inference he felt could
not be fully confirmed, unless he discovered at the contact of granite and
other strata a repetition of the phenomena exhibited so constantly by the
trap-rocks. Resolved to try his theory by this test, he went to the Grampians
and surveyed the line of junction of the granite and superincumbent
stratified masses, and found in Glen Tilt in 1785 the most clear and
unequivocal proofs in support of his views. Veins of red granite are there
seen branching out from the principal mass, and traversing the black
micaceous schist and primary limestone. The intersected stratified rocks
are so distinct in colour and appearance as to render the example in that
locality most striking, and the alteration of the limestone in contact was
very analogous to that produced by trap veins on calcareous strata. This
verification of his system filled him with delight, and called forth such
marks of joy and exultation, that the guides who accompanied him, says
his biographer, were convinced that he must have discovered a vein of
silver or gold. He was aware that the same theory would not explain the
origin of the primary schists, but these he called primary, rejecting the
term primitive, and was disposed to consider them as sedimentary rocks
altered by heat, and that they originated in some other form from the
waste of previously existing rocks.

By this important discovery of granite veins to which he had been led
by fair induction from an independent class of facts, Hutton prepared the
way for the greatest innovation on the systems of his predecessors. Vallis-
neri had pointed out the general fact, that there were certain fundamental
rocks which contained no organic remains, and which he supposed to
have been formed before the creation of living beings. Moro, Generelli,

and other Italian writers embraced the same doctrine, and Lehman regarded the mountains called by him primitive, as parts of the original nucleus of the globe. The same tenet was an article of faith in the school of Freyberg; and if any one ventured to doubt the possibility of our being enabled to carry back our researches to the creation of the present order of things, the granitic rocks were triumphantly appealed to. On them seemed written in legible characters, the memorable inscription

> Before me things create were none, save things
> Eternal,[6]

and no small sensation was excited when Hutton seemed, with unhallowed hand, desirous to erase characters already regarded by many as sacred. 'In the economy of the world,' said the Scotch geologist, 'I can find no traces of a beginning, no prospect of an end;'[7] and the declaration was the more startling when coupled with the doctrine, that all past changes on the globe had been brought about by the slow agency of existing causes. The imagination was first fatigued and overpowered by endeavouring to conceive the immensity of time required for the annihilation of whole continents by so insensible a process. Yet when the thoughts had wandered through these interminable periods, no resting place was assigned in the remotest distance. The oldest rocks were represented to be of a derivative nature, the last of an antecedent series, and that perhaps one of many pre-existing worlds. Such views of the immensity of past time, like those unfolded by the Newtonian philosophy in regard to space, were too vast to awaken ideas of sublimity unmixed with a painful sense of our incapacity to conceive a plan of such infinite extent. Worlds are seen beyond worlds immeasurably distant from each other, and beyond them all innumerable other systems are faintly traced on the confines of the visible universe.

The characteristic feature of the Huttonian theory was, as before hinted, the exclusion of all causes not supposed to belong to the present order of nature. Its greatest defect consisted in the undue influence attributed to subterranean heat, which was supposed necessary for the consolidation of all submarine deposits. Hutton made no step beyond Hooke, Moro, and Raspe, in pointing out in what manner the laws now governing earthquakes, might bring about geological changes, if sufficient time be allowed. On the contrary, he seems to have fallen far short of some of their views. He imagined that the continents were first gradually destroyed, and when their ruins had furnished materials for new continents, they were upheaved by violent and paroxysmal convulsions. He therefore required alternate periods of disturbance and repose, and such he believed had been, and would for ever be, the course of nature. Generelli, in his exposition of Moro's system, had made a far nearer approximation towards reconciling geological appearances with the state of nature as known to us, for while he agreed with Hutton, that the decay and reproduction of rocks were always in progress, proceeding with the utmost uniformity, the learned Carmelitan represented the repairs of mountains by elevation from below, to be effected by an equally constant and synchronous operation. Neither of these theories considered singly, satisfies all the conditions of the great problem, which a geologist, who rejects cosmological causes, is called upon to solve; but they probably contain together the germs of a perfect system. There can be no doubt, that periods of disturbance and repose have followed each other in succession in every region of the globe, but it may be equally true, that the energy of the subterranean movements has been always uniform as regards the *whole earth*. The force of earthquakes may for a cycle of years have been invariably confined, as it is now, to large but determinate spaces, and may then have gradually shifted its position, so that another region, which had for ages been at rest, became in its turn the grand theatre of action.

Although Hutton's knowledge of mineralogy and chemistry was considerable, he possessed but little information concerning organic remains. They merely served him as they did Werner to characterize certain strata, and to prove their marine origin. The theory of former revolutions in organic life was not yet fully recognized, and without this class of proofs in support of the antiquity of the globe, the indefinite periods demanded by the Huttonian hypothesis appeared visionary to many, and some, who deemed the doctrine inconsistent with revealed truths, indulged very uncharitable suspicions of the motives of its author. They accused him of a deliberate design of reviving the heathen dogma of an 'eternal succession,' and of denying that this world ever had a beginning. Playfair, in the biography of his friend, has the following comment on this part of their theory: – 'In the planetary motions, where geometry has carried the eye so far, both into the future and the past, we discover no mark either of the commencement or termination of the present order. It is unreasonable, indeed, to suppose that such marks should anywhere exist. The Author of nature has not given laws to the universe, which, like the institutions of men, carry in themselves the elements of their own destruction. He has not permitted in His works any symptom of infancy or of old age, or any sign by which we may estimate either their future or their past duration. *He may put an end, as he no doubt gave a beginning,* to the present system at some determinate period of time; but we may rest assured that this great catastrophe will not be brought about by the laws now existing, and that it is not indicated by any thing which we perceive.'[8]

The party feeling excited against the Huttonian doctrines, and the open disregard of candour and temper in the controversy, will hardly be credited by our readers, unless we recall to their recollection that the mind of the English public was at that time in a state of feverish excitement. A class of writers in France had been labouring industriously for many years, to diminish the influence of the clergy, by sapping the foundation of the Christian faith, and their success, and the consequences of the Revolution, had alarmed the most resolute minds, while the imagination of the more timid was continually haunted by dread of innovation, as by the phantom of some fearful dream.

Voltaire had used the modern discoveries in physics as one of the numerous weapons of attack and ridicule directed by him against the Scriptures. He found that the most popular systems of geology were accommodated to the sacred writings, and that much ingenuity had been employed to make every fact coincide exactly with the Mosaic account of the creation and deluge. It was, therefore, with no friendly feelings, that he contemplated the cultivators of geology in general, regarding the science as one which had been successfully enlisted by theologians as an ally in their cause. He knew that the majority of those who were aware of the abundance of fossil shells in the interior of continents, were still per-suaded that they were proofs of the universal deluge; and as the readiest way of shaking this article of faith, he endeavoured to inculcate scepticism, as to the real nature of such shells, and to recall from contempt the exploded dogma of the sixteenth century, that they were sports of nature. He also pretended that vegetable impressions were not those of real plants. Yet he was perfectly convinced that the shells had really belonged to living testacea, as may be seen in his essay, 'On the formation of Mountains.' He would sometimes, in defiance of all consistency, shift his ground when addressing the vulgar; and admitting the true nature of the shells collected in the Alps, and other places, pretend that they were eastern species, which had fallen from the hats of pilgrims coming from Syria. The numerous essays written by him on geological subjects were all calculated to strengthen prejudices, partly because he was ignorant of the real state of the science, and partly from his bad faith. On the other hand, they who knew that his attacks were directed by a desire to invalidate scripture, and who were unacquainted with the true merits of the question, might well deem the old diluvian hypothesis incontrovertible, if Voltaire could adduce no better argument against it, than to deny the true nature of organic remains.

It is only by careful attention to impediments originating in extrinsic causes, that we can explain the slow and reluctant adoption of the simplest truths in geology. First, we find many able naturalists adducing the fossil remains of marine animals, as proofs of an event related in Scripture. The evidence is deemed conclusive by the multitude for a century or more; for it favours opinions which they entertained before, and they are gratified by supposing them confirmed by fresh and unexpected proofs. Many, who see through the fallacy, have no wish to undeceive those who are influenced by it, approving the effect of the delusion, and conniving at it as a pious fraud; until finally, an opposite party, who are hostile to the sacred writings, labour to explode the erroneous opinion, by substituting for it another dogma which they know to be equally unsound.

The heretical vulcanists were now openly assailed in England, by imputations of the most illiberal kind. We cannot estimate the malevolence of such a persecution, by the pain which similar insinuations might now inflict; for although charges of infidelity and atheism must always be odious, they were injurious in the extreme at that moment of political excitement: and it was better perhaps for a man's good reception in society, that his moral character should have been traduced, than that he should become a mark for these poisoned weapons. We shall pass over the works of numerous divines, who may be excused for sensitiveness on points which then excited so much uneasiness in the public mind; and we shall say nothing of the amiable poet Cowper, who could hardly be expected to have inquired into the merits of doctrines in physics. But we find in the foremost ranks of the intolerant, several laymen who had high claims to scientific reputation. Amongst these, appears Williams, a mineral surveyor of Edinburgh, who published a 'Natural History of the Mineral Kingdom' in 1789, a work of great merit for that day, and of practical utility, as containing the best account of the coal strata. In his preface he misrepresents Hutton's theory altogether, and charges him with considering all rocks to be lavas of different colours and structure; and also with 'warping every thing to support the eternity of the world.' He descants on the pernicious influence of such sceptical notions, as leading to downright infidelity and atheism, 'and as being nothing less than to depose the Almighty Creator of the universe from his office.'[9]

Kirwan, president of the Royal Academy of Dublin, a chemist and mineralogist of some merit, but who possessed much greater authority in the scientific world than he was entitled by his talents to enjoy, in the introduction to his 'Geological Essays, 1799,' said 'that *sound* geology *graduated* into religion, and was required to dispel certain systems of atheism or infidelity, of which they had had recent experience.'[10] He was an uncompromising defender of the aqueous theory of all rocks, and was scarcely surpassed by Burnet and Whiston, in his desire to adduce the Mosaic writings in confirmation of his opinions.

De Luc, in the preliminary discourse to his Treatise on Geology, says, 'the weapons have been changed by which revealed religion is attacked; it is now assailed by geology, and this science has become essential to theologians.' He imputes the failure of former geological systems to their having been anti-mosaical, and directed against a 'sublime tradition.'[11] These and similar imputations, reiterated in the works of De Luc, seem to have been taken for granted by some modern writers: it is therefore necessary to state, in justice to the numerous geologists of different nations, whose works we have considered, that none of them were guilty of endeavouring, by arguments drawn from physics, to invalidate scriptural tenets. On the contrary, the majority of them, who were fortunate enough 'to discover the true causes of things,' did not deserve another part of the poet's panegyric, 'Fearless of Fortune, and resigned to Fate.'[12] The caution, and even timid reserve, of many eminent Italian authors of the earlier period is very apparent; and there can hardly be a doubt that they subscribed to certain dogmas, and particularly to the first diluvian theory, out of deference to popular prejudices, rather than from conviction. If they were guilty of dissimulation, we must not blame their want of moral courage, but reserve our condemnation for the intolerance of the times, and that inquisitorial power which forced Galileo to abjure, and the two Jesuits to disclaim the theory of Newton.*

---

* I observe that, in a most able and interesting article 'the Life of Galileo,' recently published in the 'Library of Useful Knowledge,' it is asserted that both Galileo's work, and the book of Copernicus 'Nisi corrigatur,' were still to be seen on the forbidden list of the Index at Rome in 1828. But I was assured in the same year, by Professor Scarpellini, at Rome, that Pius VII., a pontiff distinguished for his love of science, procured in 1818 a repeal of the

Hutton answered Kirwan's attacks with great warmth, and with the indignation excited by unmerited reproach. He had always displayed, says Playfair, 'the utmost disposition to admire the beneficent design manifested in the structure of the world, and he contemplated with delight those parts of his theory which made the greatest additions to our knowledge of final causes.'[13] We may say with equal truth, that in no scientific works in our language can more eloquent passages be found, concerning the fitness, harmony, and grandeur of all parts of the creation, than in those of Playfair. They are evidently the unaffected expressions of a mind, which contemplated the study of nature, as best calculated to elevate our conceptions of the attributes of the First Cause. At any other time the force and elegance of Playfair's style must have insured popularity to the Huttonian doctrines; but, by a singular coincidence, neptunianism and orthodoxy were now associated in the same creed; and the tide of prejudice ran so strong, that the majority were carried far away into the chaotic fluid, and other cosmological inventions of Werner. These fictions the Saxon Professor had borrowed with little modification, and without any improvement, from his predecessors. They had not the smallest foundation, either in Scripture, or in common sense, but were perhaps approved of by many as being so ideal and unsubstantial, that they could never come into violent collision with any preconceived opinions.

The great object of De Luc's writings was to disprove the high antiquity attributed by Hutton to our present continents, and particularly to seek out some cause for the excavation of valleys more speedy and violent than the action of ordinary rivers. Hutton had said, that the erosion of rivers, and such floods as occur in the usual course of nature, might progressively, if time be allowed, hollow out great valleys, but he had also observed, 'that on our continents there is no spot on which a river may not formerly have run.'[14] De Luc generally reasoned against him as if he had said, that the existing rivers flowing *at their present levels* had caused all these inequalities of the earth's surface; and Playfair, in his zeal to prove how much De Luc

---

edicts against Galileo and the Copernican system. He assembled the Congregation, and the late cardinal Toriozzi, assessor of the Sacred Office, proposed 'that they should wipe off this scandal from the church.' The repeal was carried, with the dissentient voice of one Dominican only. Long before this time the Newtonian theory had been taught in the Sapienza, and all catholic universities in Europe (with the exception, I am told, of Salamanca); but it was always required of professors, in deference to the decrees of the church, to use the term *hypothesis*, instead of theory. They now speak of the Copernican *theory*.

underrated the force of running water, did not sufficiently expose his misstatement of the Huttonian proposition. But we must defer the full consideration of this controverted question for the present.

While the tenets of the rival schools of Freyberg and Edinburgh were warmly espoused by devoted partisans, the labours of an individual, unassisted by the advantages of wealth or station in society, were almost unheeded. Mr. William Smith, an English surveyor, published his 'Tabular View of the British Strata' in 1790, wherein he proposed a classification of the secondary formations in the west of England. Although he had not communicated with Werner, it appeared by this work that he had arrived at the same views respecting the laws of superposition of stratified rocks; that he was aware that the order of succession of different groups was never inverted; and that they might be identified at very distant points by their peculiar organized fossils.

From the time of the appearance of the 'Tabular View,' he laboured to construct a geological map of the whole of England, and, with the greatest disinterestedness of mind, communicated the results of his investigations to all who desired information, giving such publicity to his original views, as to enable his contemporaries almost to compete with him in the race. The execution of his map was completed in 1815, and remains a lasting monument of original talent and extraordinary perseverance, for he had explored the whole country on foot without the guidance of previous observers, or the aid of fellow-labourers, and had succeeded in throwing into natural divisions the whole complicated series of British rocks.* D'Aubuisson, a distinguished pupil of Werner, paid a just tribute of praise to this remarkable performance, observing that 'what many celebrated mineralogists had only accomplished for a small part of Germany in the course of half a century, had been effected by a single individual for the whole of England.'[15]

We have now arrived at the era of living authors, and shall bring to a conclusion our sketch of the progress of opinion in geology. The contention of the rival factions of the Vulcanists and Neptunists had been carried to

---

* Werner invented a new language to express his divisions of rocks, and some of his technical terms, such as grauwacke, gneiss, and others, passed current in every country in Europe. Smith adopted for the most part English provincial terms, often of barbarous sound, such as gault, cornbrash, clunch clay, &c., and affixed them to subdivisions of the British series. Many of these still retain their place in our scientific classifications, and attest his priority of arrangement.

such a height, that these names had become terms of reproach, and the two parties had been less occupied in searching for truth, than for such arguments as might strengthen their own cause, or serve to annoy their antagonists. A new school at last arose who professed the strictest neutrality, and the utmost indifference to the systems of Werner and Hutton, and who were resolved diligently to devote their labours to observation. The reaction, provoked by the intemperance of the conflicting parties, now produced a tendency to extreme caution. Speculative views were discountenanced, and through fear of exposing themselves to the suspicion of a bias towards the dogmas of a party, some geologists became anxious to entertain no opinion whatever on the causes of phenomena, and were inclined to scepticism even where the conclusions deducible from observed facts scarcely admitted of reasonable doubt. But although the reluctance to theorize was carried somewhat to excess, no measure could be more salutary at such a moment than a suspension of all attempts to form what were termed 'theories of the earth.' A great body of new data were required, and the Geological Society of London, founded in 1807, conduced greatly to the attainment of this desirable end. To multiply and record observations, and patiently to await the result at some future period, was the object proposed by them, and it was their favourite maxim that the time was not yet come for a general system of geology, but that all must be content for many years to be exclusively engaged in furnishing materials for future generalizations. By acting up to these principles with consistency, they in a few years disarmed all prejudice, and rescued the science from the imputation of being a dangerous, or at best but a visionary pursuit.

Inquiries were at the same time prosecuted with great success by the French naturalists, who devoted their attention especially to the study of organic remains. They shewed that the specific characters of fossil shells and vertebrated animals might be determined with the utmost precision, and by their exertions a degree of accuracy was introduced into this department of science, of which it had never before been deemed susceptible. It was found that, by the careful discrimination of the fossil contents of strata, the contemporary origin of different groups could often be established, even where all identity of mineralogical character was wanting, and where no light could be derived from the order of superposition. The minute investigation, moreover, of the relics of the animate creation of former ages, had a powerful effect in dispelling the illusion which had long prevailed concerning the absence of analogy between the ancient and modern state of our planet. A close comparison of the recent and fossil species, and the inferences drawn in regard to their habits, accustomed the geologist to contemplate the earth as having been at successive periods the dwelling place of animals and plants of different races, some of which were discovered to have been terrestrial and others aquatic – some fitted to live in seas, others in the waters of lakes and rivers. By the consideration of these topics, the mind was slowly and insensibly withdrawn from imaginary pictures of catastrophes and chaotic confusion, such as haunted the imagination of the early cosmogonists. Numerous proofs were discovered of the tranquil deposition of sedimentary matter and the slow development of organic life. If many still continued to maintain, that 'the thread of induction was broken,'[16] yet in reasoning by the strict rules of induction from recent to fossil species, they virtually disclaimed the dogma which in theory they professed. The adoption of the same generic, and, in some cases, even the same specific names for the exuviæ of fossil animals, and their living analogues, was an important step towards familiarizing the mind with the idea of the identity and unity of the system in distant eras. It was an acknowledgment, as it were, that a considerable part of the ancient memorials of nature were written in a living language. The growing importance then of the natural history of organic remains, and its general application to geology, may be pointed out as the characteristic feature of the progress of the science during the present century. This branch of knowledge has already become an instrument of great power in the discovery of truths in geology, and is continuing daily to unfold new data for grand and enlarged views respecting the former changes of the earth.

When we compare the result of observations in the last thirty years with those of the three preceding centuries, we cannot but look forward with the most sanguine expectations to the degree of excellence to which geology may be carried, even by the labours of the present generation. Never, perhaps, did any science, with the exception of astronomy, unfold, in an equally brief period, so many novel and unexpected truths, and overturn so many preconceived opinions. The senses had for ages declared the earth to be at rest, until the astronomer taught that it was carried through space with inconceivable rapidity. In like manner was the surface of this planet regarded as having remained unaltered since its creation, until the geologist proved that it had been the theatre of reiterated change, and was still the subject of slow but never ending fluctuations. The discovery of other systems in the boundless regions of space was the triumph of astronomy – to trace the same system through various transformations – to behold it at successive eras adorned with different hills and valleys, lakes and seas, and peopled with new inhabitants, was the delightful meed of geological research. By the geometer were measured the regions of space, and the relative distances of the heavenly bodies – by the geologist myriads of ages were reckoned, not by arithmetical computation, but by a train of physical events – a succession of phenomena in the animate and inanimate worlds – signs which convey to our minds more definite ideas than figures can do, of the immensity of time.

Whether our investigation of the earth's history and structure will eventually be productive of as great practical benefits to mankind, as a knowledge of the distant heavens, must remain for the decision of posterity. It was not till astronomy had been enriched by the observations of many centuries, and had made its way against popular prejudices to the establishment of a sound theory, that its application to the useful arts was most conspicuous. The cultivation of geology began at a later period; and in every step which it has hitherto made towards sound theoretical principles, it has had to contend against more violent prepossessions. The practical advantages already derived from it have not been inconsiderable: but our generalizations are yet imperfect, and they who follow may be expected to reap the most valuable fruits of our labour. Meanwhile the charm of first discovery is our own, and as we explore this magnificent field of inquiry, the sentiment of a great historian of our times may continually be present to our minds, that 'he who calls what has vanished back again into being, enjoys a bliss like that of creating.'[17]

CHAPTER 5

Theoretical Errors which have Retarded the Progress of Geology

We have seen that, during the progress of geology, there have been great fluctuations of opinion respecting the nature of the causes to which all former changes of the earth's surface are referrible. The first observers conceived that the monuments which the geologist endeavours to decipher, relate to a period when the physical constitution of the earth differed entirely from the present, and that, even after the creation of living beings, there have been causes in action distinct in kind or degree from those now forming part of the economy of nature. These views have been gradually modified, and some of them entirely abandoned in proportion as observations have been multiplied, and the signs of former mutations more skilfully interpreted. Many appearances, which for a long time were regarded as indicating mysterious and extraordinary agency, are finally recognized as the necessary result of the laws now governing the material world; and the discovery of this unlooked for conformity has induced some geologists to infer that there has never been any interruption to the same uniform order of physical events. The same assemblage of general causes, they conceive, may have been sufficient to produce, by their various combinations, the endless diversity of effects, of which the shell of the earth has preserved the memorials, and, consistently with these principles, the recurrence of analogous changes is expected by them in time to come.

Whether we coincide or not in this doctrine, we must admit that the gradual progress of opinion concerning the succession of phenomena in remote eras, resembles in a singular manner that which accompanies the growing intelligence of every people, in regard to the economy of nature in modern times. In an early stage of advancement, when a great number of natural appearances are unintelligible, an eclipse, an earthquake, a flood, or the approach of a comet, with many other occurrences afterwards found to belong to the regular course of events, are regarded as prodigies.

The same delusion prevails as to moral phenomena, and many of these are ascribed to the intervention of demons, ghosts, witches, and other immaterial and supernatural agents. By degrees, many of the enigmas of the moral and physical world are explained, and, instead of being due to extrinsic and irregular causes, they are found to depend on fixed and invariable laws. The philosopher at last becomes convinced of the undeviating uniformity of secondary causes, and, guided by his faith in this principle, he determines the probability of accounts transmitted to him of former occurrences, and often rejects the fabulous tales of former ages, on the ground of their being irreconcilable with the experience of more enlightened ages.

As a belief in want of conformity in the physical constitution of the earth, in ancient and modern times, was for a long time universally prevalent, and that too amongst men who were convinced that the order of nature is *now* uniform, and has continued so for several thousand years; every circumstance which could have influenced their minds and given an undue bias to their opinions deserves particular attention. Now the reader may easily satisfy himself, that, however undeviating the course of nature may have been from the earliest epochs, it was impossible for the first cultivators of geology to come to such a conclusion, so long as they were under a delusion as to the age of the world, and the date of the first creation of animate beings. However fantastical some theories of the sixteenth century may now appear to us, – however unworthy of men of great talent and sound judgment, we may rest assured that, if the same misconceptions now prevailed in regard to the materials of human trans-actions, it would give rise to a similar train of absurdities. Let us imagine, for example, that Champollion, and the French and Tuscan literati now engaged in exploring the antiquities of Egypt, had visited that country with a firm belief that the banks of the Nile were never peopled by the human race before the beginning of the nineteenth century, and that their faith in this dogma was as difficult to shake as the opinion of our ancestors, that the earth was never the abode of living beings until the creation of the present continents, and of the species now existing, – it is easy to perceive what extravagant systems they would frame, while under the influence of this delusion, to account for the monuments discovered in Egypt. The sight of the pyramids, obelisks, colossal statues, and ruined temples, would fill them with such astonishment, that for a time they would be as men spell-bound – wholly incapacitated to reason with sobriety. They might incline at first to refer the construction of such stupendous works to some superhuman powers of a primeval world. A system might be invented resembling that so gravely advanced by Manetho, who relates that a dynasty of gods originally ruled in Egypt, of whom Vulcan, the first monarch, reigned nine thousand years. After them came Hercules and other demi-gods, who were at last succeeded by human kings. When some fanciful speculations of this kind had amused the imagination for a time, some vast repository of mummies would be discovered and would immediately undeceive those antiquaries who enjoyed an opportunity of personally examining them, but the prejudices of others at a distance, who were not eye-witnesses of the whole phenomena, would not be so easily overcome. The concurrent report of many travellers would indeed render it necessary for them to accommodate ancient theories to some of the new facts, and much wit and ingenuity would be required to modify and defend their old positions. Each new invention would violate a greater number of known analogies; for if a theory be required to embrace some false principle, it becomes more visionary in proportion as facts are multiplied, as would be the case if geometers were now required to form an astronomical system on the assumption of the immobility of the earth.

Amongst other fanciful conjectures concerning the history of Egypt, we may suppose some of the following to be started. 'As the banks of the Nile have been so recently colonized, the curious substances called mummies could never in reality have belonged to men. They may have been generated by some *plastic virtue* residing in the interior of the earth, or they may be abortions of nature produced by her incipient efforts in the work of creation. For if deformed beings are sometimes born even now, when the scheme of the universe is fully developed, many more may have been "sent before their time, scarce half made up,"[18] when the planet itself was in the embryo state. But if these notions appear to derogate from the perfection of the Divine attributes, and if these mummies be in all their parts true representations of the human form, may we not refer them to the future rather than the past? May we not be looking into the womb of nature, and not her grave? may not these images be like the shades of the unborn in Virgil's Elysium – the archetypes of men not yet called into existence?'

These speculations, if advocated by eloquent writers, would not fail to attract many zealous votaries, for they would relieve men from the painful necessity of renouncing preconceived opinions. Incredible as such scepticism may appear, it would be rivalled by many systems of the sixteenth and seventeenth centuries, and among others by that of the learned Falloppio, who regarded the tusks of fossil elephants as earthy concretions, and the vases of Monte Testaceo, near Rome, as works of nature, and not of art. But when one generation had passed away, and another not compromised to the support of antiquated dogmas had succeeded, they would review the evidence afforded by mummies more impartially, and would no longer controvert the preliminary question, that human beings had lived in Egypt before the nineteenth century: so that when a hundred years perhaps had been lost, the industry and talents of the philosopher would be at last directed to the elucidation of points of real historical importance.

But we have adverted to one only of many prejudices with which the earlier geologists had to contend. Even when they conceded that the earth had been peopled with animate beings at an earlier period than was at first supposed, they had no conception that the quantity of time bore so great a proportion to the historical era as is now generally conceded. How fatal every error as to the quantity of time must prove to the introduction of rational views concerning the state of things in former ages, may be conceived by supposing that the annals of the civil and military transactions of a great nation were perused under the impression that they occurred in a period of one hundred instead of two thousand years. Such a portion of history would immediately assume the air of a romance; the events would seem devoid of credibility, and inconsistent with the present course of human affairs. A crowd of incidents would follow each other in thick succession. Armies and fleets would appear to be assembled only to be destroyed, and cities built merely to fall in ruins. There would be the most violent transitions from foreign or intestine war to periods of profound peace, and the works effected during the years of disorder or tranquillity would be alike superhuman in magnitude.

He who should study the monuments of the natural world under the influence of a similar infatuation, must draw a no less exaggerated picture of the energy and violence of causes, and must experience the same insurmountable difficulty in reconciling the former and present state of nature. If we could behold in one view all the volcanic cones thrown up in Iceland, Italy, Sicily, and other parts of Europe, during the last five thousand years, and could see the lavas which have flowed during the same period; the dislocations, subsidences and elevations caused by earthquakes; the lands added to various deltas, or devoured by the sea, together with the effects of devastation by floods, and imagine that all these events had happened in one year, we must form most exalted ideas of the activity of the agents, and the suddenness of the revolutions. Were an equal amount of change to pass before our eyes in the next year, could we avoid the conclusion that some great crisis of nature was at hand? If geologists, therefore, have misinterpreted the signs of a succession of events, so as to conclude that centuries were implied where the characters imported thousands of years, and thousands of years where the language of nature signified millions, they could not, if they reasoned logically from such false premises, come to any other conclusion, than that the system of the natural world had undergone a complete revolution.

We should be warranted in ascribing the erection of the great pyramid to superhuman power, if we were convinced that it was raised in one day; and if we imagine, in the same manner, a mountain chain to have been elevated, during an equally small fraction of the time which was really occupied in upheaving it, we might then be justified in inferring, that the subterranean movements were once far more energetic than in our own times. We know that one earthquake may raise the coast of Chili for a hundred miles to the average height of about five feet. A repetition of two thousand shocks of equal violence might produce a mountain chain one hundred miles long, and ten thousand feet high. Now, should one only of these convulsions happen in a century, it would be consistent with the order of events experienced by the Chilians from the earliest times; but if the whole of them were to occur in the next hundred years, the entire district must be depopulated, scarcely any animals or plants could survive, and the surface would be one confused heap of ruin and desolation.

One consequence of undervaluing greatly the quantity of past time is the apparent coincidence which it occasions of events necessarily disconnected, or which are so unusual, that it would be inconsistent with all calculation of chances to suppose them to happen at one and the same time. When the unlooked for association of such rare phenomena is witnessed in the present course of nature, it scarcely ever fails to excite a suspicion of the preternatural in those minds which are not firmly convinced of the uniform agency of secondary causes; − as if the death of some individual in whose fate they are interested, happens to be accompanied by the appearance of a luminous meteor, or a comet, or the shock of an earthquake. It would be only necessary to multiply such coincidences indefinitely, and the mind of every philosopher would be disturbed. Now it would be difficult to exaggerate the number of physical events, many of them most rare and unconnected in their nature, which were imagined by the Woodwardian hypothesis to have happened in the course of a few months; and numerous other examples might be found of popular geological theories, which require us to imagine that a long succession of events happened in a brief and almost momentary period.

The sources of prejudice hitherto considered may be deemed as in a great degree peculiar to the infancy of the science, but others are common to the first cultivators of geology and to ourselves, and are all singularly calculated to produce the same deception, and to strengthen our belief that the course of nature in the earlier ages differed widely from that now established. Although we cannot fully explain all these circumstances, without assuming some things as proved, which it will be the object of another part of this work to demonstrate, we must briefly allude to them in this place.

The first and greatest difficulty, then, consists in our habitual unconsciousness that our position as observers is essentially unfavourable, when we endeavour to estimate the magnitude of the changes now in progress. In consequence of our inattention to this subject, we are liable to the greatest mistakes in contrasting the present with former states of the globe. We inhabit about a fourth part of the surface; and that portion is almost exclusively the theatre of decay and not of reproduction. We know, indeed, that new deposits are annually formed in seas and lakes, and that every year some new igneous rocks are produced in the bowels of the earth, but we cannot watch the progress of their formation; and, as they are only present to our minds by the aid of reflection, it requires an effort both of the reason and the imagination to appreciate duly their importance. It is, therefore, not surprising that we imperfectly estimate the result of operations invisible to us; and that, when analogous results of some former epoch are presented to our inspection, we cannot recognise the analogy. He who has observed the quarrying of stone from a rock, and has seen it shipped for some distant port, and then endeavours to conceive what kind of edifice will be raised by the materials, is in the same predicament as a geologist, who, while he is confined to the land, sees the decomposition of rocks, and the transportation of matter by rivers to the sea, and then endeavours to picture to himself the new strata which Nature is building beneath the waters. Nor is his position less unfavourable when, beholding a volcanic eruption, he tries to conceive what changes the column of lava has produced, in its passage upwards, on the intersected strata; or what form the melted matter may assume at great depths on cooling down; or what may be the extent of the subterranean rivers and reservoirs of liquid matter far beneath the surface. It should, therefore, be remembered, that the task imposed on those who study the earth's history requires no ordinary share of discretion, for we are precluded from collating the corresponding parts of a system existing at two different periods. If we were inhabitants of another element – if the great ocean were our domain, instead of the narrow limits of the land, our difficulties would be considerably lessened; while, on the other hand, there can be little doubt, although the reader may, perhaps, smile at the bare suggestion of such an idea, that an amphibious being, who should possess our faculties, would still more easily arrive at sound theoretical opinions in geology, since he might behold, on the one hand, the decomposition of rocks in the atmosphere, and the transportation of matter by running water; and, on the other, examine the deposition of sediment in the sea, and the imbedding of animal remains in new strata. He might ascertain, by direct observation, the action of a mountain torrent, as well as of a marine current; might compare the products of volcanos on the land with those poured out beneath the waters; and might mark, on the one hand, the growth of the forest, and on the other that of the coral reef. Yet, even with these advantages, he would be liable to fall into the greatest errors when endeavouring to reason on rocks of subterranean origin. He would seek in vain, within the sphere of his observation, for any direct analogy to the process of their formation, and would therefore be in danger of attributing them, wherever they are upraised to view, to some 'primeval state of nature.' But if we may be allowed so far to indulge the imagination, as to suppose a being, entirely confined to the nether world – some 'dusky melancholy sprite,' like Umbriel, who could 'flit on sooty pinions to the central earth,' but who was never permitted to 'sully the fair face of light,'[19] and emerge into the regions of water and of air; and if this being should busy himself in investigating the structure of the globe, he might frame theories the exact converse of those usually adopted by human philosophers. He might infer that the stratified rocks, containing shells and other organic remains, were the oldest of created things, belonging to some original and nascent state of the planet. 'Of these masses,' he might say, 'whether they consist of loose incoherent sand, soft clay, or solid rock, none have been formed in modern times. Every year some part of them are broken and shattered by earthquakes, or melted up by volcanic fire; and, when they cool down slowly from a state of fusion, they assume a crystalline form, perfectly distinct from those inexplicable rocks which are so regularly bedded, and contain stones full of curious impressions and fantastic markings. This process cannot have been carried on for an indefinite time, for in that case all the stratified rocks would long ere this have been fused and crystallized. It is therefore probable that the whole planet once consisted of these curiously-bedded formations, at a time when the volcanic fire had not yet been brought into activity. Since that period there seems to have been a gradual development of heat, and this augmentation we may expect to continue till the whole globe shall be in a state of fluidity and incandescence.'

Such might be the system of the Gnome at the very same time that the followers of Leibnitz, reasoning on what they saw on the outer surface, would be teaching the doctrine of gradual refrigeration, and averring that the earth had begun its career as a fiery comet, and would hereafter become a frozen icy mass. The tenets of the schools of the nether and of the upper world would be directly opposed to each other, for both would partake of the prejudices inevitably resulting from the continual contemplation of one class of phenomena to the exclusion of another. Man observes the annual decomposition of crystalline and igneous rocks, and may sometimes see their conversion into stratified deposits; but he cannot witness the reconversion of the sedimentary into the crystalline by subterranean fire. He is in the habit of regarding all the sedimentary rocks as more recent than the unstratified, for the same reason that we may suppose him to fall into the opposite error if he saw the origin of the igneous class only.

It is only by becoming sensible of our natural disadvantages that we shall be roused to exertion, and prompted to seek out opportunities of discovering the operations now in progress, such as do not present themselves readily to view. We are called upon, in our researches into the state of the earth, as in our endeavours to comprehend the mechanism of the heavens, to invent means for overcoming the limited range of our vision. We are perpetually required to bring, as far as possible, within the sphere of observation, things to which the eye, unassisted by art, could never obtain access. It was not an impossible contingency that astronomers might have been placed, at some period, in a situation much resembling that in which the geologist seems to stand at present. If the Italians, for example, in the early part of the twelfth century, had discovered at Amalphi, instead of the pandects of Justinian, some ancient manuscripts filled with astronomical observations relating to a period of three thousand years, and made by some ancient geometers who possessed optical instruments as perfect as any in modern Europe, they would probably, on consulting these memorials, have come to a conclusion that there had been a great revolution in the solar and sidereal systems. 'Many primary and secondary planets,' they might say, 'are enumerated in these tables, which exist no longer. Their positions are assigned with such precision, that we may assure ourselves that there is nothing in their place at present but the blue ether. Where one star is visible to us, these documents represent several thousands. Some of those which are now single, consisted then of two separate bodies, often distinguished by different colours, and revolving periodically round a common centre of gravity. There is no analogy to them in the universe at present, for they were neither fixed stars nor planets, but stood in the mutual relation of sun and planet to each other. We must conclude, therefore, that there has occurred, at no distant period, a tremendous catastrophe, whereby thousands of worlds have been annihilated at once, and some heavenly bodies absorbed into the substance of others.' When such doctrines had prevailed for ages, the discovery of one of the worlds, supposed to have been lost, by aid of the first rude telescope, would not dissipate the delusion, for the whole burden of proof would now be thrown on those who insisted on the stability of the system from the beginning of time, and these philosophers would be required to demonstrate the existence of *all* the worlds said to have been annihilated. Such popular prejudices would be most unfavourable to the advancement of astronomy; for, instead of persevering in the attempt to improve their instruments, and laboriously to make and record observations, the greater number would despair of verifying the continued existence of the heavenly bodies not visible to the naked eye. Instead of confessing the extent of their ignorance, and striving to remove it by bringing to light new facts, they would be engaged in the indolent employment of framing imaginary theories concerning catastrophes and mighty revolutions in the system of the universe.

For more than two centuries the shelly strata of the Subapennine hills afforded matter of speculation to the early geologists of Italy, and few of them had any suspicion that similar deposits were then forming in the neighbouring sea. They were as unconscious of the continued action of causes still producing similar effects, as the astronomers, in the case supposed by us, of the existence of certain heavenly bodies still giving and reflecting light, and performing their movements as in the olden time. Some imagined that the strata, so rich in organic remains, instead of being due to secondary agents, had been so created in the beginning of things by the fiat of the Almighty; and others ascribed the imbedded fossil bodies to some plastic power which resided in the earth in the early ages of the world. At length Donati explored the bed of the Adriatic, and found the closest resemblance between the new deposits there forming, and those which constituted hills above a thousand feet high in various parts of the peninsula. He ascertained that certain genera of living testacea were grouped together at the bottom of the sea in precisely the same manner as were their fossil analogues in the strata of the hills, and that some species were common to the recent and fossil world. Beds of shells, moreover, in the Adriatic, were becoming incrusted with calcareous rock; and others were recently enclosed in deposits of sand and clay, precisely as fossil shells were found in the hills. This splendid discovery of the identity of modern and ancient submarine operations was not made without the aid of artificial instruments, which, like the telescope, brought phenomena into view not otherwise within the sphere of human observation.

In like manner, in the Vicentin, a great series of volcanic and marine sedimentary rocks were examined in the early part of the last century; but no geologist suspected, before the time of Arduino, that these were partly composed of ancient submarine lavas. If, when these enquiries were first made, geologists had been told that the mode of formation of such rocks might be fully elucidated by the study of processes then going on in certain parts of the Mediterranean, they would have been as incredulous as geometers would have been before the time of Newton, if any one had informed them that, by making experiments on the motion of bodies on the earth, they might discover the laws which regulated the movements of distant planets.

The establishment, from time to time, of numerous points of identification, drew at length from geologists a reluctant admission, that there was more correspondence between the physical constitution of the globe, and more uniformity in the laws regulating the changes of its surface, from the most remote eras to the present, than they at first imagined. If, in this state of the science, they still despaired of reconciling every class of geological phenomena to the operations of ordinary causes, even by straining analogy to the utmost limits of credibility, we might have expected, that the balance of probability at least would now have been presumed to incline towards the identity of the causes. But, after repeated experience of the failure of attempts to speculate on different classes of geological phenomena, as belonging to a distinct order of things, each new sect persevered systematically in the principles adopted by their predecessors. They invariably began, as each new problem presented itself, whether relating to the animate or inanimate world, to assume in their theories, that the economy of nature was formerly governed by rules quite independent of those now established. Whether they endeavoured to account for the origin of certain igneous rocks, or to explain the forces which elevated hills or excavated valleys, or the causes which led to the extinction of certain races of animals, they first presupposed an original and dissimilar order of nature; and when at length they approximated, or entirely came round to an opposite opinion, it was always with the feeling, that they conceded what they were justified *à priori* in deeming improbable. In a word, the same men who, as natural philosophers, would have been greatly surprised to find any deviation from the usual course of Nature *in their own time*, were equally surprised, as geologists, not to find such deviations at every period of the past.

The Huttonians were conscious that no check could be given to the utmost licence of conjecture in speculating on the causes of geological phenomena, unless we can assume invariable constancy in the order of Nature. But when they asserted this uniformity without any limitation as to time, they were considered, by the majority of their contemporaries, to have been carried too far, especially as they applied the same principle to the laws of the organic, as well as of the inanimate world.

We shall first advert briefly to many difficulties which formerly appeared insurmountable, but which, in the last forty years, have been partially or entirely removed by the progress of science; and shall afterwards consider the objections that still remain to the doctrine of absolute uniformity.

In the first place, it was necessary for the supporters of this doctrine to take for granted incalculable periods of time, in order to explain the formation of sedimentary strata by causes now in diurnal action. The time which they required theoretically, is now granted, as it were, or has become absolutely requisite, to account for another class of phenomena brought to light by more recent investigations. It must always have been evident to unbiassed minds, that successive strata, containing, in regular order of superposition, distinct beds of shells and corals, arranged in families as they grow at the bottom of the sea, could only have been formed by slow and insensible degrees in a great lapse of ages; yet, until organic remains were minutely examined and specifically determined, it was rarely possible to prove that the series of deposits met with in one country was not formed simultaneously with that found in another. But we are now able to determine, in numerous instances, the relative dates of sedimentary rocks in distant regions, and to show, by their organic remains, that they were not of contemporary origin, but formed in succession. We often find, that where an interruption in the consecutive formation in one district is indicated by a sudden transition from one assemblage of fossil species to another, the chasm is filled up, in some other district, by other important groups of strata. The more attentively we study the European continent, the greater we find the extension of the whole series of geological formations. No sooner does the calendar appear to be completed, and the signs of a succession of physical events arranged in chronological order, than we are called upon to intercalate, as it were, some new periods of vast duration. A geologist, whose observations have been confined to England, is accustomed to consider the superior and newer groups of marine strata in our island as modern, and such they are, comparatively speaking; but when he has travelled through the Italian peninsula and in Sicily, and has seen strata of more recent origin forming mountains several thousand feet high, and has marked a long series both of volcanic and submarine operations, all newer than any of the regular strata which enter largely into the physical structure of Great Britain, he returns with more exalted conceptions of the antiquity of some of those modern deposits, than he before entertained of the oldest of the British series. We cannot reflect on the concessions thus extorted from us, in regard to the duration of past time, without foreseeing that the period may arrive when part of the Huttonian theory will be combated on the ground of its departing too far from the assumption of uniformity in the order of nature. On a closer investigation of extinct volcanos, we find proofs that they broke out at successive eras, and that the eruptions of one group were often concluded long before others had commenced their activity. Some were burning when one class of organic beings were in existence, others came into action when different races of animals and plants existed, – it follows, therefore, that the convulsions caused by subterranean movements, which are merely another portion of the volcanic phenomena, occurred also in succession, and their effects must be divided into separate sums, and assigned to separate periods of time; and this is not all: – when we examine the volcanic products, whether they be lavas which flowed out under water or upon dry land, we find that intervals of time, often of great length, intervened between their formation, and that the effects of one eruption were not greater in amount than that which now results during ordinary volcanic convulsions. The accompanying or preceding earthquakes, therefore, may be considered to have been also successive, and to have been in like manner interrupted by intervals of time, and not to have exceeded in violence those now experienced in the ordinary course of nature. Already, therefore, may we regard the doctrine of the sudden elevation of whole continents by paroxysmal eruptions as invalidated; and there was the greatest inconsistency in the adoption of such a tenet by the Huttonians, who were anxious to reconcile former changes to the present economy of the world. It was contrary to analogy to suppose, that Nature had been at any former epoch parsimonious of time and prodigal of violence – to imagine that one district was not at rest while another was convulsed – that the disturbing forces were not kept under subjection, so as never to carry simultaneous havoc and desolation over the whole earth, or even over one great region. If it could have been shown, that a certain combination of circumstances would at some future period produce a crisis in the subterranean action, we should certainly have had no right to oppose our experience for the last three thousand years as an argument against the probability of such occurrences in past ages; but it is not pretended that such a combination can be foreseen. In speculating on catastrophes by water, we may certainly anticipate great floods in future, and we may therefore presume that they have happened again and again in past times. The existence of enormous seas of fresh-water, such as the North American lakes, the largest of which is elevated more than six hundred feet above the level of the ocean, and is in parts twelve hundred feet deep, is alone sufficient to assure us, that the time will come, however distant, when a deluge will lay waste a considerable part of the American continent. No hypothetical agency is required to cause the sudden escape of the confined waters. Such changes of level, and opening of fissures, as have accompanied earthquakes since the commencement of the present century, or such excavation of ravines as the receding cataract of Niagara is now effecting, might breach the barriers. Notwithstanding, therefore, that we have not witnessed within the last three thousand years the devastation by deluge of a large continent, yet, as we may predict the future occurrence of such catastrophes, we are authorized to regard them as part of the present order of Nature, and they may be introduced into geological speculations respecting the past, provided we do not imagine them to have been more frequent or general than we expect them to be in time to come.

The great contrast in the aspect of the older and newer rocks, in their texture, structure, and in the derangement of the strata, appeared formerly one of the strongest grounds for presuming that the causes to which they owed their origin were perfectly dissimilar from those now in operation. But this incongruity may now be regarded as the natural result of subsequent modifications, since the difference of relative age is demonstrated to have been so immense, that, however slow and insensible the change, it must have become important in the course of so many ages. In addition to volcanic heat, to which the Vulcanists formerly attributed too much influence, we must allow for the effect of mechanical pressure, of chemical affinity, of percolation by mineral waters, or permeation by elastic fluids, and the action, perhaps, of many other forces less understood, such as electricity and magnetism. In regard to the signs of upraising and sinking, of fracture and contortion in rocks, it is evident that newer strata cannot be shaken by earthquakes, unless the subjacent rocks are also affected; so that the contrast in the relative degree of disturbance in the more ancient and the newer strata, is one of many proofs that the convulsions have happened in different eras, and the fact confirms the uniformity of the action of subterranean forces, instead of their greater violence in the primeval ages.

The popular doctrine of universal formations, or the unlimited geographical extent of strata, distinguished by similar mineral characters, appeared for a long time to present insurmountable objections to the supposition, that the earth's crust had been formed by causes now acting. If it had merely been assumed, that rocks originating from fusion by subterranean fire presented in all parts of the globe a perfect correspondence in their mineral composition, the assumption would not have been extravagant; for, as the elementary substances that enter largely into the composition of rocks are few in number, they may be expected to arrange themselves invariably in the same forms, whenever the elementary particles are freely exposed to the action of chemical affinities. But when it was imagined that sedimentary mixtures, including animal and vegetable remains, and evidently formed in the beds of ancient seas, were of homogeneous nature throughout a whole hemisphere, or even farther, the dogma precluded at once all hope of recognizing the slightest analogy between the ancient and modern causes of decay and reproduction. For we know that existing rivers carry down from different mountain-chains sediment of distinct colours and composition; where the chains are near the sea, coarse sand and gravel is swept in; where they are distant, the finest mud. We know, also, that the matter introduced by springs into lakes and seas is very diversified in mineral composition; in short, contemporaneous strata now in the progress of formation are greatly varied in their composition, and could never afford formations of homogeneous mineral ingredients co-extensive with the greater part of the earth's surface. This theory, however, is as inapplicable to the effects of those operations to which the formation of the earth's crust is due, as to the effects of existing causes. The first investigators of sedimentary rocks had never reflected on the great areas occupied by modern deltas of large rivers; still less on the much greater areas over which marine currents, preying alike on river-deltas, and continuous lines of sea-coast, might be diffusing homogeneous mixtures. They were ignorant of the vast spaces over which calcareous and other mineral springs abound upon the land and in the sea, especially in and near volcanic regions, and of the quantity of matter discharged by them. When, therefore, they ascertained the extent of the geographical distribution of certain groups of ancient strata – when they traced them continuously from one extremity of Europe to the other, and found them flanking, throughout their entire range, great mountain-chains, they were astonished at so unexpected a discovery; and, considering themselves at liberty to disregard all modern analogy, they indulged in the sweeping generalization, that the law of continuity prevailed throughout strata of contemporaneous origin over the whole planet. The difficulty of dissipating this delusion was extreme, because some rocks, formed under similar circumstances at different epochs, present the same external characters, and often the same internal composition; and all these were assumed to be contemporaneous until the contrary could be shown, which, in the absence of evidence derived from direct superposition, and in the scarcity of organic remains, was often impossible.

Innumerable other false generalizations have been derived from the same source; such, for instance, as the former universality of the ocean, now disproved by the discovery of the remains of terrestrial vegetation, contemporary with every successive race of marine animals; but we shall dwell no longer on exploded errors, but proceed at once to contend against weightier objections, which will require more attentive consideration.

CHAPTER 6

Assumed Discordance of the Ancient and Existing Causes of Change Controverted – Climate

That the climate of the northern hemisphere has undergone an important change, and that its mean annual temperature must once have resembled that now experienced within the tropics, was the opinion of some of the first naturalists who investigated the contents of ancient strata. Their conjecture became more probable when the shells and corals of the secondary rocks were more carefully examined, for these organic remains were found to be intimately connected by generic affinity with species now living in warmer latitudes. At a later period, many reptiles, such as turtles, tortoises, and large saurian animals, were discovered in the European strata in great abundance; and they supplied new and powerful arguments, from analogy, in support of the doctrine, that the heat of the climate had been great when our secondary formations were deposited. Lastly, when the botanist turned his attention to the specific determination of fossil plants, the evidence acquired the fullest confirmation, for the flora of a country is peculiarly influenced by temperature; and the ancient vegetation of the earth might, more readily than the forms of animals, have afforded conflicting proofs, had the popular theory been without foundation. When the examination of animal and vegetable remains was extended to rocks in the most northern parts of Europe and North America, and even to the Arctic regions, indications of the same revolution in climate were discovered.

It cannot be said, that in this, as in many other departments of geology, we have investigated the phenomena of former eras, and neglected those of the present state of things. On the contrary, since the first agitation of this interesting question, the accessions to our knowledge of living animals and plants have been immense, and have far surpassed all the data previously obtained for generalizing, concerning the relation of certain types of organization to particular climates. The tropical and temperate zones of South America and of Australia have been explored; and, on close comparison, it has been found, that scarcely any of the species of the animate creation in these extensive continents are identical with those inhabiting the old world. Yet the zoologist and botanist, well acquainted with the geographical distribution of organic beings in other parts of the globe, would have been able, if distinct groups of species had been presented to them from these regions, to recognise those which had been collected from latitudes within, and those which were brought from without the tropics.

Before we attempt to explain the probable causes of great vicissitudes of temperature on the earth's surface, we shall take a rapid view of some of the principal data which appear to warrant, to the utmost extent, the popular opinions now entertained on the subject. To insist on the soundness of the inference, is the more necessary, because some zoologists have of late undertaken to vindicate the uniformity of the laws of nature, not by accounting for former fluctuations in climate, but by denying the value of the evidence on this subject.[20]

It is not merely by reasoning from analogy that we are led to infer a diminution of temperature in the climate of Europe; there are direct proofs in confirmation of the same doctrine, in the only countries hitherto investigated by expert geologists where we could expect to meet with direct proofs. It is not in England or Northern France, but around the borders of the Mediterranean, from the South of Spain to Calabria, and in the islands of the Mediterranean, that we must look for conclusive evidence on this question; for it is not in strata, where the organic remains belong to extinct species, but where living species abound in a fossil state, that a theory of climate can be subjected to the experimentum crucis. In Sicily, Ischia, and Calabria, where the fossil testacea of the more recent strata belong almost entirely to species now known to inhabit the Mediterranean, the conchologist remarks, that individuals in the inland deposits exceed in their average size their living analogues.* Yet no doubt can be

---

* I collected several hundred species of shells in Sicily, some from an elevation of several thousand feet, and forty species or more in Ischia, partly from an elevation of above one thousand feet, and these were carefully compared with recent shells procured by Professor O. G. Costa, from the Neapolitan seas. Not only were the fossil species for the most part identical with those now living, but the relative abundance in which different species occur in the strata and in the sea corresponds in a remarkable manner. Yet the larger average size of the fossil individuals of many species was very striking. A comparison of the fossil shells of the more modern strata of Calabria and Otranto, in the collection of Professor Costa, afforded similar results.

entertained, on the ground of such difference in their dimensions, of their specific identity, because the living individuals attain sometimes, though rarely, the average size of the fossils; and so perfect is the preservation of the latter, that they retain, in some instances, their colour, which affords an additional element of comparison.

As we proceed northwards in the Italian peninsula, and pass from the region of active, to that of extinct volcanos – from districts now violently convulsed from time to time, to those which are comparatively undisturbed by earthquakes, we find the assemblage of fossil shells, in the modern (Subapennine) strata, to depart somewhat more widely from the type of the neighbouring seas. The proportion of species, identifiable with those now living in the Mediterranean, is still considerable; but it no longer predominates, as in the South of Italy, over the unknown species. Although occurring in localities which are removed several degrees farther from the equator (as at Sienna, Parma, Asti, &c.), the shells yield clear indications of a hotter climate. Many of them are common to the Subapennine hills, to the Mediterranean, and to the Indian Ocean. Those in the fossil state, and their living analogues from the tropics, correspond in size; whereas the individuals of the same species from the Mediterranean are dwarfish and degenerate, and stunted in their growth, for want of conditions which the Indian Ocean still supplies.

This evidence amounts to demonstration, and is not neutralized by any facts of a conflicting character; such, for instance, as the association, in the same group, of individuals referrible to species now confined to arctic regions. On the contrary, whenever any of the fossils shells are identified with living species foreign to the Mediterranean, it is not in the Northern Ocean, but between the tropics, that they must be sought. On the other hand, the associated unknown species belong, for the most part, to genera which are either exclusively limited to equinoctial regions, or are now most largely developed there. Of the former, the genus Pleurotoma is a remarkable example; of the latter, the genus Cypræa.

When we proceed to the central and northern parts of Europe, far from the modern theatres of volcanic action, and where there is no evidence of considerable inequalities of the earth's surface having been produced since the present species were in existence, our opportunities are necessarily more limited of procuring evidence from the contents of marine strata. It is only in lacustrine deposits, or in ancient river-beds, or in the sand and gravel of land-floods, or the stalagmite of ancient caverns once inhabited by wild beasts, that we can obtain access to proofs of the changes which animal life underwent during those periods when the marine strata already adverted to were deposited farther to the south. As far, however, as proofs from analogy can be depended upon, nothing can be more striking than the harmony of the testimony derived from the last-mentioned sources. We often find, in such situations, the remains of extinct species of quadrupeds, such as the elephant, rhinoceros, hippopotamus, hyæna, and tiger, which belong to genera now confined to warmer regions. Some of the accompanying fossil species, which are identifiable with those now living, belong to animals which inhabit the same latitudes at present. It seems, therefore, fair to infer, that the same change of climate which has caused certain Indian species of testacea to become rare, or to degenerate in size, or to disappear from the Mediterranean, and certain genera of the Subapennine hills, now exclusively tropical, to retain no longer any representatives in the adjoining seas, has also contributed to the annihilation of certain genera of land-mammifera, which inhabited the continents at about the same epoch. The mammoth (*Elephas primigenius*), and other extinct animals of the same era, may not have required the same temperature as their living congeners within the tropics; but we may infer, that the climate was milder than that now experienced in some of the regions once inhabited by them, because, in Northern Russia, where their bones are found in immense numbers, it would be difficult, if not impossible, for such animals to obtain subsistence at present, during an arctic winter.* It has been said, that as the modern northern animals migrate, the Siberian elephant may also have shifted his place during the inclemency of the season, but this conjecture seems forced, even in regard to the elephant, and still more so, when applied to the Siberian rhinoceros, found in the frozen gravel of that country; as animals of this genus are heavy and slow in their motions, and can hardly be supposed to have

---

* I fully agree with Dr. Fleming, that the kind of food which the existing species of elephant prefers will not enable us to determine, or even to offer a feasible conjecture, concerning that of the extinct species. No one, as he observes, acquainted with the gramineous character of the food of our fallow-deer, stag, or roe, would have assigned a lichen to the rein-deer. But, admitting that the trees and herbage on which the fossil elephants and rhinoceroses may have fed were not of a tropical character, but such perhaps as now grow in the temperate zone, it is still highly improbable that the vegetation which nourished these great quadrupeds was as scanty as that of our arctic regions, or that it was covered during the greater part of every year by snow.

accomplished great periodical migrations to southern latitudes. That the mammoth, however, continued for a long time to exist in Siberia after the winters had become extremely cold, is demonstrable, since their bones are found in icebergs, and in the frozen gravel, in such abundance as could only have been supplied by many successive generations. So many skeletons could not have belonged to herds which lived at one time in the district, even if those northern countries had once been clothed with vegetation as luxuriant as that of an Indian jungle. But, if we suppose the change to have been extremely slow, and to have consisted, not so much in a diminution of the mean annual temperature, as in an alteration from what has been termed an 'insular' to an 'excessive' climate, from one in which the temperature of winter and summer were nearly equalized to one wherein the seasons were violently contrasted, we may, perhaps, explain the phenomenon. Siberia and other arctic regions, after having possessed for ages a more uniform temperature, may, after certain changes in the form of the arctic land, have become occasionally exposed to extremely severe winters. When these first occurred at distant intervals, the drift snow would fill the valleys, and herds of herbivorous quadrupeds would be surprised and buried in a frozen mass, as often happens to cattle and human beings, overwhelmed, in the Alpine valleys of Switzerland, by avalanches. When valleys have become filled with ice, as those of Spitzbergen, the contraction of the mass causes innumerable deep rents, such as are seen in the mer de glace on Mont Blanc. These deep crevices usually become filled with loose snow, but sometimes a thin covering is drifted across the mouth of the chasm, capable of sustaining a certain weight. Such treacherous bridges are liable to give way when heavy animals are crossing, which are then precipitated at once into the body of a glacier, which slowly descends to the sea, and becomes a floating iceberg. As bears, foxes, and deer now abound in Spitzbergen, we may confidently assume that the imbedding of animal remains in the glaciers of that island must be an event of almost annual occurrence. The conversion of drift snow into permanent glaciers and icebergs, when it happens to become covered over with alluvial matter, transported by torrents and floods, is by no means a rare phenomenon in the arctic regions. During a series of milder seasons intervening between the severe winters, the mammoths may have recovered their numbers, and the rhinoceroses may have multiplied again, so that the repetition of such catastrophes may have been indefinite. The increasing cold, and greater frequency of inclement winters, would at last thin their numbers, and their final extirpation would be consummated by the rapid augmentation of other herbivorous quadrupeds, more fitted for the new climate.

That the greater part of the elephants lived in Siberia after it had become subject to intense cold, is confirmed, among other reasons, by the state of the ivory, which has been so largely exported in commerce. Its perfect preservation indicates, that from the period when the individuals died, their remains were either buried in a frozen soil, or at least were not exposed to decay in a warm atmosphere. The same conclusion may be deduced from the clothing of the mammoth, of which the entire carcase was discovered by Mr. Adams on the shores of the frozen ocean, near the mouth of the river Lena, inclosed in a mass of ice. The skin of that individual was covered with long hair and with thick wool, about an inch in length. Bishop Heber informs us, that along the lower range of the Himalaya mountains, in the north-eastern borders of the Delhi territory, between lat. 29° and 30°, he saw an Indian elephant covered with shaggy hair. In that region, where, within a short space, a nearly tropical, and a cold climate meet, dogs and horses become covered, in the course of a winter or two, with shaggy hair, and many other species become, in as short a time, clothed with the same fine short shawl-wool, which distinguishes the indigenous species of the country. Lions, tigers, hyænas, are there found with elks, chamois, and other species of genera usually abundant in colder latitudes.

If we pass from the consideration of these more modern deposits, whether of marine or continental origin, in which existing species are intermixed with the extinct, to strata of somewhat higher antiquity, (older tertiary strata, Calcaire Grossier, London clay, fresh-water formations of Paris and Isle of Wight, &c.,) we can only reason from analogy, since the species, whether of mammalia, reptiles, or testacea, are scarcely in any instance identifiable with any now in being. In these strata, whether they were formed in seas or lakes, we find the remains of many animals, analogous to those of hot climates, such as the crocodile, turtle, and tortoise, and many large shells of the genus nautilus, and plants indicating such a temperature as is now found along the borders of the Mediterranean. A great interval of time appears to have elapsed between the deposition of the last mentioned (tertiary) strata, and the *secondary* formations, which constitute the principal portion of the more elevated land in Europe. In these secondary rocks a very distinct assemblage of organized fossils are entombed, all of unknown species, and many of them referrible to genera, and families now most abundant between the tropics. Among the most remarkable, are many gigantic reptiles, some of them herbivorous, others carnivorous, and far exceeding in size any now known even in the torrid zone. The genera are for the most part extinct, but some of them, as the crocodile and monitor, have still representatives in the warmest parts of the earth. Coral reefs also were evidently numerous in the seas of the same period, and composed of species belonging to genera now characteristic of a tropical climate. The number of immense chambered shells also leads us to infer an elevated temperature; and the associated fossil plants, although imperfectly known, tend to the same conclusion, the Cycadeæ constituting the most numerous family. But the study of the fossil flora of the coal deposits of still higher antiquity, has yielded the most extraordinary evidence of an extremely hot climate, for it consisted almost exclusively of large vascular cryptogamic plants. We learn, from the labours of M. Ad. Brongniart, that there existed, at that epoch, Equiseta upwards of ten feet high, and from five to six inches in diameter; tree ferns of from forty to fifty feet in height, and aborescent Lycopodiaceæ, of from sixty to seventy feet high. Of the above classes of vegetables, the species are all small at present in cold climates; while in tropical regions, there occur, together with small species, many of a much greater size, but their development at present, even in the hottest parts of the globe, is inferior to that indicated by the petrified forms of the coal formation. An elevated and uniform temperature, and great humidity in the air, are the causes most favourable for the numerical predominance, and the great size of these plants within the torrid zone at present.* If the gigantic size and form of these fossil plants are remarkable, still more so is the extent of

* Humboldt, in speaking of the vegetation of the present era, considers the laws which govern the distribution of vegetable forms to be sufficiently constant to enable a botanist, who is informed of the number of one class of plants, to conjecture, with tolerable accuracy, the relative number of all others. It is premature, perhaps, to apply this law of proportion to the fossil botany of strata, between the coal formation and the chalk, as M. Adolphe Brongniart has attempted, as the number of species hitherto procured is so inconsiderable, that the quotient would be materially altered by the addition of one or two species. It may also be objected, that the fossil flora consists of such plants as may accidentally have been floated into seas, lakes, or estuaries, and may often, perhaps always, give a false representation of the numerical relations of families, then living on the land. Yet, after allowing for all liability to error on these grounds, the argument founded on the comparative numbers of the fossil plants of the carboniferous strata is very strong.

Martius informs us, that on seeing the tesselated surface of the stems of arborescent ferns their geographical distribution; for impressions of arborescent ferns, such as characterize our English carboniferous strata, have been brought from Melville island, in latitude 75°. The corals and chambered shells, which occur in beds interstratified with the coal (as in mountain limestone), afford also indications of a warm climate, – the gigantic orthocerata of this era being, to recent multilocular shells, what the fossil ferns, equiseta, and other plants of the coal strata, are in comparison with plants now growing within the tropics. These shells also, like the vegetable impressions, have been brought from rocks in very high latitudes in North America.

In vain should we attempt to explain away the phenomena of the carboniferous and other secondary formations, by supposing that the plants were drifted from equatorial seas. During the accumulation, and consolidation of so many sedimentary deposits, and the various movements and dislocations to which they were subjected at different periods, rivers and currents must often have changed their direction, and wood might as often be floated from the arctic towards tropical seas, as in an opposite direction. It is undeniable, that the materials for future beds of lignite and coal are now amassed in high latitudes far from the districts where the forests grew, and on shores where scarcely a stunted shrub can now exist. The Mackenzie, and other rivers of North America, carry pines with their roots attached for many hundred miles towards the north, into the arctic sea, where they are imbedded in deltas, and some of them drifted still farther, by currents towards the pole. But such agency, although it might account for some partial anomalies in the admixture of vegetable remains of different climes, can by no means weaken the arguments deduced from the general character of fossil vegetable remains. We cannot suppose the leaves of tree ferns to be transported by water for thousands of miles, without being injured; nor, if this were possible, would the same hypothesis explain the presence of uninjured corals and multilocular shells of contemporary origin, for these must have lived in the same latitudes where they are now inclosed in rocks. The plants, moreover, whose remains have given rise to the coal beds, must be supposed to have grown upon the same land, the destruction of which provided materials for the sandstones and conglomerates of that group of strata. The coarseness of the particles of many of these rocks attests that they were not borne from very remote

---

in Brazil, he was reminded of their prototypes, in the impressions which he had seen in the coal-mines of Germany.

localities, but were most probably derived from islands in a vast sea, which was continuous, at that time, over a great part of the northern hemisphere, as is demonstrated by the great extent of the mountain and transition limestone formations. The same observation is applicable to many secondary strata of a later epoch. There must have been dry land in these latitudes, to provide materials by its disintegration for sandstones, – to afford a beach whereon the oviparous reptiles deposited their eggs, – to furnish an habitation for the opossum of Stonesfield, and the insects of Solenhofen. The vegetation of the same lands, therefore, must in general have imparted to fossil floras their prevailing character.

From the considerations above enumerated, we must infer, that the remains both of the animal and vegetable kingdom preserved in strata of different ages, indicate that there has been a great diminution of temperature throughout the northern hemisphere, in the latitudes now occupied by Europe, Asia, and America. The change has extended to the arctic circle, as well as to the temperate zone. The heat and humidity of the air, and the uniformity of climate, appear to have been most remarkable when the oldest strata hitherto discovered were formed. The approximation to a climate similar to that now enjoyed in these latitudes, does not commence till the era of the formations termed tertiary, and while the different tertiary rocks were deposited in succession; the temperature seems to have been still farther lowered, and to have continued to diminish gradually, even after the appearance of a great portion of existing species upon the earth.

CHAPTER 7

Climate, continued

As the proofs enumerated in the last chapter indicate that the earth's surface has experienced great changes of climate since the deposition of the older sedimentary strata, we have next to inquire, how such vicissitudes can be reconciled with the existing order of nature. The cosmogonist has availed himself of this, as of every obscure problem in geology, to confirm his views concerning a period when the laws of the animate and inanimate world were wholly distinct from those now established; and he has in this, as in all other cases, succeeded so far, as to divert attention from that class of facts, which, if fully understood, might probably lead to an explanation of the phenomenon. At first, it was imagined that the earth's axis had been for ages perpendicular to the plane of the ecliptic, so that there was a perpetual equinox, and unity of seasons throughout the year: – that the planet enjoyed this *'paradisiacal'* state until the era of the great flood; but in that catastrophe, whether by the shock of a comet, or some other convulsion, it lost its equal poize, and hence the obliquity of its axis, and with that the varied seasons of the temperate zone, and the long nights and days of the polar circles. When the advancement of astronomical science had exploded this theory, it was assumed, that the earth at its creation was in a state of fluidity, and red hot, and that ever since that era it had been cooling down, contracting its dimensions, and acquiring a solid crust, – an hypothesis equally arbitrary, but more calculated for lasting popularity, because, by referring the mind directly to the beginning of things, it requires no support from observations, nor from any ulterior hypothesis. They who are satisfied with this solution are relieved from all necessity of inquiry into the present laws which regulate the diffusion of heat over the surface, for however well these may be ascertained, they cannot possibly afford a full and exact elucidation of the internal changes of an embryo world. As well might an ornithologist study the plumage and external form of a full-fledged bird, in the hope of divining the colour of its egg, or the mysterious metamorphoses of the yolk during incubation.

But if, instead of vague conjectures as to what might have been the state of the planet at the era of its creation, we fix our thoughts steadily on the connection at present between climate and the distribution of land and sea; and if we then consider what influence former fluctuations in the physical geography of the earth must have had on superficial temperature, we may perhaps approximate to a true theory. If doubt still remain, it should be ascribed to our ignorance of the laws of Nature, not to revolutions in her economy; – it should stimulate us to farther research, not tempt us to indulge our fancies in framing imaginary systems for the government of infant worlds.

In considering the laws which regulate the diffusion of heat over the globe, says Humboldt, we must beware not to regard the climate of Europe as a type of the temperature which all countries placed under the same latitudes enjoy. The physical sciences, observes this philosopher, always bear the impress of the places where they began to be cultivated; and, as in geology, an attempt was at first made to refer all the volcanic phenomena to those of the volcanos in Italy, so in meteorology, a small part of the old world, the centre of the primitive civilization of Europe, was for a long time considered a type to which the climate of all corresponding latitudes might be referred. But this region, constituting only one-seventh of the whole globe, proved eventually to be the exception to the general rule; and for the same reason we may warn the geologist to be on his guard, and not hastily to assume that the temperature of the earth in the present era is a type of that which most usually obtains, since he contemplates far mightier alterations in the position of land and sea, than those which now cause the climate of Europe to differ from that of other countries in the same parallels.

It is now well ascertained that zones of equal warmth, both in the atmosphere and in the waters of the ocean, are neither parallel to the equator nor to each other.* It is also discovered that the same mean annual temperature may exist in two places which enjoy very different

* We are indebted to Baron Alex. Humboldt for collecting together in a beautiful essay, the scattered data on which some approximation to a true theory of the distribution of heat over the globe may be founded. Many of these data are derived from the author's own observations, and many from the works of M. Prevost on the radiation of heat, and other writers. See Humboldt on Isothermal Lines, Mémoires d'Arcueil, tom. iii. translated in the Edin. Phil. Journ. vol. iii. July, 1820.

climates, for the seasons may be nearly equalized or violently contrasted. Thus the lines of equal winter temperature do not coincide with the lines of equal annual heat, or the isothermal lines. The deviations of all these lines from the same parallel of latitude, are determined by a multitude of circumstances, among the principal of which are the position, direction, and elevation of the continents and islands, the position and depth of the sea, and the direction of currents and of winds.

It is necessary to go northwards in Europe in order to find the same mean quantity of annual heat as in North America. On comparing these two continents, it is found that places situated in the same latitudes, have sometimes a mean difference of temperature amounting to 11° or even sometimes 17° of Fahrenheit; and places on the two continents which have the same mean temperature, have sometimes a difference in latitude of from 7° to 13°. The principal cause of greater intensity of cold in corresponding latitudes of North America and Europe, is the connexion of the former country with the polar circle, by a large tract of land, some of which is from three to five thousand feet in height, and, on the other hand, the separation of Europe from the arctic circle by an ocean. The ocean has a tendency to preserve every where a mean temperature, which it communicates to the contiguous land, so that it tempers the climate, moderating alike an excess of heat or cold. The elevated land, on the other hand, rising to the colder regions of the atmosphere, becomes a great reservoir of ice and snow, attracts, condenses, and congeals vapour, and communicates its cold to the adjoining country. For this reason, Greenland, forming part of a continent which stretches northward to the 82nd degree of latitude, experiences under the 60th parallel a more rigorous climate than Lapland under the 72nd parallel.

But if land be situated between the 40th parallel and the equator, it produces exactly the opposite effect, unless it be of extreme height, for it then warms the tracts of land or sea that intervene between it and the polar circle. For the surface being in this case exposed to the vertical, or nearly vertical rays of the sun, absorbs a large quantity of heat, which it diffuses by radiation into the atmosphere. For this reason, the western parts of the old continent derive warmth from Africa, 'which, like an immense furnace,' says Malte-Brun, 'distributes its heat to Arabia, to Turkey in Asia, and to Europe.'[21] On the contrary, Asia in its north-eastern extremity, experiences in the same latitude extreme cold, for it has land on the north between the 60th and 70th parallel, while to the south it is separated from the equator by the North Pacific.

In consequence of the more equal temperature of the waters of the ocean, the climate of islands and coasts differs essentially from that of the interior of continents, the former being characterized by mild winters and less temperate summers; for the sea breezes moderate the cold of winter, as well as the summer heat. When, therefore, we trace round the globe those belts in which the mean annual temperature is the same, we often find great differences in climate; for there are *insular* climates where the seasons are nearly equalized, and *excessive* climates as they have been termed, where the temperature of winter and summer is strongly contrasted. The whole of Europe, compared with the eastern parts of America and Asia, has an insular climate. The northern part of China, and the Atlantic region of the United States, exhibit 'excessive climates.' We find at New York, says Humboldt, the summer of Rome and the winter of Copenhagen; at Quebec, the summer of Paris and the winter of Petersburgh. At Pekin, in China, where the mean temperature of the year is that of the coasts of Brittany, the scorching heats of summer are greater than at Cairo, and the winters as rigorous as at Upsal.

If lines be drawn round the globe through all those places which have the same winter temperature, they are found to deviate from the terrestrial parallels much farther than the lines of equal mean annual heat. For the lines of equal winter in Europe are often curved so as to reach parallels of latitude 9° or 10° distant from each other, whereas the isothermal lines only differ from 4° to 5°.

Among other influential causes, both of remarkable diversity in the mean annual heat, and of unequal division of heat in the different seasons, are the direction of currents and the accumulation and drifting of ice in high latitudes. That most powerful current, the Gulf stream, after doubling the Cape of Good Hope, flows to the northward along the western coast of Africa, then crosses the Atlantic, and accumulates in the Gulf of Mexico. It then issues through the Straits of Bahama, running northwards at the rate of four miles an hour, and retains in the parallel of 38°, nearly one thousand miles from the above strait, a temperature 10° Fahr. warmer than the air. The general climate of Europe is materially affected by the volume of warmer water thus borne northwards, for it maintains an open sea free from ice in the meridian of East Greenland and Spitzbergen, and thus moderates the cold of all the lands lying to the south. Until the waters of the great current reach the circumpolar sea, their specific gravity is less than that of the lower strata of water; but when they arrive near Spitzbergen, they meet with the water of melted ice which is still lighter, for it is a well known law of this fluid, that it passes the point of greatest density when cooled down below 40°, and between that and the freezing point expands again. The warmer current, therefore, being now the heavier, sinks below the surface, so that in the lower regions it is found to be from 16° to 20° Fahrenheit, above the mean temperature of the climate. The movements of the sea, however, cause this under current sometimes to appear at the surface, and greatly to moderate the cold.

The great glaciers generated in the valleys of Spitzbergen, in the 79° of north latitude, are almost all cut off at the beach, being melted by the feeble remnant of heat retained by the Gulf stream. In Baffin's Bay, on the contrary, on the east coast of old Greenland, where the temperature of the sea is not mitigated by the same cause, and where there is no warmer under-current, the glaciers stretch out from the shore, and furnish repeated crops of mountainous masses of ice which float off into the ocean. The number and dimensions of these bergs is prodigious. Capt. Ross saw several of them together in Baffin's Bay aground in water fifteen hundred feet deep! Many of them are driven down into Hudson's Bay, and, accumulating there, diffuse excessive cold over the neighbouring continent, so that Captain Franklin reports, that at the mouth of Hayes river, which lies in the same latitude as the north of Prussia or the south of Scotland, ice is found every where in digging wells at the depth of four feet!

When we compare the climate of the northern and southern hemispheres, we obtain still more instruction in regard to the influence of the distribution of land and sea on climate. The dry land in the southern hemisphere, is to that of the northern in the ratio only of one to three, excluding from our consideration that part which lies between the pole and the 74° of south latitude, which has hitherto proved inaccessible. The predominance of ice in the antarctic over the arctic zone is very great; for that which encircles the southern pole, extends to lower latitudes by ten degrees than that around the north pole. It is probable that this remarkable difference is partly attributable, as Cook conjectured, to the existence of a considerable tract of high land between the 70th parallel of south latitude and the pole. There is, however, another reason suggested by Humboldt, to which great weight is due, – the small quantity of land in the tropical and temperate zones south of the line. If Africa and New Holland extended farther to the south, a diminution of ice would take place in consequence of the radiation of heat from these continents during summer, which would warm the contiguous sea and rarefy the air. The heated aërial currents would then ascend and flow more rapidly towards the south pole, and moderate the winter. In confirmation of these views, it is stated that the cap of ice, which extends as far as the 68° and 71° of south latitude, advances more towards the equator whenever it meets a free sea; that is, wherever the extremities of the present continents are not opposite to it; and this circumstance seems explicable only on the principle above alluded to, of the radiation of heat from the lands so situated.

Before the amount of difference between the temperature of the two hemispheres was ascertained, it was referred by astronomers to the acceleration of the earth's motion in its perihelium; in consequence of which the spring and summer of the southern hemisphere are shorter, by nearly eight days, than those seasons north of the equator. A sensible effect is probably produced by this source of disturbance, but it is quite inadequate to explain the whole phenomenon. It is, however, of importance to the geologist to bear in mind, that in consequence of the procession of the equinoxes the two hemispheres receive alternately, each for a period of upwards of 10,000 years, a greater share of solar light and heat. This cause may sometimes tend to counterbalance inequalities resulting from other circumstances of a far more influential nature; but, on the other hand, it must sometimes tend to increase the extreme of deviation which certain combinations of causes produce at distant epochs. But, whatever may now be the inferiority of heat in the temperate and arctic zones south of the line, it is quite evident that the cold would be far more intense if there happened, instead of open sea, to be tracts of elevated land between the 55th and 70th parallel; for, in Sandwich land, in 54° and 58° of south latitude, the perpetual snow and ice reach to the sea beach; and what is still more astonishing, in the island of Georgia, which is in the 53° south latitude, or the same parallel as the central counties of England, the perpetual snow descends to the level of the ocean. When we consider this fact, and then recollect that the highest mountains in Scotland do not attain the limit of perpetual snow on this side of the equator, we learn that latitude is one only of many powerful causes, which determine the climate of particular regions of the globe. The permanence of the snow, in this instance, is partly due to the floating ice, which chills the atmosphere and condenses the vapour, so that in summer the sun cannot pierce through the foggy air. The distance to which icebergs float from the polar regions on the opposite sides of the line, is, as might have been anticipated, very different. Their extreme limit in the northern hemisphere appears to be the Azores (north latitude 42°), to which isles they are sometimes drifted from Baffin's Bay. But in the other hemisphere they have been seen, within the last two years, at different points off the Cape of Good Hope, between latitude 36° and 39°. One of these was two miles in circumference, and 150 feet high. Others rose from 250 to 300 feet above the level of the sea, and were, therefore, of great volume below, since it is ascertained, by experiments on the buoyancy of ice floating in sea-water, that for every solid foot seen above, there must at least be eight feet below water. If ice islands from the north polar regions floated as far, they might reach Cape St. Vincent, and, then being drawn by the current that always sets in from the Atlantic through the Straits of Gibraltar, be drifted into the Mediterranean, where clouds and mists would immediately deform the serene sky of spring and summer.

The great extent of sea gives a particular character to climates south of the equator, the winters being mild, and the summers cold. Thus, in Van Dieman's land, corresponding nearly in latitude to Rome, the winters are more mild than at Naples, and the summers not warmer than those at Paris, which is 7° farther from the equator. The effect on vegetation is very remarkable: – tree ferns, for instance, which require abundance of moisture, and an equalization of the seasons, are found in Van Dieman's land in latitude 42°, and in New Zealand in south latitude 45°. The orchideous parasites also advance towards the 38° and 42° of south latitude.*

Having offered these brief remarks on the diffusion of heat over the globe in the present state of the surface, we shall now proceed to speculate

---

* These forms of vegetation might perhaps be developed in still higher latitudes, if the ice in the antarctic circle did not extend farther from the pole than in the arctic. Humboldt observes, that it is in the *mountainous, temperate, humid*, and *shady* parts of the equatorial regions, that the family of ferns produces the greatest number of species. As we know, therefore, that elevation often compensates the effect of latitude in plants, we may easily understand that a class of vegetables which grow at a certain height in the torrid zone, would flourish on the plains far from the equator, provided the temperature throughout the year was equally uniform.

on the vicissitudes of climate, which must attend those endless variations in the geographical features of our planet, which are contemplated in geology. In order to confine ourselves within the strict limits of analogy, we shall assume, 1st, That the proportion of dry land to sea continues always the same. 2dly, That the volume of the land rising above the level of the sea, is a constant quantity; and not only that its mean, but that its extreme height, are only liable to trifling variations. 3dly, That both the mean and extreme depth of the sea are equal at every epoch; and, 4thly, It will be consistent, with due caution, to assume, that the grouping together of the land in great continents is a necessary part of the economy of nature; for it is possible, that the laws which govern the subterranean forces, and which act simultaneously along certain lines, cannot but produce, at every epoch, continuous mountain-chains; so that the subdivision of the whole land into innumerable islands may be precluded. If it be objected, that the maximum of elevation of land and depth of sea are probably not constant, nor the gathering together of all the lands in certain parts, nor even perhaps the relative extent of land and water; we reply, that the arguments which we shall adduce will be greatly strengthened, if, in these peculiarities of the surface, there be considerable deviations from the present type. If, for example, all other circumstances being the same, the land is at one time more divided into islands than at another, a greater uniformity of climate might be produced, the mean temperature remaining unaltered; or if, at another era, there were mountains higher than the Himalaya, these, when placed in high latitudes would cause a greater excess of cold. So if we suppose, that at certain periods no chain of hills in the world rose beyond the height of 10,000 feet, a greater heat might then have prevailed than is compatible with the existence of mountains thrice that elevation.

However constant we believe the relative proportion of sea and land to continue, we know that there is annually some small variation in their respective geographical positions, and that in every century the land is in some parts raised, and in others depressed by earthquakes, and so likewise is the bed of the sea. By these and other ceaseless changes, the configuration of the earth's surface has been remodelled again and again since it was the habitation of organic beings, and the bed of the ocean has been lifted up to the height of some of the loftiest mountains. The imagination is apt to take alarm, when called upon to admit the formation of such irregularities of the crust of the earth, after it had become the habitation of living creatures; but if time be allowed, the operation need not subvert the ordinary repose of nature, and the result is insignificant, if we consider how slightly the highest mountain chains cause our globe to differ from a perfect sphere. Chimborazo, although it rises to more than 21,000 feet above the surface of the sea, would only be represented on an artificial globe, of about six feet in diameter, by a grain of sand less than one-twentieth of an inch in thickness. The superficial inequalities of the earth, then, may be deemed minute in quantity, and their distribution at any particular epoch must be regarded in geology as temporary peculiarities, like the height and outline of the cone of Vesuvius in the interval between two eruptions. But, although the unevenness of the surface is so unimportant, in reference to the magnitude of the globe, it is on the position and direction of these small inequalities that the state of the atmosphere and both the local and general climate are mainly dependent.

Before we consider the effect which a material change in the distribution of land and sea must occasion, it may be well to remark, how greatly organic life may be affected by those minor mutations, which need not in the least degree alter the general temperature. Thus, for example, if we suppose, by a series of convulsions, a certain part of Greenland to become sea, and, in compensation, a tract of land to rise and connect Spitzbergen with Lapland, – an accession not greater in amount than one which the geologist can prove to have occurred in certain districts bordering the Mediterranean, within a comparatively modern period, – this altered form of the land might occasion an interchange between the climate of certain parts of North America and of Europe, which lie in corresponding latitudes. Many European species would probably perish in consequence, because the mean temperature would be greatly lowered; and others would fail in America because it would there be raised. On the other hand, in places where the mean annual heat remained unaltered, some species which flourish in Europe, where the seasons are more uniform, would be unable to resist the great heat of the North American summer, or the intense cold of the winter; while others, now fitted by their habits for the great contrast of the American seasons, would not be fitted for the *insular* climate of Europe. Many plants, for instance, will endure a severe frost, but cannot ripen their seeds without a certain intensity of summer heat and a certain quantity of light; others cannot endure the same intensity of heat or cold. It is now established, that many species of animals, which are at present the contemporaries of man, have survived great changes in the physical geography of the globe. If such species be termed modern, in comparison to races which preceded them, their remains, nevertheless, enter into submarine deposits many hundred miles in length, and which have since been raised from the deep to no inconsiderable altitude. When, therefore, it is shewn that changes of the temperature of the atmosphere may be the consequence of such physical revolutions of the surface, we ought no longer to wonder that we find the distribution of existing species to be *local*, in regard to *longitude* as well as latitude. If all species were now, by an exertion of creative power, to be diffused uniformly throughout those zones where there is an equal degree of heat, and in all respects a similar climate, they would begin from this moment to depart more and more from their original distribution. Aquatic and terrestrial species would be displaced, as Hooke long ago observed, so often as land and water exchanged places; and there would also, by the formation of new mountains and other changes, be transpositions of climate, contributing, in the manner before alluded to, to the local extermination of species.

If we now proceed to consider the circumstances required for a *general* change of temperature, it will appear, from the facts and principles already laid down, that whenever a greater extent of high land is collected in the polar regions, the cold will augment; and the same result will be produced when there shall be more sea between or near the tropics; while, on the contrary, so often as the above conditions are reversed, the heat will be greater. If this be admitted, it will follow as a corollary, that unless the superficial inequalities of the earth be fixed and permanent, there must be never-ending fluctuations in the mean temperature of every zone, and that the climate of one era can no more be a type of every other, than is one of our four seasons of all the rest. It has been well said, that the earth is covered by an ocean, and in the midst of this ocean there are two great islands, and many smaller ones; for the whole of the continents and islands occupy an area scarcely exceeding one-fourth of the whole superficies of the spheroid. Now, on a fair calculation, we may expect that at any given epoch, there will not be more than about one-fourth dry land in a particular region; such, for example, as the arctic and antarctic circles. If, therefore, at present there should happen in the only one of these regions which we can explore, to be much more than this average proportion of land, and some of it above five thousand feet in height, this alone affords ground for concluding, that in the present state of things, the mean heat of the climate is below that which the earth's surface, in its more ordinary state, would enjoy. This presumption is heightened, when we remember that the mean depth of the Atlantic ocean is calculated to be about three miles, and that of the Pacific four miles; so that we might look not only for more than two-thirds sea in the frigid zones, but for water of great depth, which could not readily be reduced to the freezing point. The same opinion is farther confirmed, when we compare the quantity of land lying between the poles and the 30th parallels of north and south latitude, and the quantity placed between those parallels and the equator; for it is clear, that at present we must have not only more than the usual degree of cold in the polar regions, but also less than the average quantity of heat generated in the intertropical zone.

In order to simplify our view of the various changes in climate, which different combinations of geographical circumstances may produce, we shall first consider the conditions necessary for bringing about the extreme of cold, or what may be termed the winter of the 'great year,' or geological cycle, and afterwards, the conditions requisite for producing the maximum of heat, or the summer of the same year.

To begin with the northern hemisphere. Let us suppose those hills of the Italian peninsula and of Sicily, which are of comparatively modern origin, and contain many fossil shells identical with living species, to subside again into the sea, from which they have been raised, and that an extent of land of equal area and height (varying from one to three thousand feet) should rise up in the Arctic ocean, between Siberia and the north pole. In speaking of such changes, we need not allude to the manner in which we conceive it possible that they may be brought about, nor of the time required for their accomplishment, – reserving for a future occasion, not only the proofs that revolutions of equal magnitude have taken place, but that analogous mutations are still in gradual progress. The alteration now supposed in the physical geography of the northern regions would cause additional snow and ice to accumulate where now there is usually an open sea; and the temperature of the greater part of Europe would be somewhat lowered, so as to resemble more nearly that of corresponding latitudes of North America; or, in other words, it might be necessary to travel about 10° farther south, in order to meet with the same climate which we now enjoy. There would be no compensation derived from the disappearance of land in the Mediterranean countries; for, on the contrary, the mean heat of the soil so situated, is probably far above that which would belong to the sea, by which we imagine it to be replaced. But let the configuration of the surface be still further varied, and let some large district within or near the tropics, such as Mexico for example, with its mountains rising to the height of twelve thousand feet and upwards, be converted into sea, while lands of equal elevation and extent are transferred to the arctic circle. From this change there would, in the first place, result a sensible diminution of temperature near the tropic, for the soil of Mexico would no longer be heated by the sun; so that the atmosphere would be less warm, as also the Atlantic, and the Gulf stream. On the other hand, the whole of Europe, Northern Asia, and North America, would feel the influence of the enormous quantity of ice and snow, now generated at vast heights on the new arctic continent. If, as we have already seen, there are some points in the southern hemisphere where snow is perpetual to the level of the sea, in latitudes as low as central England, such might now assuredly be the case throughout a great part of Europe. If at present the extreme limits of drifted icebergs are the Azores, they might easily reach the equator after the changes above supposed. To pursue the subject still farther, let the Himalaya mountains, with the whole of Hindostan, sink down, and their place be occupied by the Indian ocean, and then let an equal extent of territory and mountains, of the same vast height, stretch from North Greenland to the Orkney islands. It seems difficult to exaggerate the amount to which the climate of the northern hemisphere would now be cooled down. But, notwithstanding the great refrigeration which would thus be produced, it is probable that the difference of mean temperature between the arctic and equatorial latitudes would not be increased in a very high ratio, for no great disturbance can be brought about in the climate of a particular region, without immediately affecting all other latitudes, however remote. The heat and cold which surround the globe are in a state of constant and universal flux and reflux. The heated and rarefied air is always rising and flowing from the equator towards the poles in the higher regions of the atmosphere, and, in the lower, the colder air is flowing back to restore the equilibrium. That this circulation is constantly going on in the aërial currents is not disputed, and that a corresponding interchange takes place in the seas, is demonstrated, according to Humboldt, by the cold which is found to exist at great depths between the tropics; and, among other proofs, may be mentioned the great volume of water which the Gulf stream is constantly bearing northwards, while another current flows *from* the north along the coast of Greenland and Labrador, and helps to restore the equilibrium.*

Currents of heavier and colder water pass from the poles towards the equator, which cool the inferior parts of the ocean; so that the heat of the torrid zone, and the cold of the polar circle, balance each other. The refrigeration, therefore, of the polar regions, resulting from the supposed alteration in the distribution of land and sea, would be immediately communicated to the tropics, and from them would extend to the antarctic circle, where the atmosphere and the ocean would be cooled, so that ice and snow would augment. Although the mean temperature of higher latitudes in the southern hemisphere is, as we have stated, for the most part lower than that of the same parallels in the northern, yet for a considerable space on each side of the line, the mean annual heat of the waters is found to be the same in corresponding parallels. When, therefore, by the new position of the land, the generating of icebergs had become of frequent occurrence in the temperate zone, and when they were frequently drifted as far as the equator, the same degree of cold would immediately be communicated as far as the tropic of Capricorn, and from thence to the lands or ocean to the south. The freedom, then, of the circulation of heat and cold from pole to pole being duly considered, it will be evident that the mean quantity of heat which at two different periods visits the same point, may differ far more widely than the mean quantity which any two points receive in the same parallels of latitude, at one and the same period. For the range of temperature in a given zone, or in other words, the curves of the isothermal lines, must always be circumscribed within narrow limits, the climate of each place in that zone being controlled by the combined influence of the geographical peculiarities of all other parts of the earth. But, when we compare the state of things as existing at two distinct epochs, a particular zone may at one time be under the influence of one class of disturbing causes, as for example those of a refrigerating nature, and at another time may be affected by a combination of opposite circumstances. The lands to the north of Greenland cause the present climate of North America to be

---

* In speaking of the circulation of air and water in this chapter, no allusion is made to the trade winds, or to irregularities in the direction of currents, caused by the rotatory motion of the earth. These causes prevent the movements from being direct from north to south, or from south to north, but they do not affect the theory of a constant circulation.

colder than that of Europe in the same latitudes, but they also affect, to a certain extent, the temperature of the atmosphere in Europe; and the entire removal from the northern hemisphere of that great source of refrigeration would not assimilate the mean temperature of America to that now experienced in Europe, but would render the continents on both sides of the Atlantic much warmer.

To return to the state of the earth, after the changes before supposed by us, we must not omit to dwell on the important effects to which a wide expanse of perpetual snow would give rise. It is probable that nearly the whole sea, from the poles to the parallels of 45°, would be frozen over, for it is well known that the immediate proximity of land, is not essential to the formation and increase of field ice, provided there be in some part of the same zone a sufficient quantity of glaciers generated on or near the land, to cool down the sea. Field ice is almost always covered with snow, through which the sun's rays are unable to penetrate, and thus not only land as extensive as our existing continents, but immense tracts of sea in the frigid and temperate zones, would now present a solid surface covered with snow, and reflecting the sun's rays for the greater part of the year. Within the tropics, moreover, where we suppose the ocean to predominate, the sky would no longer be serene and clear, as in the present era; but the melting of floating ice would cause quick condensations of vapour, and fogs and clouds would deprive the vertical rays of the sun of half their power. The whole planet, therefore, would receive annually a smaller proportion of solar influence, and the external crust would part, by radiation, with some of the heat which had been accumulated in it, during a different state of the surface. This heat would be dissipated into the spaces surrounding our atmosphere, which, according to the calculations of M. Fourier, have a temperature much inferior to that of freezing water.

At this period, the climate of equinoctial lands might resemble that of the present temperate zone, or perhaps be far more wintery. They who should then inhabit the small isles and coral reefs, which are now seen in the Indian ocean and South Pacific, would wonder that zoophytes of such large dimensions had once been so prolific in those seas; or if, perchance, they found the wood and fruit of the cocoa-nut tree or the palm silicified by the waters of some mineral spring, or incrusted with calcareous matter, they would muse on the revolutions that had annihilated such genera, and replaced them by the oak, the chestnut, and the pine. With equal admiration would they compare the skeletons of their small lizards with the bones of fossil alligators and crocodiles more than twenty feet in length, which, at a former epoch, had multiplied between the tropics; and when they saw a pine included in an iceberg, drifted from latitudes which we now call temperate, they would be astonished at the proof thus afforded, that forests had once grown where nothing could be seen in their own times but a wilderness of snow.

As we have not yet supposed any mutations to have taken place in the relative position of land and sea in the southern hemisphere, we might still increase greatly the intensity of cold, by transferring the land still remaining in the equatorial and contiguous regions, to higher southern latitudes; but it is unnecessary to pursue the subject farther, as we are too ignorant of the laws governing the direction of subterranean forces, to determine whether such a crisis be within the limits of possibility. At the same time we may observe, that the distribution of land at present is so remarkably irregular, and appears so capricious, if we may so express ourselves, that the two extremes of terrestrial heat and cold are probably separated very widely from each other. The globe may now be equally divided, so that one hemisphere shall be entirely covered with water, with the exception of some promontories and islands, while the other shall contain less water than land; and what is still more extraordinary, on comparing the extra-tropical lands in the northern and southern hemispheres, the former are found to be to the latter in the proportion of thirteen to one! To imagine all the lands, therefore, in high, and all the sea in low latitudes, would scarcely be a more anomalous state of the surface.

Let us now turn from the contemplation of the winter of the 'great year,' and consider the opposite train of circumstances, which would bring on the spring and summer. That some part of the vast ocean which forms the Atlantic and Pacific, should at certain periods occupy entirely one or both of the polar regions, and should extend, interspersed with islands, only to the parallels of 40°, and even 30°, is an event that may be supposed in the highest degree probable, in the course of many great geological revolutions. In order to estimate the degree to which the general temperature would then be elevated, we should begin by considering separately the effect of the diminution of certain portions of land, in high northern latitudes, which might cause the sea to be as open in every direction, as it is at present towards the north pole, in the meridian of Spitzbergen. By transferring the same lands to the torrid zone, we might gain farther accessions of heat, and cause the ice towards the south pole to diminish.

We might first continue these geographical mutations, until we had produced as mild a climate in high latitudes as exists at those points in the same parallel where the mean annual heat is now greatest. We should then endeavour to calculate what farther alterations would be required to double the amount of change; and the great deviation of isothermal lines at present seems to authorize us to infer, that without an entire revolution of the surface, we might cause the mean temperature to vary to an extent equivalent to 20° or even 30° of latitude, – in other words, we might transfer the temperature of the torrid zone, to the mean parallel, and of the latter, to the arctic regions. By additional transpositions, therefore, of land and sea, we might bring about a still greater variation, so that, throughout the year, all signs of frost should disappear from the earth.

The plane of congelation would rise in the atmosphere in all latitudes; and as our hypothesis would place all the highest mountains in the torrid zone, they would be clothed with rich vegetation to their summits. We must recollect that even now it is necessary to ascend to the height of 15,000 feet in the Andes under the line; and in the Himalaya mountains, which are without the tropic, to 17,000 feet before we reach the limit of perpetual snow. When the absorption of the solar rays was unimpeded, even in winter, by a coat of snow, the mean heat of the earth's crust would augment to considerable depths, and springs, which we know to be an index of the mean temperature of the climate, would be warmer in all latitudes. The waters of lakes, therefore, and rivers, would be much hotter in winter, and would be never chilled in summer by the melting of snow. A remarkable uniformity of climate would prevail amid the numerous archipelagos of the polar ocean, amongst which the tepid waters of equatorial currents would freely circulate. The general humidity of the atmosphere would far exceed that of the present period, for increased heat would promote evaporation in all parts of the globe. The winds would be first heated in their passage over the tropical plains, and would then gather moisture from the surface of the deep, till, charged with vapour, they would arrive at northern regions, and, encountering a cooler atmosphere, would discharge their burden in warm rain. If, during the long night of a polar winter, the snows should whiten the summit of some arctic islands, and ice collect in the bays of the remotest Thule, they would be dissolved as rapidly by the returning sun, as are the snows of Etna by the blasts of the sirocco.

We learn from those who have studied the geographical distribution of plants, that in very low latitudes, at present, the vegetation of small islands remote from continents has a peculiar character, and the ferns and allied families, in particular, bear a great proportion to the total number of other vegetables. Other circumstances being the same, the more remote the isles are from the continents, the greater does this proportion become. Thus, in the continent of India, and the tropical parts of New Holland, the proportion of ferns to the phanerogamic plants is only as one to twenty-six; whereas, in the South Sea Islands, it is as one to four, or even as one to three. We might expect, therefore, in the summer of the 'great year,' which we are now considering, that there would be a great predominance of tree-ferns and plants allied to palms and arborescent grasses in the isles of the wide ocean, while the dicotyledonous plants and other forms now most common in temperate regions would almost disappear from the earth. Then might those genera of animals return, of which the memorials are preserved in the ancient rocks of our continents. The huge iguanodon might reappear in the woods, and the ichthyosaur in the sea, while the pterodactyle might flit again through umbrageous groves of tree-ferns. Coral reefs might be prolonged beyond the arctic circle, where the whale and the narwal now abound. Turtles might deposit their eggs in the sand of the sea beach, where now the walrus sleeps, and where the seal is drifted on the ice-floe.

But, not to indulge these speculations farther, we may observe, in conclusion, that however great, in the lapse of ages, may be the vicissitudes of temperature in every zone, it accords with our theory that the general climate should not experience any sensible change in the course of a few thousand years, because that period is insufficient to affect the leading features of the physical geography of the globe. Notwithstanding the apparent uncertainty of the seasons, it is found that the mean temperature of particular localities is very constant, provided we compare observations made at different periods for a series of years. Yet, there must be exceptions to this rule, and even the labours of man have, by the drainage of lakes and marshes, and the felling of extensive forests, caused such changes in the atmosphere as raise our conception of the important influence of those forces to which even the existence in certain latitudes of land or water, hill or valley, lake or sea, must be ascribed. If we possessed accurate information of the amount of local fluctuation in climate in the course of twenty centuries, it would often, undoubtedly, be considerable. Certain tracts, for example, on the coast of Holland and of England, consisted of cultivated land in the time of the Romans, which the sea, by gradual encroachments, has at length occupied. Here an alteration has been effected; for neither the division of heat in the different seasons, nor the mean annual heat of the atmosphere investing the sea is precisely the same as that which rests on the land. In those countries also where the earthquake and volcano are in full activity, a much shorter period may produce a sensible variation. The climate of the once fertile plain of Malpais in Mexico must differ materially from that which prevailed before the middle of the last century; for, since that time, six mountains, the highest of them rising 1700 feet above the plateau, have been thrown up by volcanic eruptions. It is by the repetition of an indefinite number of local revolutions due to volcanic and various other causes, that a general change of climate is finally brought about.

CHAPTER 8

Climate, continued

We stated, in the sixth chapter, our reasons for concluding that the mean annual temperature of the northern hemisphere was considerably more elevated when the old carboniferous strata were deposited; as also that the climate had been modified more than once since that epoch, and that it approximated by successive changes more and more nearly to that now prevailing in the same latitudes. Further, we endeavoured, in the last chapter, to prove that vicissitudes in climate of no less importance may be expected to recur in future, if it be admitted that causes now active in nature have power, in the lapse of ages, to vary to an unlimited extent the relative position of land and sea. It next remains for us to inquire whether the alterations, which the geologist can prove to have actually taken place at former periods, in the geographical features of the northern hemisphere, coincide in their nature, and in the time of their occurrence, with such revolutions in climate as would naturally have followed, according to the meteorological principles already explained.

We may select the great carboniferous series, including the transition and mountain limestones, and the coal, as the oldest system of rocks of which the organic remains furnish any decisive evidence as to climate. We have already insisted on the indications which they afford of great heat and uniformity of temperature, extending over a vast area, from about 45° to 60°, or perhaps, if we include Melville Island, to near 75° north latitude.

When we attempt to restore in imagination the distribution of land and sea, as they existed at that remote epoch, we discover that our information is at present limited to latitudes north of the tropic of cancer, and we can only hope, therefore, to point out that the condition of the earth, so far as relates to our temperate and arctic zones, was such as the theory before offered would have led us to anticipate. Now there is scarcely any land hitherto examined in Europe, Northern Asia, or North America, which has not been raised from the bosom of the deep, since the origin of the carboniferous rocks, or which, if previously raised, has not subsequently acquired additional altitude. If we were to submerge again all the marine strata, from the transition limestone to the most recent shelly beds, the summits of some primary mountains alone would remain above the waters. These facts, it is true, considered singly, are not conclusive as to the universality of the ancient ocean in the northern hemisphere, because the movements of earthquakes occasion the subsidence as well as the upraising of the surface, and by the alternate rising and sinking of particular spaces, at successive periods, a great area may become entirely covered with marine deposits, although the whole has never been beneath the waters at one time, nay, even though the relative proportion of land and sea may have continued unaltered throughout the whole period. There is, however, the highest presumption against such an hypothesis, because the land in the northern hemisphere is now in great excess, and this circumstance alone should induce us to suppose that, amidst the repeated changes which the surface has undergone, the sea has usually predominated in a much greater degree. But when we study the mineral composition and fossil contents of the older strata, we find evidence of a more positive and unequivocal kind in confirmation of the same opinion.

Calcareous rocks, containing the same class of organic remains as our transition and mountain limestones, extend over a great part of the central and northern parts of Europe, are found in the lake district of North America, and even appear to occur in great abundance as far as the border of the Arctic sea. The organic remains of these rocks consist principally of marine shells, corals, and the teeth and bones of fish; and their nature, as well as the continuity of the calcareous beds of homogeneous mineral composition, concur to prove that the whole series was formed in a deep and expansive ocean, in the midst of which, however, there were many isles. These isles were composed partly of primary and partly of volcanic rocks, which being exposed to the erosive action of torrents, to the undermining power of the waves beating against the cliffs, and to atmospheric decomposition, supplied materials for pebbles, sand, and shale, which, together with substances introduced by mineral springs and volcanos in frequent eruption, contributed the inorganic parts of the carboniferous strata. The disposition of the beds in that portion of this group which is of mechanical origin, and which incloses the coal, has been truly described to be such as would result from the waste of small islands placed in rows and forming the highest points of ridges of submarine mountains. The disintegration of such clusters of isles would produce around and between them detached deposits of various dimensions, which, when subsequently raised above the waters, would resemble the strata formed in a chain of lakes. The insular masses of primary rock would preserve their original relative superiority of height, and would often surround the newer strata on several sides, like the boundary heights of lake basins.

As might have been expected, the zoophytic, and shelly limestones of the same era, (as the mountain limestone,) sometimes alternate with the rocks of mechanical origin, but appear to have been, in ordinary cases, diffused far and wide over the bottom of the sea, remote from any islands, and where no grains of sand were transported by currents. The associated volcanic rocks, resemble the products of submarine eruptions, the tuffs being sometimes interstratified with calcareous shelly beds, or with sandstones, just as might be expected if the sand and ejected matter of which they are probably composed had been intermixed with the waters of the sea, and had then subsided like other sediment. The lavas also often extend in spreading sheets, and must have been poured out on a surface rendered horizontal by sedimentary depositions. There is, moreover, a compactness and general absence of porosity in these igneous rocks which distinguishes them from most of those which are produced on the sides of Etna or Vesuvius, and other land-volcanos. The modern submarine lavas of Sicily, which alternate with beds of shells specifically identical with those now living in the Mediterranean, have almost all their cavities filled with calcareous and other ingredients, and have been converted into amygdaloids, and this same change we must suppose such parts of the Etnean lava currents as enter the sea to be undergoing at present, because we know the water on the adjoining coast to be copiously charged with carbonate of lime in solution. It is, therefore, one among many reasons for inferring the submarine origin of our ancient trap rocks, that there are scarcely any instances, in which the cellular hollows, left by bubbles of elastic fluid, have not subsequently been filled by calcareous, siliceous, or other mineral ingredients, such as now abound in the hot springs of volcanic countries.

If, on the other hand, we examine the fossil remains in these strata, we find the vegetation of the coal strata declared by botanists to possess the characters of an insular, not a continental flora, and we may suppose the carbonaceous matter to have been derived partly from trees swept from the rock by torrents into the sea, and partly from such peaty matter as often discolours and blackens the rills flowing through marshy grounds in our temperate climate, where the vegetation is probably less rank, and its decomposition less rapid than in the moist and hot climate of the era under consideration. There is only one instance yet on record of the remains of a saurian animal having been found in a member of the carboniferous series. The larger oviparous reptiles usually inhabit rivers of considerable size in warm latitudes, and had crocodiles and other animals of that class been as abundant as in some secondary formations, we must have inferred the existence of many rivers, which could only have drained large tracts of land. Nor have the bones of any terrestrial mammalia rewarded our investigations. Had any of these, belonging to quadrupeds of large size, occurred, they would have supplied an argument against the resemblance of the ancient northern archipelagos to those of the modern Pacific, since in the latter no great indigenous quadrupeds have been met with. It is, indeed, a general character of small islands situated at a remote distance from continents, to be altogether destitute of land quadrupeds, except such as appear to have been conveyed to them by man. Kerguelen's land, which is of no inconsiderable size, placed in a latitude corresponding to that of the Scilly islands, may be cited as an example, as may all the groups of fertile islands in the Pacific ocean between the tropics, where no quadrupeds have been found, except the dog, the hog, and the rat, which have probably been brought to them by the natives, and also bats, which may have made their way along the chain of islands which extend from the shores of New Guinea far into the southern Pacific. Even the isles of New Zealand, which may be compared to Ireland and Scotland in dimensions, appear to possess no indigenous quadrupeds, except the bat; and this is rendered the more striking, when we recollect that the northern extremity of New Zealand stretches to latitude 34°, where the warmth of the climate must greatly favour the prolific development of organic life. Lastly, no instance has yet been discovered of a pure lacustrine formation of the carboniferous era; although there are some instances of shells, apparently fresh-water, which may have been washed in by small streams, and do not by any means imply a considerable extent of dry land. All circumstances, therefore, point to one conclusion; – the subaqueous character of the igneous products – the continuity of the calcareous strata over vast spaces – the marine nature of their organic remains – the basin-shaped disposition of the mechanical rocks – the absence of large fluviatile and of land quadrupeds – the non-existence of pure lacustrine strata – the insular character of the flora, – all concur with wonderful harmony to establish the prevalence throughout the northern hemisphere of a great ocean, interspersed with small isles. If we seek for points of analogy to this state of things, we must either turn to the north Pacific, and its numerous submarine or insular volcanos between Kamtschatka and New Guinea, or, in order to obtain a more perfect counterpart to the coralline and shelly limestones, we may explore the archipelagos of the south Pacific, between Australia and South America, where volcanos are not wanting, and where coral reefs, consisting in great part of compact limestone, are spread over an area not inferior, perhaps, to that of our ancient calcareous rocks, though we suppose these to be prolonged from the lakes of North America to central Europe.

No geologists have ever denied, that when our oldest conchiferous rocks were produced, great continents were wanting in the temperate and arctic zones north of the equator; but they have even gone farther, and have been disposed to speculate on the universality of what they termed the primeval ocean. As well might a new Zealander, who had surveyed and measured the quantity of land between the south pole and the tropic of Capricorn, assume that the same proportion would be found to exist between the tropic of Cancer and the north pole. By this generalization, he would imagine twelve out of thirteen parts of the land of our temperate and arctic zones to be submerged. Such theorists should be reminded, that if the ocean was ever universal, its mean depth must have been inferior, and if so, the probability of deep water within the arctic circle is much lessened, and the likelihood of a preponderance of ice increased, and the heat of the ancient climate rendered more marvellous. To this objection, however, they will answer, that they do not profess to restrict themselves to existing analogies, and they may suppose the volume of water in the primeval ocean to have been greater. Besides, the high temperature, say they, was caused by heat which emanated from the interior of the new-born planet. In vain should we suggest to such reasoners, that when the ocean was in excess in high latitudes, the land in all probability predominated within the tropics, where, being exposed to the direct rays of the sun, it may have heated the winds and currents which flowed from lower to higher latitudes. In vain should we contend that a greater expanse of ocean, if general throughout the globe, would imply a comparative evenness of the superficial crust of the earth, and such an hypothesis would oblige us to conclude that the disturbances caused by subterranean movements in ancient times were inferior to those of later date. Will these arguments be met by the assumption, that earthquakes were feebler in the earlier ages, or wholly unknown, – as, according to Werner, there were no volcanos? Such a doctrine would be inconsistent with other popular prejudices respecting the extraordinary violence of the operations of nature in the olden time; and it is probable, therefore, that refuge will be taken in the old dogma of Lazzoro Moro, who imagined that the bed of the first ocean was as regular as its surface, and if so, it may be contended that sufficient time did not elapse between the creation of the world and the origin of the carboniferous strata, to allow the derangement necessary to produce great continents and Alpine chains.

But it would be idle to controvert, by reference to modern analogies, the conjectures of those who think they can ascend in their retrospect to the origin of our system. Let us, therefore, consider what changes the crust of the globe suffered after the consolidation of that ancient series of rocks to which we have adverted. Now, there is evidence that, before our secondary strata were formed, those of older date (from the old red sandstone to the coal inclusive) were fractured and contorted, and often thrown into vertical positions. We cannot enter here into the geological details by which it is demonstrable, that at an epoch extremely remote, some parts of the carboniferous series were lifted above the level of the sea, others sunk to greater depths beneath it, and the former, being no longer protected by a covering of water, were partially destroyed by torrents and the waves of the sea, and supplied matter for newer horizontal beds. These were arranged on the truncated edges of the submarine portions of the more ancient series, and the fragments included in the more modern conglomerates still retain their fossil shells and corals, so as to enable us to determine the parent rocks from whence they were derived. By such remodelling of the surface the small islands of the first period increased in size, and new land was introduced into northern regions, consisting partly of primary and volcanic rocks and partly of the newly raised carboniferous strata. Among other proofs that earthquakes were then governed by the same laws which now regulate the subterranean forces, we find that they were restrained within limited areas, so that the site of Germany was not agitated, while that of some parts of England was convulsed. The older rocks, therefore, remained in some cases undisturbed at the bottom of the ancient ocean, and in this case the strata of the succeeding epoch were deposited upon them in conformable position. By reference to groups largely developed on the continent, but which are some of them entirely wanting, and others feebly represented in our own country, we find that the apparent interruption in the chain of events between the formation of our coal and the lias arises merely from local deficiency in the suite of geological monuments. During the great interval which separated the formation of these groups, new species of animals and plants made their appearance, and in their turn became extinct; volcanos broke out, and were at length exhausted; rocks were destroyed in one region, and others accumulated elsewhere, while, in the mean time, the geographical condition of the northern hemisphere suffered material modifications. Yet the sea still extended over the greater part of the area now occupied by the lands which we inhabit, and was even of considerable depth in many localities where our highest mountain-chains now rise. The vegetation, during a part at least of this new period (from the lias to the chalk inclusive), appears to have approached to that of the larger islands of the equatorial zone. These islands appear to have been drained by rivers of considerable size, which were inhabited by crocodiles and gigantic oviparous reptiles, both herbivorous and carnivorous, belonging for the most part to extinct genera. Of the contemporary inhabitants of the land we have as yet acquired but scanty information, but we know that there were flying reptiles, insects, and small insectivorous mammifera, allied to the opossum. In farther confirmation of the opinion that countries of considerable extent now rose above the sea in the temperate zone, we may mention the discovery of a large estuary formation in the south-west of England of higher antiquity than the chalk, containing terrestrial plants and fresh-water testacea, tortoises, and large reptiles, – in a word, such an assemblage as the delta of the Ganges, or a large river in a hot climate might be expected to produce.

In the present state of our knowledge, we cannot pretend to institute a close comparison between the climate which prevailed during the gradual deposition of our secondary formations and that of the older carboniferous rocks, for the general temperature of the surface must at both epochs have been so dissimilar to that now experienced in the same, or perhaps in any latitudes, that proofs from analogy lose much of their value, and a larger body of facts is required to support theoretical conclusions. If the signs of intense heat diminish, as some suppose, in the newer groups of this great series, there are nevertheless indications in the animal forms of the continued prevalence of a climate which we might consider as tropical in its character.

We may now turn our attention to the phenomena of the tertiary strata, which afford evidence of an abrupt transition from one description of climate to another. If this remarkable break in the regular sequence of physical events is merely apparent, arising from the present imperfect state of our knowledge, it nevertheless serves to set in a clearer point of view the intimate connexion between great changes in the physical geography of the earth, and revolutions in the mean temperature of the air and water. We have already shewn that when the climate was hottest, the northern hemisphere was for the most part occupied by the ocean, and it remains for us to point out, that the refrigeration did not become considerable, until a very large portion of that ocean was converted into land, nor even until it was in some parts replaced by high mountain chains. Nor did the cold reach its maximum until these chains attained their full height, and the lands their full extension. A glance at the best geological maps now constructed of various countries in the northern hemisphere, whether in North America or Europe, will satisfy the inquirer that the greater part of the present land has been raised from the deep, either between the period of the deposition of the chalk and that of the strata termed tertiary, or at subsequent periods, during which, various tertiary groups were formed in succession. For, as the secondary rocks from the lias to the chalk inclusive, are, with a few unimportant exceptions, marine, it follows that every district now occupied by them has been converted into land since they originated. We may prove, by reference to the relative altitudes of the secondary and tertiary groups, and several other circumstances, that a considerable part of the elevation of the older series was accomplished before the newer was formed. The Apennines, for example, as the Italian geologists hinted long before the time of Brocchi, and as that naturalist more clearly demonstrated, rose several thousand feet above the level of the Mediterranean, before the deposition of the recent Subapennine beds which flank them on either side. What now constitutes the central calcareous chain of the Apennines, must for a long time have been a narrow ridgy peninsula, branching off at its northern extremity from the Alps near Savonna. A line of volcanos afterwards burst out in the sea, parallel to the axis of the older ridge. These igneous vents were extremely numerous, and the ruins of some of their cones and craters (as those in Tuscany, for example) indicate such a continued series of eruptions, almost all subsequent to the deposition of the Subapennine strata, that we cannot wonder at the vast changes in the relative level of land and sea which were produced. However minute the effect of each earthquake which preceded or intervened between such countless eruptions, the aggregate result of their elevating or depressing operation may well be expected to display itself in seas of great depth, and hills of considerable altitude. Accordingly, the more recent shelly beds, which often contain rounded pebbles derived from the waste of contiguous parts of the older Apennine rocks, have been raised from one to two thousand feet; but they never attain the loftier eminences of the Apennines, nor penetrate far into the higher and more ancient valleys; for the whole peninsula was evidently subjected to the action of the same subterranean movements, and the older and newer groups of strata changed their level, in relation to the sea, but not to each other.

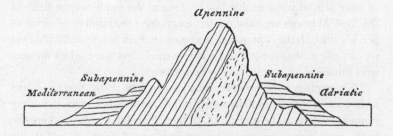

1. Transverse section of the Italian peninsula

In the above diagram, exhibiting a transverse section of the Italian peninsula, the superior elevation of the more ancient group, and its unconformable stratification in relation to the more recent beds is expressed. The latter, however, are often much more disturbed at the point of contact than is here represented, and in some cases they have suffered such derangement as to dip towards, instead of from, the more ancient chain. There is usually, moreover, a valley at the junction of the Apennine and Subapennine strata, owing to the greater degradation which the newer and softer beds have undergone; but this intervening depression is not universal.

These phenomena are exhibited in the Alps on a much grander scale;

those mountains being encircled by a great zone of tertiary rocks of different ages, both on their southern flank towards the plains of the Po, and on the side of Switzerland and Austria, and at their eastern termination towards Styria and Hungary. This tertiary zone marks the position of former seas or gulfs, like the Adriatic, which were many thousand feet deep, and wherein strata accumulated, some single groups of which are not inferior in thickness to the whole of our secondary formations in England. These marine tertiary strata rise to the height of from two to four thousand feet and upwards, and consist of formations of different ages, characterized by different assemblages of organized fossils. The older tertiary groups generally rise to greater heights, and form interior zones nearest to the Alps. We may imagine some future convulsion once more to upraise this stupendous chain, together with the adjoining bed of the sea, so that the greatest mountains of Europe might rival the Andes in elevation, in which case the deltas of the Po, Adige, and Brenta now encroaching upon the Adriatic, might be uplifted so as to form another exterior belt, of considerable height, around the south-eastern flank of the Alps. Although we have not yet ascertained the number of different periods at which the Alps gained accessions to their height and width, yet we can affirm, that the last series of movements occurred when the seas were inhabited by *many existing species of animals*.

There appears to be no sedimentary formations in the Alps so ancient as the rocks of our carboniferous series; while, on the other hand, secondary strata as modern as the green sand of English geologists, and perhaps the chalk, enter into some of the higher and central ridges. Down to the period, therefore, when the rocks, from our lias to the chalk inclusive, were deposited, there was sea where now the principal chain of Europe extends, and that chain attained more than half its present elevation and breadth between the eras when our newer secondary and oldest tertiary rocks originated. The remainder of its growth, if we may so speak, is of much more recent date, some of the latest changes, as we have stated, having been coeval with the existence of many animals belonging to species now contemporary with man. The Pyrenees, also, have acquired the whole of their present altitude, which in Mont Perdu exceeds eleven thousand feet, since the origin of some of the newer members of our secondary series. The granitic axis of that chain does not rise so high as a ridge formed by marine calcareous beds, the organic remains of which shew them to be the equivalents of our lower chalk, or a formation of about that age. The tertiary strata at the base of this great chain are only slightly raised above the sea, and retain a horizontal position, without partaking of any of the disturbances to which the older series has been subjected, so that the great barrier between France and Spain was almost entirely upheaved in the interval between the deposition of the secondary and tertiary strata. The Jura, also, owe the greatest part of their present elevation to subterranean convulsions which happened after the deposition of certain tertiary groups; at which time that portion which had been previously raised above the level of the sea underwent an entire alteration of form. In other parts of the continent, as in France and England, where the newer rocks lie in basins surrounded by gently-rising hills, we find evidence that considerable spaces were redeemed from the original ocean and converted into dry land after the chalk was formed, and before the origin of the tertiary deposits. In these cases, the secondary strata were not raised into lofty mountain chains, like the Alps, Apennines, and Pyrenees, but the proofs are not less clear of their partial conversion into land anterior to the tertiary era. The chalk, for example, must have originated in the sea in the form of sediment from tranquil water; but before the tertiary rocks of the Paris and London basins were deposited, large portions of it had been so raised as to be exposed to the destroying power of the elements. The layers of flint had been washed out by torrents and rivers from their cretaceous matrix, rounded by attrition, and transported to the sea, where oysters attached themselves, and in some localities grew to a full size, until covered by other beds of flint-pebbles or sand. These newer derivative deposits are found abundantly along the borders, and in the inferior strata of our tertiary basins, and they are often interstratified with lignite. We may fairly infer, that the various trees and plants which enter into the composition of this lignite, grew on the surface of the same chalk which was then wasting away and affording to the torrents a constant supply of flint gravel.

We cannot dwell longer on the distinct periods when the secondary and various tertiary groups were upraised, without anticipating details which belong to other parts of this treatise; but we may observe, that although geologists have neglected to point out the relation of changes in the configuration of the earth's surface with fluctuations in general temperature, they do not dispute the fact, that the sea covered the regions where a great part of the land in Europe is now placed, until after the period when the newer groups of secondary rocks were formed. There is, therefore, confessedly a marked coincidence in point of time between the greatest alteration in climate and the principal revolution in the physical geography of the northern hemisphere. It is very probable that the abruptness of the transition from the organic remains of the secondary to those of the tertiary epoch, may not be wholly ascribable to the present deficiency of our information. We shall doubtless hereafter discover many intermediate gradations, (and one of these may be recognized in the calcareous beds of Maestricht,) by which a passage was effected from one state of things to another; but it is not impossible that the interval between the chalk and tertiary formations constituted an era in the earth's history, when the passage from one class of organic beings to another was, comparatively speaking, rapid. For if the doctrines explained by us in regard to vicissitudes of temperature are sound, it will follow that changes of equal magnitude in the geographical features of the globe, may at different periods produce very unequal effects on climate, and, so far as the existence of certain animals and plants depends on climate, the duration of species may often be shortened or protracted, according to the rate at which the change in temperature proceeded.

Let us suppose that the laws which regulate the subterranean forces are constant and uniform, (which we are entitled to assume, until some convincing proofs can be adduced to the contrary;) we may then infer, that a given amount of alteration in the superficial inequalities of the surface of the planet always requires for its consummation nearly equal periods of time. Let us then imagine the quantity of land between the equator and the tropic in one hemisphere to be to that in the other as thirteen to one, which, as we before stated, represents the unequal proportion of the extra-tropical lands in the two hemispheres at present. Then let the first geographical change consist in the shifting of this preponderance of land from one side of the line to the other, from the southern hemisphere, for example, to the northern. Now this would not affect the *general* temperature of the earth. But if, at another epoch, we suppose a continuance of the same agency to transfer an equal volume of land from the torrid zone to the temperate and arctic regions of the northern hemisphere, there might be so great a refrigeration of the mean temperature *in all latitudes*, that scarcely any of the pre-existing races of animals would survive, and, unless it pleased the Author of Nature that the planet should be uninhabited, new species would be substituted in the room of the extinct. We ought not, therefore, to infer, that equal periods of time are always attended by an equal amount of change in organic life, since a great fluctuation in the mean temperature of the earth, the most influential cause which can be conceived in exterminating whole races of animals and plants, must, in different epochs, require unequal portions of time for its completion.

The only geological monument yet discovered, which throws light on the period immediately succeeding the deposition of the chalk, is the series of calcareous beds in St. Peter's Mount at Maestricht. The turtles and gigantic reptiles there found, seem to indicate that the hot climate of the secondary era had not then been greatly modified; but as it seems that but a small proportion of the fossil species hitherto discovered are identical with known chalk fossils, there may perhaps have been a considerable lapse of ages between the consolidation of our upper chalk, and the completion of the Maestricht group. During these ages, part of the gradual rise of the Alps and Pyrenees may have been accomplished; for we know that earthquakes may work mighty changes during what we may call a small portion of one zoological era, since there are hills in Sicily which have gained more than three thousand feet in height, while the assemblage of testacea and zoophytes inhabiting the Mediterranean has only suffered slight alterations, and a large part of the countries bordering the Mediterranean have been remodelled since about one-third of the existing species were in being.

Before we conclude this chapter, we may be expected to offer some remarks on the gradual diminution of the supposed central heat of the globe, a doctrine which appears of late years to have increased in popu- larity. Baron Fourier, after making a curious series of experiments on the cooling of incandescent bodies, has endeavoured by profound mathemat- ical calculations to prove that the actual distribution of heat in the earth's envelope is precisely that which would have taken place if the globe had been formed in a medium of a very high temperature, and had afterwards been constantly cooled. He supposes that the matter of our planet, as Leibnitz formerly conjectured, was in an intensely heated state at the era of its creation, and that the incandescent fluid nucleus has been parting ever since with portions of its original heat, thereby contracting its dimensions, – a process which has not yet entirely ceased. But it is admitted, that there are no positive facts in support of this contraction; on the contrary, La Place has shewn, by reference to astronomical observations made in the time of Hipparchus, that in the last two thousand years there has been no sensible contraction of the globe by cooling down, for had this been the case, even to an extremely small amount, the day would have been shortened in an appreciable degree. The reader will bear in mind, that the question as to the existence of a central heat is very different from that of the gradual refrigeration of the interior of the earth. Many observations and experiments appear to countenance the idea, that in descending from the surface to those slight depths to which man can penetrate, there is a progressive increase of heat; but if this be established, and if, as some are not afraid to infer, we dwell on a thin crust which covers a central ocean of liquid incandescent lava, we ought still to be very reluctant to concede on slight evidence that the internal heat is *variable in quantity*.

In our ignorance of the sources and nature of volcanic fire, it seems more consistent with philosophical caution, to assume that there is no instability in this part of the terrestrial system. We know that different regions have been subject in succession to a series of violent subterranean convulsions, and that fissures have opened from which hot vapours, thermal springs, and at some points red hot liquid lavas have issued to the surface. This evolution of heat often continues for ages after the extinction of volcanos and after the cessation of earthquakes, as in Central France, for example, and it seems perfectly natural, that each part of the earth's crust should, as M. Fourier states to be the fact, present the appearance of a heated body slowly cooling down. This may be owing chiefly to the shifting of the volcanic foci; but some effect may perhaps be due to that unequal absorption of the solar rays to which we have alluded, when speaking of the different temperature of the earth, according to the varying distribution of its superficial inequalities. M. Cordier announces as the result of his experiments and observations on the temperature of the interior of the earth, that the heat increases rapidly with the depth, but the increase does not follow the same law over the whole earth, being twice or three times as much in one country as in another, and these differences not being in constant relation either with the latitudes or longitudes of places. All this is precisely what we should have expected to arise from variations in the intensity of volcanic heat, and from that change of position, which the principal theatres of volcanic action have undergone at different periods, as the geologist can distinctly prove. But M. Cordier conjectures that there is a connexion between such phenomena and the secular refrigeration and contraction of the internal fluid mass, and that the changes of climate, of which there are geological proofs, favour this hypothesis.

We cannot help suspecting that if it had appeared that *the same species of animals and plants* had continued to inhabit the seas, lakes, and continents, before and after the great physical mutations which the northern hemisphere has undergone since the secondary strata were formed, the difficulty of explaining the ancient climate of the globe would have appeared far more insurmountable than at present. It would have been so contrary to the elementary truths of meteorology to suppose no refrigeration to have followed from the rising of so many new mountain-chains in northern latitudes, that recourse would probably have been had in that case also to cosmological speculations. It might have been argued with much plausibility, that as the accession of high ridges covered with perpetual snow and glaciers had not occasioned any perceptible increase of cold, so as to affect the state of organic life, there must have been some new source of heat which counterbalanced that refrigerating cause. This, it might have been said, was the increased development of *central fire* issuing from innumerable fissures opened in the crust of the earth, when it was shaken by convulsions which raised the Alps and other colossal chains.

But, without entering into farther discussion on the merits of the hypothesis of gradual refrigeration, let us hope that experiments will continue to be made, to ascertain whether there be internal heat in the globe, and what laws may govern its distribution. When its existence has been incontrovertibly established, it will be time to enquire whether it be subject to secular variations. Should these also be confirmed, we may begin to indulge speculations respecting the cause, but let us not hastily assume that it has reference to the original formation of the planet, with which it might be as unconnected as with its final dissolution. In the mean time we know that great changes in the external configuration of the earth's crust have at various times taken place, and we may affirm that they *must* have produced *some effect* on climate. The extent of their influence ought, therefore, to form a primary object of enquiry, more especially as there seems an obvious coincidence between the eras at which the principal accessions of land in high latitudes were made, and the successive periods when the diminution of temperature was most decided.

CHAPTER 9

Theory of the Progressive Development of Organic Life

We have considered, in the preceding chapters, many of the most popular grounds of opposition to the doctrine, that all former changes of the organic and inorganic creation are referrible to one uninterrupted succession of physical events, governed by the laws now in operation.

As the principles of the science must always remain unsettled so long as no fixed opinions are entertained on this fundamental question, we shall proceed to examine other objections which have been urged against the assumption of uniformity in the order of nature. We shall cite the words of a late distinguished writer, who has formally advanced some of the weightiest of these objections. 'It is impossible,' he affirms, 'to defend the proposition, that the present order of things is the ancient and constant order of nature, only modified by existing laws – in those strata which are deepest, and which must, consequently, be supposed to be the earliest deposited forms, even of vegetable life, are rare; shells and vegetable remains are found in the next order; the bones of fishes and oviparous reptiles exist in the following class; the remains of birds, with those of the same genera mentioned before, in the next order; those of quadrupeds of extinct species in a still more recent class; and it is only in the loose and slightly-consolidated strata of gravel and sand, and which are usually called diluvian formations, that the remains of animals such as now people the globe are found, with others belonging to extinct species. But, in none of these formations, whether called secondary, tertiary, or diluvial, have the remains of man, or any of his works, been discovered; and whoever dwells upon this subject must be convinced, that the present order of things, and the comparatively recent existence of man as the master of the globe, is as certain as the destruction of a former and a different order, and the extinction of a number of living forms which have no types in being. In the oldest secondary strata there are no remains of such animals as now belong to the surface; and in the rocks, which may be regarded as more recently deposited, these remains occur but rarely, and with abundance of extinct species; – there seems, as it were, a gradual approach to the present system of things, and a succession of destructions and creations preparatory to the existence of man.'[22]

In the above passages, the author deduces two important conclusions from geological data; first, that in the successive groups of strata, from the oldest to the most recent, there is a progressive development of organic life, from the simplest to the most complicated forms; – secondly, that man is of comparatively recent origin. It will be easy to shew that the first of these propositions, though very generally received, has no foundation in fact. The second, on the contrary, is indisputable, and it is important, therefore, to consider how far its admission is inconsistent with the assumption, that the system of the natural world has been uniform from the beginning, or rather from the era when the oldest rocks hitherto discovered were formed.

We shall first examine the geological proofs appealed to in support of the theory of the successive development of animal and vegetable life, and their progressive advancement to a more perfect state. No geologists, who are in possession of all the data now established respecting fossil remains, will for a moment contend for the doctrine in all its detail, as laid down by the great chemist to whose opinions we have referred. But naturalists, who are not unacquainted with recent discoveries, continue to defend the ancient doctrine in a somewhat modified form. They say that, in the first period of the world, (by which they mean the earliest of which we have yet procured any memorials,) the vegetation consisted almost entirely of cryptogamic plants, while the animals which co-existed were almost entirely confined to zoophytes, testacea, and a few fish. Plants of a less simple structure succeeded in the next epoch, when oviparous reptiles began also to abound. Lastly, the terrestrial flora became most diversified and most perfect when the highest orders of animals, the mammifera and birds, were called into existence.

Now, in the first place, we may observe, that many naturalists have been guilty of no small inconsistency in endeavouring to connect the phenomena of the earliest vegetation with a nascent condition of organic life, and at the same time to deduce, from the numerical predominance of certain types of form, the greater heat of the ancient climate. The arguments in favour of the latter conclusion are without any force, unless we can assume that the rules followed by the Author of Nature in the creation and distribution of organic beings were the same formerly as now; and that as certain families of animals and plants are now most abundant, or exclusively confined to regions where there is a certain temperature, a certain degree of humidity, intensity of light, and other conditions, so also the same phenomena were exhibited at every former era. If this postulate be denied, and the prevalence of particular families be declared to depend on a certain order of precedence in the introduction of different classes into the earth, and if it be maintained that the standard of organization was raised successively, we must then ascribe the numerical preponderance in the earlier ages of plants of simpler structure, *not to the heat*, but to those different laws which regulate organic life in newly created worlds. If, according to the laws of progressive development, cryptogamic plants always flourish for ages before the dicotyledonous order can be established, then is the small proportion of the latter fully explained; for in this case, whatever may have been the mildness or severity of the climate, they could not make their appearance. Before we can infer an elevated temperature in high latitudes, from the presence of arborescent Ferns, Lycopodiaceæ, and other allied families, we must be permitted to assume, that at all times, past and future, a heated and moist atmosphere pervading the northern hemisphere has a tendency to produce in the vegetation a predominance of analogous types of form. We grant, indeed, that there may be a connexion between an extraordinary profusion of monocotyledonous plants, and a youthful condition of the world, if the dogma of certain cosmogonists be true, that planets, like certain projectiles, are always red hot when they are first cast; but to this arbitrary hypothesis we need not again revert.

Between two and three hundred species of plants are now enumerated as belonging to the carboniferous era, and, with very few exceptions, not one of them are dicotyledonous. But these exceptions are as fatal to the doctrine of successive development as if there were a thousand, although they do not by any means invalidate the conclusion in regard to the heat of the ancient climate, for that depends on the numerical relations of the different classes.

The animal remains in the most ancient series of European sedimentary rocks (from the graywacke to the coal inclusive), consist chiefly of corals and testacea. Some estimate may generally be formed of the comparative extent of our information concerning the fossil remains of a particular era, by reference to the number of species of shells obtained from a particular group of strata. Some of the rarest species cannot be discovered, unless the more abundant kinds have been found again and again; and if the variety brought to light be very considerable, it proves not only great diligence of research, but a good state of preservation of the organic contents of that formation. In the older rocks, many causes of destruction have operated, of which the influence has been rendered considerable by the immense lapse of ages during which they have acted. Mechanical pressure, derangement by subterranean movements, the action of chemical affinity, the percolation of acidulous waters and other agencies, have obliterated, in a greater or less degree, all traces of organization in fossil bodies. Sometimes only obscure or unintelligible impressions are left, and the lapidifying process has often effaced not only the characters by which the species, but even those whereby the class might be determined. The number of organic forms which have disappeared from the oldest strata, may be conjectured from the fact, that their former existence is in many cases merely revealed to us by the unequal weathering of an exposed face of rock, by which certain parts are made to stand out in relief. As the number of species of shells found in the English series, from the graywacke to the coal inclusive, after attentive examination, amounts only to between one and two hundred species, we cannot be surprised that so few examples of vertebrated animals have as yet occurred. The remains of fish, however, appear in one of the lowest members of the group, which entirely destroys the theory of the precedence of the simplest forms of animals. The vertebra also of a saurian, as we before stated, has been met with in the mountain limestone of Northumberland, so that the only negative fact remaining in support of the doctrine of the imperfect development of the higher orders of animals in remote ages, is the absence of birds and mammalia. The former are generally wanting in deposits of all ages, even where the highest order of animals occurs in abundance. Land mammifera could not, as we have before suggested, be looked for in strata formed in an ocean interspersed with isles, such as we must suppose to have existed in the northern hemisphere, when the carboniferous rocks were formed.

As all are agreed that the ancient strata in question were subaqueous, and for the most part submarine, from what data we may ask do naturalists infer the non-existence or even the rarity of warm-blooded quadrupeds in the earlier ages? Have they dredged the bottom of the ocean throughout an area co-extensive with that now occupied by the carboniferous rocks, and have they found that with the number of between one and two hundred species of shells they always obtain the remains of at least one land quadruped? Suppose our mariners were to report that on sounding in the Indian ocean near some coral reefs, and at some distance from the land, they drew up on hooks attached to their line portions of a leopard, elephant, or tapir; should we not be sceptical as to the accuracy of their statements; and if we had no doubt of their veracity, might we not suspect them to be unskilful naturalists? or, if the fact were unquestioned, should we not be disposed to believe that some vessel had been wrecked on the spot? The casualties must be rare indeed whereby land quadrupeds are swept by rivers and torrents into the sea, and still rarer must be the contingency of such a floating body not being devoured by sharks or other predaceous fish, such as were those of which we find the teeth preserved in some of the carboniferous strata. But if the carcase should escape and should happen to sink where sediment was in the act of accumulating, and if the numerous causes of subsequent disintegration should not efface all traces of the body included for countless ages in solid rock, is it not contrary to all calculation of chances that we should hit upon the exact spot, – that mere point in the bed of the ancient ocean, where the precious relic was entombed? Can we expect for a moment that when we have only succeeded amidst several thousand fragments of corals and shells, in finding a few bones of *aquatic* or *amphibious* animals, that we should meet with a single skeleton of an inhabitant of the land?

Clarence, in his dream, saw 'in the slimy bottom of the deep,'

> —a thousand fearful wrecks;
> A thousand men, that fishes gnaw'd upon;
> Wedges of gold, great anchors, heaps of pearl.

Had he also beheld amid 'the dead bones that lay scatter'd by,'[23] the carcasses of lions, deer, and the other wild tenants of the forest and the plain, the fiction would have been deemed unworthy of the genius of Shakspeare. So daring a disregard of probability, so avowed a violation of analogy, would have been condemned as unpardonable even where the poet was painting those incongruous images which present themselves to a disturbed imagination during the visions of the night. But the cosmogonist is not amenable, even in his waking hours, to these laws of criticism; for he assumes either that the order of nature was formerly distinct, or that the globe was in a condition to which it can never again be reduced by changes which the existing laws of nature can bring about. This assumption being once admitted, inexplicable anomalies and violations of analogy, instead of offending his judgment, give greater consistency to his reveries.

The organic contents of the secondary strata in general consist of corals and marine shells. Of the latter, the British strata (from the inferior oolite to the chalk inclusive) have yielded about six hundred species. Vertebrated animals are very abundant, but they are almost entirely confined to fish and reptiles. But some remains of cetacea have also been met with in the oolitic series of England, and the bones of two species of warm-blooded quadrupeds of extinct genera allied to the Opossum. The occurrence of one individual of the higher classes of mammalia, whether marine or terrestrial, in these ancient strata, is as fatal to the theory of successive development, as if several hundreds had been discovered.

The tertiary strata, as will appear from what we have already stated, were deposited when the physical geography of the northern hemisphere had been entirely altered. Large inland lakes had become numerous as in central France and many other countries. There were gulfs of the sea into which large rivers emptied themselves, where strata were formed like those of the Paris basin. There were then also littoral formations in progress, such as are indicated by the English *Crag*, and the *Faluns* of the Loire. The state of preservation of the organic remains of this period is very different from that of fossils in the older rocks, the colours of the shells, and even the cartilaginous ligaments uniting the valves being in some cases retained. No less than twelve hundred species of testacea have been found in the beds of the Paris basin, and an equal number in the more modern formations of the Subapennine hills; and it is a most curious fact in natural history, that the zoologist has already acquired more extensive information concerning the testacea which inhabited the ancient seas of northern latitudes at that era, than of those now living in the same parallels in Europe. The strata of the Paris basin are partly of fresh-water origin, and filled with the spoils of the land. They have afforded a great number of skeletons of land quadrupeds, but these relics are confined almost entirely to one small member of the group, and their conservation may be considered as having arisen from some local and accidental combination of circumstances. On the other hand, the scarcity of terrestrial mammalia in submarine sediment is elucidated, in a striking manner, by the extremely small number of such remains hitherto procured from the Subapennine hills. The facilities of investigation in these strata, which undergo rapid disintegration, are perhaps unexampled in the rest of Europe, and they have been examined by collectors for three hundred years. But, although they have already yielded twelve hundred species of testacea, the authenticated examples of associated remains of terrestrial mammalia are extremely scanty; and several of those which have been cited by earlier writers as belonging to the elephant or rhinoceros, have since been declared, by able anatomists, to be the bones of whales and other cetacea. In about five or ten instances, perhaps, bones of the mastodon, rhinoceros, and some other animals, have been observed in this formation with marine shells attached. These must have been washed into the bed of the ancient sea when the strata were forming, and they serve to attest the contiguity of land inhabited by large herbivora, which renders the rarity of such exceptions more worthy of attention. On the contrary, the number of skeletons of existing animals in the upper Val d'Arno, which are usually considered to be referrible to the same age as the Subapennine beds, occur in a deposit which was formed entirely in an inland lake, surrounded by lofty mountains.

The inferior member of our oldest tertiary formations in England, usually termed the plastic clay, has hitherto proved as destitute of mammiferous remains, as our ancient coal strata; and this point of resemblance between these deposits is the more worthy of observation, because the lignite, in the one case, and the coal in the other, are exclusively composed of terrestrial plants. From the London clay we have procured three or four hundred species of testacea, but the only bones of vertebrated animals are those of reptiles and fish. On comparing, therefore, the contents of these strata with those of our oolitic series, we find the supposed order of precedence inverted. In the more ancient system of rocks, mammalia, both of the land and sea, have been recognized, whereas in the newer, if negative evidence is to be our criterion, nature has made a retrograde, instead of a progressive, movement, and no animals more exalted in the scale of organization than reptiles are discoverable.

Not a single bone of a quadrumanous animal has ever yet been discovered in a fossil state, and their absence has appeared, to some geologists, to countenance the idea that the type of organization most nearly resembling the human came last in the order of creation, and was scarcely perhaps anterior to that of man. But the evidence on this point is quite inconclusive, for we know nothing, as yet, of the details of the various classes of the animal kingdom which inhabited the land up to the consolidation of the newest of the secondary strata; and when a large part of the tertiary formations were in progress, the climate does not appear to have been of such a tropical character as seems necessary for the development of the tribe of apes, monkeys, and allied genera. Besides, it must not be forgotten, that almost all the animals which occur in subaqueous deposits are such as frequent marshes, rivers, or the borders of lakes, as the rhinoceros, tapir, hippopotamus, ox, deer, pig, and others. On the other hand, species which live in trees are extremely rare in a fossil state, and we have no data as yet for determining how great a number of the one kind we ought to find, before we have a right to expect a single individual of the other. If, therefore, we are led to infer, from the presence of crocodiles and turtles in the London clay, and from the cocoa-nuts and spices found in the isle of Sheppey, that at the period when our older tertiary strata were formed, the climate was hot enough for the quadrumanous tribe, we nevertheless could not hope to discover any of their skeletons until we had made considerable progress in ascertaining what were the contemporary Pachydermata; and not one of these, as we have already remarked, has been discovered as yet in any strata of this epoch in England.*

It is, therefore, clear, that there is no foundation in geological facts, for the popular theory of the successive development of the animal and vegetable world, from the simplest to the most perfect forms; and we shall now proceed to consider another question, whether the recent origin of man lends any support to the same doctrine, or how far the influence of man may be considered as such a deviation from the analogy of the order of things previously established, as to weaken our confidence in the uniformity of the course of nature. We need not dwell on the proofs of the low antiquity of our species, for it is not controverted by any geologist; indeed, the real difficulty which we experience consists in tracing back the signs of man's existence on the earth to that comparatively modern period when species, now his contemporaries, began to predominate. If

* The only exception of which I have heard is the tooth of an Anoplotherium, mentioned by Dr. Buckland as having been found in the collection of Mr. Allan, labelled 'Binstead, Isle of Wight.' The quarries of Binstead are entirely in the lower fresh-water formation, and such is undoubtedly the geological position in which we might look for the bones of such an animal. My friend Mr. Allan has shewn me this tooth, to which, unfortunately, none of the matrix is attached, so that it is still open to a captious sceptic to suspect that a Parisian fossil was so ticketed by mistake.

there be a difference of opinion respecting the occurrence in certain deposits of the remains of man and his works, it is always in reference to strata confessedly of the most modern order; and it is never pretended that our race co-existed with assemblages of animals and plants, of which *all the species* are extinct. From the concurrent testimony of history and tradition, we learn that parts of Europe, now the most fertile and most completely subjected to the dominion of man, were, within less than three thousand years, covered with forests, and the abode of wild beasts. The archives of nature are in perfect accordance with historical records; and when we lay open the most superficial covering of peat, we sometimes find therein the canoes of the savage, together with huge antlers of the wild stag, or horns of the wild bull. In caves now open to the day in various parts of Europe, the bones of large beasts of prey occur in abundance; and they indicate, that at periods extremely modern in the history of the globe, the ascendancy of man, if he existed at all, had scarcely been felt by the brutes.

No inhabitant of the land exposes himself to so many dangers on the waters as man, whether in a savage or a civilized state, and there is no animal, therefore, whose skeleton is so liable to become imbedded in lacustrine or submarine deposits; nor can it be said, that his remains are more perishable than those of other animals, for in ancient fields of battle, as Cuvier has observed, the bones of men have suffered as little decomposition as those of horses which were buried in the same grave. But even if the more solid parts of our species had disappeared, the impression would have remained engraven on the rocks as have the traces of the tenderest leaves of plants, and the integuments of many animals. Works of art, moreover, composed of the most indestructible materials, would have outlasted almost all the organic contents of sedimentary rocks; edifices, and even entire cities have, within the times of history, been buried under volcanic ejections, or submerged beneath the sea, or engulphed by earthquakes; and had these catastrophes been repeated throughout an indefinite lapse of ages, the high antiquity of man would have been inscribed in far more legible characters on the frame-work of the globe, than are the forms of the ancient vegetation which once covered the isles of the northern ocean, or of those gigantic reptiles, which at later periods peopled the seas and rivers of the northern hemisphere.

Assuming, then, that man is, comparatively speaking, of modern origin, can his introduction be considered as one step in a progressive system by which, as some suppose, the organic world advanced slowly from a more simple to a more perfect state? To this question we may reply, that the superiority of man depends not on those faculties and attributes which he shares in common with the inferior animals, but on his reason by which he is distinguished from them.

If the organization of man were such as would confer a decided pre-eminence upon him, even if he were deprived of his reasoning powers, and provided only with such instincts as are possessed by the lower animals, he might then be supposed to be a link in a progressive chain, especially if it could be shewn that the successive development of the animal creation had always proceeded from the more simple to the more compound, from species most remote from the human type to those most nearly approaching to it. But this is an hypothesis which, as we have seen, is wholly unsupported by geological evidence. On the other hand, we may admit, that man is of higher dignity than were any pre-existing beings on the earth, and yet question whether his coming was a step in the gradual advancement of the organic world: for the most highly civilized people may sometimes degenerate in strength and stature, and become inferior in their physical attributes to the stock of rude hunters from which they descended. If then the physical organization of man may remain stationary, or even become deteriorated, while the race makes the greatest progress to higher rank and power in the scale of rational being, the animal creation also may be supposed to have made no progress by the addition to it of the human species, regarded merely as a part of the organic world. But, if this reasoning appear too metaphysical, let us waive the argument altogether, and grant that the animal nature of man, even considered apart from the intellectual, is of higher dignity than that of any other species; still the introduction at a certain period of our race upon the earth, raises no presumption whatever that each former exertion of creative power was characterized by the successive development of *irrational* animals of higher orders. The comparison here instituted is between things so dissimilar, that when we attempt to draw such inferences, we strain analogy beyond all reasonable bounds. We may easily conceive that there was a considerable departure from the succession of phenomena previously exhibited in the organic world, when so new and extraordinary a circumstance arose, as the union, for the first time, of moral and intellectual faculties capable of indefinite improvement, with the animal nature. But we have no right to expect that there were any similar deviations from analogy – any corresponding steps in a progressive scheme, at former periods, when no similar circumstances occurred.

But another, and a far more difficult question may arise out of the admission that man is comparatively of modern origin. Is not the interference of the human species, it may be asked, such a deviation from the antecedent course of physical events, that the knowledge of such a fact tends to destroy all our confidence in the uniformity of the order of nature, both in regard to time past and future? If such an innovation could take place after the earth had been exclusively inhabited for thousands of ages by inferior animals, why should not other changes as extraordinary and unprecedented happen from time to time? If one new cause was permitted to supervene, differing in kind and energy from any before in operation, why may not others have come into action at different epochs? Or what security have we that they may not arise hereafter? If such be the case, how can the experience of one period, even though we are acquainted with all the possible effects of the then existing causes, be a standard to which we can refer all natural phenomena of other periods?

Now these objections would be unanswerable, if adduced against one, who was contending for the absolute uniformity throughout all time of the succession of sublunary events – if, for example, he was disposed to indulge in the philosophical reveries of some Egyptian and Greek sects, who represented all the changes both of the moral and material world as repeated at distant intervals, so as to follow each other in their former connexion of place and time. For they compared the course of events on our globe to astronomical cycles, and not only did they consider all sublunary affairs to be under the influence of the celestial bodies, but they taught that on the earth, as well as in the heavens, the same identical phenomena recurred again and again in a perpetual vicissitude. The same individual men were doomed to be re-born, and to perform the same actions as before; the same arts were to be invented, and the same cities built and destroyed. The Argonautic expedition was destined to sail again with the same heroes, and Achilles with his Myrmidons, to renew the combat before the walls of Troy.

> Another *Tiphys* shall new Seas explore,
> Another *Argos* land the Chiefs,
> upon th' *Iberian* Shore:
> Another *Helen* other Wars create,
> And great *Achilles* urge the Trojan Fate.[24]

94

The geologist, however, may condemn these tenets as absurd, without running into the opposite extreme, and denying that the order of nature has, from the earliest periods, been uniform in the same sense in which we believe it to be uniform at present. We have no reason to suppose, that when man first became master of a small part of the globe, a greater change took place in its physical condition than is now experienced when districts, never before inhabited, become successively occupied by new settlers. When a powerful European colony lands on the shores of Australia, and introduces at once those arts which it has required many centuries to mature; when it imports a multitude of plants and large animals from the opposite extremity of the earth, and begins rapidly to extirpate many of the indigenous species, a mightier revolution is effected in a brief period, than the first entrance of a savage horde, or their continued occupation of the country for many centuries, can possibly be imagined to have produced. If there be no impropriety in assuming that the system is uniform when disturbances so unprecedented occur in certain localities, we can with much greater confidence apply the same language to those primeval ages when the aggregate number and power of the human race, or the rate of their advancement in civilization, must be supposed to have been far inferior.

If the barren soil around Sidney had at once become fertile upon the landing of our first settlers; if, like the happy isles whereof the poets have given us such glowing descriptions, those sandy tracts had begun to yield spontaneously an annual supply of grain, we might then, indeed, have fancied alterations still more remarkable in the economy of nature to have attended the first coming of our species into the planet. Or if, when a volcanic island like Ischia was, for the first time brought under cultivation by the enterprise and industry of a Greek colony, the internal fire had become dormant, and the earthquake had remitted its destructive violence, there would then have been some ground for speculating on the debilitation of the subterranean forces, when the earth was first placed under the dominion of man. But after a long interval of rest, the volcano bursts forth again with renewed energy, annihilates one-half of the inhabitants, and compels the remainder to emigrate. Such exiles, like the modern natives of Cumana, Calabria, Sumbawa, and other districts, habitually convulsed by earthquakes, would probably form no very exalted estimate of the sagacity of those geological theorists, who, contrasting the *human* with antecedent epochs, have characterized it as *the period of repose.*

In reasoning on the state of the globe immediately before our species was called into existence, we may assume that all the present causes were in operation, with the exception of man, until some geological arguments can be adduced to the contrary. We must be guided by the same rules of induction as when we speculate on the state of America in the interval that elapsed between the period of the introduction of man into Asia, the cradle of our race, and that of the arrival of the first adventurers on the shores of the New World. In that interval, we imagine the state of things to have gone on according to the order now observed in regions unoccupied by man. Even now, the waters of lakes, seas, and the great ocean, which teem with life, may be said to have no immediate relation to the human race – to be portions of the terrestrial system of which man has never taken, nor ever can take, possession, so that the greater part of the inhabited surface of the planet remains still as insensible to our presence, as before any isle or continent was appointed to be our residence.

The variations in the external configuration of the earth, and the successive changes in the races of animals and plants inhabiting the land and sea, which the geologist beholds when he restores in imagination the scenes presented by certain regions at former periods, are not more full of wonderful or inexplicable phenomena, than are those which a traveller would witness who traversed the globe from pole to pole. Or if there be more to astonish and perplex us in searching the records of the past, it is because one district may, in an indefinite lapse of ages, become the theatre of a greater number of extraordinary events, than the whole face of the globe can exhibit at one time. However great the multiplicity of new appearances, and however unexpected the aspect of things in different parts of the present surface, the observer would never imagine that he was transported from one system of things to another, because there would always be too many points of resemblance, and too much connexion between the characteristic features of each country visited in succession, to permit any doubt to arise as to the continuity and identity of the whole plan.

'In our globe;' says Paley, 'new countries are continually discovered, but the old laws of nature are always found in them: new plants perhaps, or animals, but always in company with plants and animals which we already know, and always possessing many of the same general properties. We never get amongst such original, or totally different modes of existence, as to indicate that we are come into the province of a different Creator, or under the direction of a different will. In truth, the same order of things attends us wherever we go.'[25] But the geologist is in danger of drawing a contrary inference, because he has the power of passing rapidly from the events of one period to those of another – of beholding, at one glance, the effects of causes which may have happened at intervals of time incalculably remote, and during which, nevertheless, no local circumstances may have occurred to mark that there is a great chasm in the chronological series of nature's archives. In the vast interval of time which may really have elapsed between the results of operations thus compared, the physical condition of the earth may, by slow and insensible modifications, have become entirely altered, one or more races of organic beings may have passed away, and yet have left behind, in the particular region under contemplation, no trace of their existence. To a mind unconscious of these intermediate links in the chain of events, the passage from one state of things to another must appear so violent, that the idea of revolutions in the system inevitably suggests itself. The imagination is as much perplexed by such errors as to time, as it would be if we could annihilate space, and by some power, such as we read of in tales of enchantment, could transfer a person who had laid himself down to sleep in a snowy arctic wilderness, to a valley in a tropical region, where on awaking he would find himself surrounded by birds of brilliant plumage, and all the luxuriance of animal and vegetable forms of which nature is there so prodigal. The most reasonable supposition, perhaps, which a philosopher could make, if by the necromancer's art he was placed in such a situation, would be, that he was dreaming; and if a geologist forms theories under a similar delusion, we should not expect him to preserve more consistency in his speculations, than in the train of ideas in an ordinary dream.

But if, instead of inverting the natural order of inquiry, we cautiously proceed in our investigations, from the known to the unknown, and begin by studying the most modern periods of the earth's history, attempting afterwards to decipher the monuments of more ancient changes, we can never so far lose sight of analogy, as to suspect that we have arrived at a new system, governed by different physical laws. In more recent formations, consisting often of strata of great thickness, the shells of the present seas and lakes, and the remains of animals and plants now living on the land, are imbedded in great numbers. In those of more ancient date, many of the same species are found associated with others now extinct. These unknown kinds again are observed in strata of still higher antiquity, connected with a great number of others which have also no living representatives, till at length we arrive at periods of which the monuments contain exclusively the remains of species with many genera foreign to the present creation. But even in the oldest rocks which contain organic remains, some genera of marine animals are recognized, of which species still exist in our seas, and these are repeated at different intervals in all the intermediate groups of strata, attesting that, amidst the great variety of revolutions of which the earth's surface has been the theatre, there has never been a departure from the conditions necessary for the existence of certain unaltered types of organization. The uniformity of animal instinct, observes Mr. Stewart, pre-supposes a corresponding regularity in the physical laws of the universe, 'insomuch that if the established order of the material world were to be essentially disturbed, (the instincts of the brutes remaining the same,) all their various tribes would inevitably perish.'[26] Now, any naturalist will be convinced, on slight reflection, of the justice of this remark. He will also admit that the same species have always retained the same instincts, and therefore that all the strata wherein any of *their* remains occur, must have been formed when the phenomena of inanimate matter were the same as they are in the actual condition of the earth. The same conclusion must also be extended to the extinct animals with which the remains of these living species are associated; and by these means we are enabled to establish the permanence of the existing physical laws, throughout the whole period when the tertiary deposits were formed. We have already stated that, during that vast period, a large proportion of all the lands in the northern hemisphere were raised above the level of the sea.

The modifications in the system of which man is the instrument, do not, in all probability, constitute so great a deviation from analogy as we usually imagine; we often, for example, form an exaggerated estimate of the extent of the power displayed by man in extirpating some of the inferior animals, and causing others to multiply; a power which is circumscribed within certain limits, and which, in all likelihood, is by no means exclusively exerted by our species. The growth of human population cannot take place without diminishing the numbers, or causing the entire destruction of many animals. The larger carnivorous species give way before us, but other quadrupeds of smaller size, and innumerable birds, insects, and plants, which are inimical to our interests, increase in spite of us, some attacking our food, others our raiment and persons, and others interfering with our agricultural and horticultural labours. We force the ox and the horse to labour for our advantage, and we deprive the bee of his store; but, on the other hand, we raise the rich harvest with the sweat of our brow, and behold it devoured by myriads of insects, and we are often as incapable of arresting their depredations as of staying the shock of an earthquake, or the course of a stream of burning lava. The changes caused by other species, as they gradually diffuse themselves over the globe, are inferior probably in magnitude, but are yet extremely analogous to those which we occasion. The lion, for example, and the migratory locust, must necessarily, when they first made their way into districts now occupied by them, have committed immense havoc amongst the animals and plants which became their prey. They may have caused many species to diminish, perhaps wholly to disappear; but they must also have enabled some others greatly to augment in number, by removing the natural enemies by which they had been previously kept down. It is probable from these, and many other considerations, that as we enlarge our knowledge of the system, we shall become more and more convinced, that the alterations caused by the interference of man deviate far less from the analogy of those effected by other animals than we usually suppose. We are often misled, when we institute such comparisons, by our knowledge of the wide distinction between the instincts of animals and the reasoning power of man; and we are apt hastily to infer, that the effects of a rational and an irrational species, considered merely *as physical agents*, will differ almost as much as the faculties by which their actions are directed. A great philosopher has observed, that we can only command nature by obeying her laws, and this principle is true even in regard to the astonishing changes which are superinduced in the qualities of certain animals and plants by domestication and garden culture. We can only effect such surprising alterations by assisting the development of certain instincts, or by availing ourselves of that mysterious law of their organization, by which individual peculiarities are transmissible from one generation to another.

We are not, however, contending that a real departure from the antecedent course of physical events cannot be traced in the introduction of man. If that latitude of action which enables the brutes to accommodate themselves in some measure to accidental circumstances, could be imagined to have been at any former period so great, that the operations of instinct were as much diversified as are those of human reason, it might perhaps be contended, that the agency of man did not constitute an anomalous deviation from the previously established order of things. It might then have been said, that the earth's becoming at a particular period the residence of human beings, was an era in the moral, not in the physical world – that our study and contemplation of the earth, and the laws which govern its animate productions, ought no more to be considered in the light of a disturbance or deviation from the system, than the discovery of the satellites of Jupiter should be regarded as a physical event in the history of those heavenly bodies, however influential they may have become from that time in advancing the progress of sound philosophy among men, and in augmenting human resources by aiding navigation and commerce. The distinctness, however, of the human, from all other species, considered merely as an efficient cause in the physical world, is real, for we stand in a relation to contemporary species of animals and plants, widely different from that which other irrational animals can ever be supposed to have held to each other. We modify their instincts, relative numbers, and geographical distribution, in a manner superior in degree, and in some respects very different in kind from that in which any other species can affect the rest. Besides, the progressive movement of each successive generation of men causes the human species to differ more from itself in power at two distant periods, than any one species of the higher order of animals differs from another. The establishment, therefore, by geological evidence of the first intervention of such a peculiar and unprecedented agency, long after other parts of the animate and inanimate world existed, affords ground for concluding that the experience during thousands of ages of all the events which may happen on this globe would not enable a philosopher to speculate with confidence concerning future contingencies. If an intelligent being, therefore, after observing the order of events for an indefinite series of ages had witnessed at last so wonderful an innovation as this, to what extent would his belief in the regularity of the system be weakened? – would he cease to assume that there was permanency in the laws of nature? – would he no longer be guided in his speculations by the strictest rules of induction? To this question we may reply, that had he previously presumed to dogmatize respecting the absolute uniformity of the order of nature, he would undoubtedly be checked by witnessing this new and unexpected event, and would form a more just estimate of the limited range of his own knowledge, and the unbounded extent of the scheme of the universe. But he would soon perceive that no one of the fixed and constant laws of the animate or inanimate world was subverted by human agency, and that the modifications produced were on the occurrence of new and extraordinary circumstances, and those not of a *physical*, but a *moral* nature. The deviation permitted, would also appear to be as slight as was consistent with the accomplishment of the new *moral* ends proposed, and to be in a great degree temporary in its nature, so that whenever the power of the new agent was withheld, even for a brief period, a relapse would take place to the ancient state of things; the domesticated animal, for example, recovering in a few generations its wild instinct, and the garden-flower and fruit-tree reverting to the likeness of the parent stock.

Now, if it would be reasonable to draw such inferences with respect to the future, we cannot but apply the same rules of induction to the past. It will scarcely be disputed that we have no right to anticipate any modifications in the results of existing causes in time to come, which are not conformable to analogy, unless they be produced by the progressive development of human power, or perhaps from some other new relations between the moral and material worlds. In the same manner we must concede, that when we speculate on the vicissitudes of the animate and inanimate creation in former ages, we have no ground for expecting any anomalous results, unless where man has interfered, or unless clear indications appear of some other *moral* source of temporary derangement. When we are unable to explain the monuments of past changes, it is always more probable that the difficulty arises from our ignorance of all the existing agents, or all their possible effects in an indefinite lapse of time, than that some cause was formerly in operation which has ceased to act; and if in any part of the globe the energy of a cause appears to have decreased, it is always probable, that the diminution of intensity in its action is merely local, and that its force is unimpaired, when the whole globe is considered. But should we ever establish by unequivocal proofs, that certain agents have, at particular periods of past time, been more potent instruments of change over the entire surface of the earth than they now are, it will be more consistent with philosophical caution to presume, that after an interval of quiescence they will recover their pristine vigour, than to regard them as worn out.

The geologist who yields implicit assent to the truth of these principles, will deem it incumbent on him to examine with minute attention all the changes now in progress on the earth, and will regard every fact collected respecting the causes in diurnal action, as affording him a key to the interpretation of some mystery in the archives of remote ages. Our estimate, indeed, of the value of all geological evidence, and the interest derived from the investigation of the earth's history, must depend entirely on the degree of confidence which we feel in regard to the permanency of the laws of nature. Their immutable constancy alone can enable us to reason from analogy, by the strict rules of induction, respecting the events of former ages, or, by a comparison of the state of things at two distinct geological epochs, to arrive at the knowledge of general principles in the economy of our terrestrial system.

The uniformity of the plan being once assumed, events which have occurred at the most distant periods in the animate and inanimate world will be acknowledged to throw light on each other, and the deficiency of our information respecting some of the most obscure parts of the present creation will be removed. For as by studying the external configuration of the existing land and its inhabitants, we may restore in imagination the appearance of the ancient continents which have passed away, so may we obtain from the deposits of ancient seas and lakes an insight into the nature of the subaqueous processes now in operation, and of many forms of organic life, which, though now existing, are veiled from our sight. Rocks, also produced by subterranean fire in former ages at great depths in the bowels of the earth, present us, when upraised by gradual movements, and exposed to the light of heaven, with an image of those changes which the deep-seated volcano may now occasion in the nether regions. Thus, although we are mere sojourners on the surface of the planet, chained to a mere point in space, enduring but for a moment of time, the human mind is not only enabled to number worlds beyond the unassisted ken of mortal eye, but to trace the events of indefinite ages before the creation of our race, and is not even withheld from penetrating into the dark secrets of the ocean, or the interior of the solid globe; free, like the spirit which the poet described as animating the universe,

> Thro' Heav'n, and Earth, and Oceans depth he throws
> His Influence round, and kindles as he goes.[27]

Aqueous Causes

We defined geology to be the science which investigates the former changes that have taken place in the organic, as well as in the inorganic kingdoms of nature; and we now proceed to inquire what changes are now in progress in both these departments. Vicissitudes in the inorganic world are most apparent, and as on them all fluctuations in the animate creation must in a great measure depend, they may claim our first consideration. We may divide the great agents of change in the inorganic world into two principal classes, the aqueous and the igneous. To the former belong Rivers, Torrents, Springs, Currents, and Tides; to the latter, Volcanos and Earthquakes. Both these classes are instruments of decay as well as of reproduction; but they may be also regarded as antagonist forces. The *aqueous* agents are incessantly labouring to reduce the inequalities of the earth's surface to a level, while the *igneous*, on the other hand, are equally active in restoring the unevenness of the external crust, partly by heaping up new matter in certain localities, and partly by depressing one portion, and forcing out another of the earth's envelope. It is difficult, in a scientific arrangement, to give an accurate view of the combined effects of so many forces in simultaneous operation; because, when we consider them separately, we cannot easily estimate either the extent of their efficacy, or the kind of results which they produce. We are in danger, therefore, when we attempt to examine the influence exerted singly by each, of overlooking the modifications which they produce on one another; and these are so complicated, that sometimes the igneous and aqueous forces co-operate to produce a joint effect, to which neither of them unaided by the other could give rise, – as when repeated earthquakes unite with running water to widen a valley. Sometimes the organic combine with the inorganic causes; as when a reef, composed of shells and corals, protects one line of coast from the destroying power of tides or currents, and turns them against some other point; or when drift timber floated into a lake, fills a hollow to which the stream would not have had sufficient velocity to convey earthy sediment.

It is necessary, however, to divide our observations on these various causes, and to classify them systematically, endeavouring as much as possible to keep in view that the effects in nature are mixed, and not simple, as they may appear in an artificial arrangement.

In treating, first, of the aqueous causes, we may consider them under two divisions: first, those which are connected with the circulation of water from the land to the sea, under which are included all the phenomena of rivers and springs; secondly, those which arise from the movements of water in lakes, seas, and the ocean, wherein are comprised the phenomena of tides and currents. In turning our first attention to the former division, we find that the effects of rivers may be subdivided into those of a destroying and those of a renovating nature. In the former are included the erosion of rocks and the transportation of matter to lower levels; in the latter, the formation of sand-bars and deltas, the shallowing of seas, &c.

Action of Running Water. – We shall begin, then, by describing the destroying and transporting power of running water, as exhibited by torrents and rivers. It is well known that the lands elevated above the sea attract in proportion to their volume and density a larger quantity of that aqueous vapour which the heated atmosphere continually absorbs from the surface of lakes and seas. By this means, the higher regions become perpetual reservoirs of water, which descend and irrigate the lower valleys and plains. In consequence of this provision, almost all the water is first carried to the highest regions, and is then made to descend by steep declivities towards the sea; so that it acquires superior velocity, and removes a greater quantity of soil than it would do if the rain had been distributed over the low plains and high mountains equally in proportion to their relative areas. Almost all the water is also made by these means to pass over the greatest distances which each region affords, before it can regain the sea. The rocks in the higher regions are particularly exposed to atmospheric influences, to frost, rain, and vapour, and to great annual alternations, of moisture and desiccation – of cold and heat. Among the most powerful agents of decay may be mentioned the mechanical action of water, which possesses the remarkable property of expanding during congelation. When water has penetrated into crevices and cavities, it rends open, on freezing, the most solid rocks with the force of a lever, and for this reason, although in cold climates the comparative quantity of rain which falls is very inferior, and although it descends more gradually than in tropical regions, yet the severity of frost, and the greater inequalities of temperature, compensate for this diminished power of degradation, and cause it to proceed with equal, if not greater rapidity than in high latitudes. The solvent power of water also is very great, and acts particularly on the calcareous and alkaline elements of stone, especially when it holds carbonic acid in solution, which is abundantly supplied to almost every large river by springs, and is collected by rain from the atmosphere. The oxygen of the atmosphere is also gradually absorbed by all animal and vegetable productions, and by almost all mineral masses exposed to the open air. It gradually destroys the equilibrium of the elements of rocks, and tends to reduce into powder, and to render fit for soils, even the hardest aggregates belonging to our globe. And as it is well known that almost every thing affected by rapid combustion may also be affected gradually by the slow absorption of oxygen, the surface of the hardest rocks exposed to the air may be said to be slowly burning away.

When earthy matter has once been intermixed with running water, a new mechanical power is obtained by the attrition of sand and pebbles, borne along with violence by a stream. Running water charged with foreign ingredients being thrown against a rock, excavates it by mechanical force, sapping and undermining till the superincumbent portion is at length precipitated into the stream. The obstruction causes a temporary increase of the water, which then sweeps down the barrier. By a repetition of these land-slips, the ravine is widened into a small, narrow valley, in which sinuosities are caused by the deflexion of the stream first to one side and then to the other. The unequal hardness of the materials through which the channel is eroded, tends also to give new directions to the lateral force of excavation. When by these, or by accidental shiftings of the alluvial matter in the channel, and numerous other causes, the current is made to cross its general line of descent, it eats out a curve in the opposite bank, or the side of the hill bounding the valley, from which curve it is turned back again at an equal angle, and recrossing the line of descent, it gradually hollows out another curve lower down, in the opposite bank, till the whole sides of the valley, or river-bed, present a succession of salient and retiring angles.

Among the causes of deviation from a straight course by which torrents and rivers tend to widen the valleys through which they flow, may be mentioned the confluence of lateral torrents, swoln irregularly at different seasons in mountainous regions by partial storms, and discharging at different times unequal quantities of debris into the main channel.

When the tortuous flexures of a river are extremely great, the aberration from the direct line of descent is often restored by the river cutting

2. Diagram explanatory of the sinuosity of river-courses

through the isthmus which separates two neighbouring curves. Thus, in the annexed diagram, the extreme sinuosity of the river has caused it to return for a brief space in a contrary direction to its main course, so that a peninsula is formed, and the isthmus (at *a*) is consumed on both sides by currents flowing in opposite directions. In this case an island is soon formed, – on either side of which a portion of the stream usually remains. These windings occur not only in the channels of rivers flowing through flat alluvial plains, but large valleys also are excavated to a great depth through solid rocks in this serpentine form. In the valley of the Moselle, between Berncastle and Roarn, which is sunk to a depth of from six to eight hundred feet through an elevated platform of transition rocks, the curves are so considerable that the river returns, after a course of seventeen miles in one instance, and nearly as much in two others, to within a distance of a few hundred yards of the spot it passed before. The valley of the Meuse, near Givet, and many others in different countries, offer similar windings. Mr. Scrope has remarked, that these tortuous flexures are decisively opposed to the hypothesis, that any violent and transient rush of water suddenly swept out such valleys; for great floods would produce straight channels in the direction of the current, not sinuous excavations, wherein rivers flow back again in an opposite direction to their general line of descent.

Our present purpose, however, relates to the force of aqueous erosion, and the transportation of materials by running water, considered separately, and not to the question so much controverted respecting the formation of valleys in general. This subject cannot be fully discussed without referring to all the powers to which the inequalities of the earth's surface, and the very existence of land above the level of the sea, are due. Nor even when we have described the influence of all the chemical and mechanical agents which operate at one period in effecting changes in the external form of the land, shall we be enabled to present the reader with a comprehensive theory of the origin of the present valleys. It will be necessary to consider the complicated effects of all these causes at distinct geological epochs, and to inquire how particular regions, after having remained for ages in a state of comparative tranquillity, and under the influence of a certain state of the atmosphere, may be subsequently remodelled by another series of subterranean movements, – how the new direction, volume, and velocity acquired by rivers and torrents may modify the former surface, – what effects an important difference in the mean temperature of the climate, or the greater intensity of heat and cold at different seasons, may produce, – what pre-existing valleys, under a new configuration of the land, may cease to give passage to large bodies of water, or may become entirely dried up, – how far the relative level of certain districts in the more modern period may become precisely the reverse of those which prevailed at the more ancient era. When these and other essential elements of the problem are all duly appreciated, the reader will not be surprised to learn, that amongst geologists who have neglected them there has prevailed a great contrariety of opinion on these topics. Some writers of distinguished talent have gone so far as to contend, that the origin of the greater number of existing valleys was simply due to the agency of one cause, and that it was consummated in a brief period of time. But without discussing the merits of the general question, we may observe that we agree with the author before cited, that the sinuosity of deep valleys is one among many proofs that they have been shaped out progressively, and not by the simultaneous action of one or many causes; and when we consider other agents of change, we shall have opportunities of pointing out a multitude of striking facts in confirmation of the gradual nature of the process to which the inequalities of hill and valley owe their origin.

In regard to the transporting power of water, we are often surprised at the facility wherewith streams of a small size, and which descend a slight declivity, bear along coarse sand and gravel; for we usually estimate the weight of rocks in air, and do not reflect sufficiently on their comparative buoyancy when submerged in a denser fluid. The specific gravity of many rocks is not more than twice that of water, and very rarely more than thrice, so that almost all the fragments propelled by a stream have lost a third, and many of them half of what we usually term their weight.

It has been proved by experiment, in contradiction to the theories of the early writers on hydrostatics, to be a universal law, regulating the motion of running water, that the velocity at the bottom of the stream is everywhere less than in any part above it, and is greatest at the surface. Also that the superficial particles in the middle of the stream move swifter than those at the sides. This retardation of the lowest and lateral currents is produced by friction, and when the velocity is sufficiently great, the soil composing the sides and bottom gives way. A velocity of three inches per second is ascertained to be sufficient to tear up fine clay, – six inches per second, fine sand, – twelve inches per second, fine gravel, – and three feet per second, stones of the size of an egg.

When this mechanical power of running water is considered, we are prepared for the transportation of large quantities of gravel, sand, and mud, by the torrents and rivers which descend with great velocity from the mountainous regions. But a question naturally arises, how the more tranquil rivers of the valleys and plains, flowing on comparatively level ground, can remove the prodigious burden which is discharged into them by their numerous tributaries, and by what means they are enabled to convey the whole mass to the sea. If they had not this power, their channels would be annually choked up, and the lower valleys and districts adjoining mountain-chains would be continually strewed over with fragments of rock and sterile sand. But this evil is prevented by a general law regulating the conduct of running water, that two equal streams do not occupy a bed of double surface. In proportion, therefore, as the whole fluid mass increases, the space which it occupies decreases relatively to the volume of water; and hence there is a smaller proportion of the whole retarded by friction against the bottom and sides of the channel. The portion thus unimpeded moves with great velocity, so that the main current is often accelerated in the lower country, notwithstanding that the slope of the channel is lessened. It not unfrequently happens, as we shall afterwards demonstrate by examples, that two large rivers, after their junction, have only the surface which one of them had previously; and even in some cases their united waters are confined in a narrower bed than each of them filled before. By this beautiful adjustment, the water which drains the interior country is made continually to occupy less room as it approaches the sea; and thus the most valuable part of our continents, the rich deltas, and great alluvial plains, are prevented from being constantly under water.

Many remarkable illustrations of the power of running water in moving stones and heavy materials were afforded by the late storm and flood which occurred on the 3rd and 4th of August 1829, in Aberdeenshire and other counties, in Scotland. The floods extended almost simultaneously and in equal violence over a space of about five thousand square miles, being that part of the north-east of Scotland which would be cut off by two lines drawn from the head of Lochrannoch, one towards Inverness, and another to Stonehaven. All the rivers within that space were flooded, and the destruction of roads, lands, buildings, and crops along the courses of the streams was very great. The elements during this storm assumed all the characters which mark the tropical hurricanes: the wind blowing in sudden gusts and whirlwinds, the lightning and thunder being such as is rarely witnessed in that climate, and heavy rain falling without intermission. The bridge over the Dee at Ballatu consisted of five arches, having upon the whole a water-way of two hundred and sixty feet. The bed of the river on which the piers rested, was composed of rolled pieces of granite and gneiss. The bridge was built of granite, and had stood uninjured for twenty years, but the different parts were swept away in succession by the flood, and the whole mass of masonry disappeared in the bed of the river. 'The river Don,' observes Mr. Farquharson, in his account of the inundations, 'has upon my own premises forced a mass of four or five hundred tons of stones, many of them two or three hundred pounds weight, up an inclined plane, rising six feet in eight or ten yards; and left them in a rectangular heap, about three feet deep on a flat ground; and, singularly enough, the heap ends abruptly at its lower extremity. A large stone, of three or four tons which I have known for many years in a deep pool of the river, has been moved about one hundred yards from its place.'[28]

The power even of a small rivulet, when swoln by rain, in removing heavy bodies, was lately exemplified in the College, a small stream which flows at a moderate declivity from the eastern water-shed of the Cheviot-Hills. Several thousand tons weight of gravel and sand were transported to the plain of the Till, and a bridge then in progress of building was carried away, some of the arch-stones of which, weighing from half to three-quarters of a ton each, were propelled two miles down the rivulet.

On the same occasion, the current tore away from the abutment of a mill-dam a large block of greenstone-porphyry, weighing nearly two tons, and transported the same to the distance of a quarter of a mile. Instances are related as occurring repeatedly, in which from one to three thousand tons of gravel are in like manner removed to great distances in one day.

In the cases above adverted to, the waters of the river and torrent were dammed back by the bridges which acted as partial barriers, and illustrate the irresistible force of a current when obstructed. Bridges are also liable to be destroyed by the tendency of rivers to shift their course, whereby the pier, or the rock on which the foundation stands, is undermined.

When we consider how insignificant are the volume and velocity of the rivers and streams in our island, when compared to those of the Alps and other lofty chains, and how, during the various changes which the levels of different districts have undergone, the various contingencies which give rise to floods, must in the lapse of ages be multiplied, we may easily conceive that the quantity of loose superficial matter distributed over Europe must be very considerable. That the position also of a great portion of these travelled materials should now appear most irregular, and should often bear no relation to the existing water-drainage of the country, is a necessary consequence, as we shall afterwards see, of the combined operations of running water and subterranean movements.

In mountainous regions and high northern latitudes, the moving of heavy stones by water is greatly assisted by the ice which adheres to them, and which forming together with the rock a mass of less specific gravity, is readily borne along. The glaciers also of alpine regions, formed of consolidated snow, bear down upon their surface a prodigious burden of rock and sand mixed with ice. These materials are generally arranged in long ridges, which sometimes in the Alps are thirty or forty feet high, running parallel to the borders of the glacier, like so many lines of intrenchment. These mounds of debris are sometimes three or more deep, and have generally been brought in by lateral glaciers: the whole accumulation is slowly conveyed to the lower valleys, where, on the melting of the glacier, it is swept away by rivers.

The rapidity with which even the smallest streams hollow out deep channels in soft and destructible soils is remarkably exemplified in volcanic countries, where the sand and half-consolidated tuffs oppose but a slight resistance to the torrents which descend the mountain side. After the heavy rains which followed the eruption of Vesuvius in 1822, the water flowing from the Atrio del Cavallo, cut in three days a new chasm through strata of tuff and volcanic ejected matter to the depth of twenty-five feet. The old mule road was seen, in 1828, intersected by this new ravine. But such facts are trifling when compared to the great gorges which are excavated in somewhat similar materials in the great plateau of Mexico, where an ancient system of valleys, originally worn out of granite and secondary rocks, has been subsequently filled with strata of tuff, pumice, lava, and trachytic conglomerate, to the thickness of several thousand feet. The rivers and torrents annually swoln by tropical rains, are now actively employed in removing these more recent deposits, and in re-excavating the ancient water-courses.

The gradual erosion of deep chasms through some of the hardest rocks, by the constant passage of running water charged with foreign matter, is another phenomenon of which striking examples may be adduced. Some of the clearest illustrations of this excavating power are presented by many valleys in Central France, where the channels of rivers have been barred up by solid currents of lava, through which the streams have re-excavated a passage to the depth of from twenty to seventy feet and upwards, and often of great width. In these cases there are decisive proofs that neither the sea nor any denuding wave, or extraordinary body of water, have passed over the spot since the melted lava was consolidated. Every hypothesis of the intervention of sudden and violent agency is entirely excluded, because the cones of *loose* scoriæ, out of which the lavas flowed, are oftentimes at no great elevation above the rivers, and have remained undisturbed during the whole period which has been sufficient for the hollowing out of such enormous ravines. But we shall reserve a more detailed account of the volcanic district of Central France for another part of this work, and at present confine ourselves to examples derived from events which have happened since the time of history.

Some lavas of Etna, produced by eruptions of which the date is known, have flowed across two of the principal rivers in Sicily; and in both cases the streams, dispossessed of their ancient beds, have opened for themselves new channels. An eruption from Mount Mojo, an insulated cone at the northern base of Etna, sent forth, in the year 396, B.C., in the reign of Dionysius I., a great lava-stream, which crossed the river Caltabianca in two places. The lowermost point of obstruction is seen on the eastern side of Etna, on the high road from Giardini to Catania, where one pier of the bridge on either bank is based upon a remnant of the solid lava, which has been breached by the river to the depth of fourteen feet. But the Caltabianca, although it has been at work for more than two and twenty centuries, has not worn through the igneous rock so as to lay open the gravel of its ancient bed. The declivity, however, of the alluvial plain is very slight; and as the extent of excavation in a given time depends on the volume and velocity of the stream, and the destructibility of the rock, we must carefully ascertain all these circumstances before we attempt to deduce from such examples a measure of the force of running water in a given period.

Recent Excavation of the Simeto. – The power of running water to hollow out compact rock is exhibited, on a larger scale, at the western base of Etna, where a great current of lava (A A, diagram 3), descending from near the

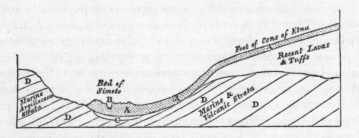

3. Diagram showing the recent excavation of lava at the foot of Etna by the river Simeto

summit of the great volcano, has flowed to the distance of five or six miles, and then reached the alluvial plain of the Simeto, the largest of the Sicilian rivers which skirts the base of Etna, and falls into the sea a few miles south of Catania. The lava entered the river about three miles above the town of Adernò, and not only occupied its channel for some distance, but, crossing to the opposite side of the valley, accumulated there in a rocky mass. Gemmellaro gives the year 1603 as the date of the eruption. The appearance of the current clearly proves that it is one of the most modern of those of Etna, for it has not been covered or crossed by subsequent streams or ejections, and the olives on its surface are all of small size, yet older than the natural wood on the same lava. In the course, therefore, of about two centuries the Simeto has eroded a passage from fifty to several hundred feet wide, and in some parts from forty to fifty feet deep.

The portion of lava cut through is in no part porous or scoriaceous, but consists of a compact homogeneous mass of hard blue rock, somewhat lighter than ordinary basalt, containing crystals of olivine and glassy felspar. The general declivity of this part of the bed of the Simeto is not considerable, but, in consequence of the unequal waste of the lava, two waterfalls occur at Passo Manzanelli, each about six feet in height. Here the chasm (B, diagram No. 3) is about forty feet deep, and only fifty broad.

The sand and pebbles in the river bed consist chiefly of a brown quartzose sandstone, derived from the upper country; but the matter derived from the volcanic rock itself must have greatly assisted the attrition. This river, like the Caltabianca, has not yet cut down to the ancient bed of which it was dispossessed, and of which we have indicated the probable position in the annexed diagram (C, No. 3).

On entering the narrow ravine where the water foams down the two cataracts, we are entirely shut out from all view of the surrounding country; and a geologist who is accustomed to associate the characteristic features of the landscape with the relative age of certain rocks, can scarcely dissuade himself from the belief that he is contemplating a scene in some rocky gorge of a primary district. The external forms of the hard blue lava are as massive as any of the most ancient trap-rocks of Scotland. The solid surface is in some parts smoothed and almost polished by attrition, and covered in others with a white lichen, which imparts to it an air of extreme antiquity, so as greatly to heighten the delusion. But the moment we reascend the cliff, the spell is broken; for we scarcely recede a few paces, before the ravine and river disappear, and we stand on the black and rugged surface of a vast current of lava, which seems unbroken, and which we can trace up nearly to the distant summit of that majestic cone which Pindar called 'the pillar of heaven,'[29] and which still continues to send forth a fleecy wreath of vapour, reminding us that its fires are not extinct, and that it may again give out a rocky stream, wherein other scenes like that now described may present themselves to future observers.

Falls of Niagara. – The falls of Niagara afford a magnificent example of the progressive excavation of a deep valley in solid rock. That river flows from Lake Erie to Lake Ontario, the former lake being three hundred and thirty feet above the latter, and the distance between them being thirty-two miles. On flowing out of the upper lake, the river is almost on a level with its banks; so that, if it should rise perpendicularly eight or ten feet, it would lay under water the adjacent flat country of Upper Canada on the West, and of the State of New York on the East. The river, where it issues, is about three quarters of a mile in width. Before reaching the falls, it is propelled with great rapidity, being a mile broad, about twenty-five feet deep, and having a descent of fifty feet in half a mile. An island at the very verge of the cataract divides it into two sheets of water; one of these, called the Horse-shoe Fall, is six hundred yards wide, and one hundred and fifty-eight feet perpendicular; the other, called the American Falls, is about two hundred yards in width, and one hundred and sixty-four feet in height. The breadth of the island is about five hundred yards. This great sheet of water is precipitated over a ledge of hard limestone, in horizontal strata, below which is a somewhat greater thickness of soft shale, which decays and crumbles away more rapidly, so that the calcareous rock forms an overhanging mass, projecting forty feet or more above the hollow space below. The blasts of wind, charged with spray, which rise out of the pool into which this enormous cascade is projected, strike against the shale beds, so that their disintegration is constant; and the superincumbent limestone, being left without a foundation, falls from time to time in rocky masses. When these enormous fragments descend, a shock is felt at some distance, accompanied by a noise like a distant clap of thunder. After the river has passed over the falls, its character, observes Captain Hall, is immediately and completely changed. It then runs furiously along the bottom of a deep wall-sided valley, or huge trench, which has been cut into the horizontal strata by the continued action of the stream during the lapse of ages. The cliffs on both sides are in most places perpendicular, and the ravine is only perceived on approaching the edge of the precipice.

The waters which expand at the falls, where they are divided by the island, are contracted again, after their union, into a stream not more than one hundred and sixty yards broad. In the narrow channel, immediately below this immense rush of water, a boat can pass across the stream with ease. The pool, it is said, into which the cataract is precipitated, being one hundred and seventy feet deep, the descending water sinks down and forms an under-current, while a superficial eddy carries the upper stratum back *towards* the main fall. This is not improbable; and we must also suppose, that the confluence of two streams, which meet at a considerable angle, tends mutually to neutralize their forces. The bed of the river below the falls is strewed over with huge fragments which have been hurled down into the abyss. By the continued destruction of the rocks, the falls have, within the last forty years, receded nearly fifty yards, or, in other words, the ravine has been prolonged to that extent. Through this deep chasm the Niagara flows for about seven miles; and then the table-land, which is almost on a level with Lake Erie, suddenly sinks down at a town called Queenstown, and the river emerges from the ravine into a plain which continues to the shores of Lake Ontario.

There seems good foundation for the general opinion, that the falls were once at Queenstown, and that they have gradually retrograded from that place to their present position, about seven miles distant. If the ratio of recession had never exceeded fifty yards in forty years, it must have required nearly ten thousand years for the excavation of the whole ravine; but no probable conjecture can be offered as to the quantity of time consumed in such an operation, because the retrograde movement may have been much more rapid when the whole current was confined within a space not exceeding a fourth or fifth of that which the falls now occupy. Should the erosive action not be accelerated in future, it will require upwards of thirty thousand years for the falls to reach Lake Erie (twenty-five miles distant), to which they seem destined to arrive in the course of time, unless some earthquake changes the relative levels of the district. The table-land, extending from Lake Erie, consists uniformly of the same geological formations as are now exposed to view at the falls. The upper stratum is an ancient alluvial sand, varying in thickness from ten to one hundred and forty feet; below which is a bed of hard limestone, about ninety feet in thickness, stretching nearly in a horizontal direction over the whole country, and forming the bed of the river *above* the falls, as do the inferior shales *below*. The lower shale is nearly of the same thickness as the limestone. Should Lake Erie remain in its present state until the period when the ravine recedes to its shores, the sudden escape of that great body of water would cause a tremendous deluge; for the ravine would be much more than sufficient to drain the whole lake, of which the average depth was found, during the late survey, to be only ten or twelve fathoms. But, in consequence of its shallowness, Lake Erie is fast filling up with sediment, and the annual growth of the deltas of many rivers and torrents which flow into it is remarkable. Long Point, for example, near the influx of Big Creek River, was observed, during the late survey, to advance three miles in as many years. A question therefore arises, whether Lake Erie may not be converted into dry land before the Falls of Niagara recede so far. In speculating on this contingency, we must not omit one important condition of the problem. As the surface of the lake is contracted in size, the loss of water by evaporation will diminish; and unless the supply shall decrease in the same ratio (which seems scarcely probable), Niagara must augment continually in volume, and by this means its retrograde movement may hereafter be much accelerated.

[*The analysis of aqueous forces continues through Chapters 11 to 17. Two chapters (11 and 12) demonstrate the erosive power of rivers such as the Po in Italy and the Mississippi in the United States. Their force can be catastrophic, and disasters involving burst lakes in the Alps are dealt with at length. Aqueous causes can also be constructive, as illustrated by the deposits being produced by mineral springs and river deltas (Chapters 13 and 14). Three chapters (15–17) discuss oceanic tides and currents, including tidal waves; the silting up of estuaries; the formation of sand dunes; and the transportation of rocks by icebergs.*]

Igneous Causes

We have hitherto considered the changes wrought, since the times of history and tradition, by the continued action of aqueous causes on the earth's surface; and we have next to examine those resulting from igneous agency. As the rivers and springs on the land, and the tides and currents in the sea, have, with some slight modifications, been fixed and constant to certain localities from the earliest periods of which we have any records, so the volcano and the earthquake have, with few exceptions, continued, during the same lapse of time, to disturb the same regions. But as there are signs, on almost every part of our continent, of great power having been exerted by running water on the surface of the land, and by tides and currents on cliffs bordering the sea, where, in modern times, no rivers have excavated, and no tidal currents undermined – so we find signs of volcanic vents and violent subterranean movements in places where the action of fire has long been dormant. We can explain why the intensity of the force of aqueous causes should be developed in succession in different districts. Currents, for example, and tides, cannot destroy our coasts, shape out or silt up estuaries, break through isthmuses, and annihilate islands, form shoals in one place and remove them from another, without the direction and position of their destroying and transporting power becoming transferred to new localities. Neither can the relative levels of the earth's crust, above and beneath the waters, vary from time to time, as they are admitted to have varied at former periods, and as we shall demonstrate that they still do, without the continents being, in the course of ages, modified, and even entirely altered, in their external configuration. Such events must clearly be accompanied by a complete change in the volume, velocity, and direction of the streams and land floods to which certain regions give passage. That we should find, therefore, cliffs where the sea once committed ravages, and from which it has now retired – estuaries where high tides once rose, but which are now dried up – valleys hollowed out by water, where no streams now flow; – all these and similar phenomena are the necessary consequences of physical causes now in operation; and we may affirm that, if there be no instability in the laws of Nature, similar fluctuations must recur again and again in time to come.

But however natural it may be that the force of running water in numerous valleys, and of tides and currents in many tracts of the sea, should now be *spent*, it is by no means so easy to explain why the violence of the earthquake and the fire of the volcano should also have become locally extinct, at successive periods. We can look back to the time when the marine strata, whereon the great mass of Etna rests, had no existence; and that time is extremely modern in the earth's history. This alone affords ground for anticipating that the eruptions of Etna will one day cease.

> Nor *Ætna* vomiting sulphureous Fire
> Will ever belsh; for Sulphur will expire,
> (The Veins exhausted of the liquid Stores)
> Time was she cast no Flame; in time will cast no more.[30]

are the memorable words which are put into the mouth of Pythagoras by the Roman poet, and they are followed by speculations as to the causes of volcanic vents shifting their position. Whatever doubts the philosopher expresses as to the nature of these causes, it is assumed, as incontrovertible, that the points of eruption will hereafter vary, *because they have formerly done so*.

We have endeavoured to show, by former chapters, how utterly this principle of reasoning is set at nought by the modern schools of geology, which not only refuse to conclude that great revolutions in the earth's surface are now in progress, or that they will take place *because* they have often been repeated in former ages, but assume the improbability of such a conclusion and throw the whole weight of proof on those by whom that doctrine is embraced.

In our view of igneous causes we shall consider, first, the volcano, and afterwards the earthquake; for although both are probably the effects of the same subterranean process, they give rise to very different phenomena on the surface of the globe. Both are confined to certain regions, but the subterranean movements are least violent in the immediate proximity of volcanic vents, especially where the discharge of aëriform fluids and melted rock is made constantly from the same crater. We say that there are certain regions to which both the points of eruption, and the movements of great earthquakes are confined; and we shall begin by tracing out the geographical boundaries of some of these, that the reader may be aware of the magnificent scale on which the agency of subterranean fire is now simultaneously developed. Over the whole of the vast tracts alluded to, active volcanic vents are distributed at intervals, and most commonly arranged in a linear direction. Throughout the intermediate spaces there is abundant evidence that the subterranean fire is at work continuously, for the ground is convulsed from time to time by earthquakes; gaseous vapours, especially carbonic acid gas, are disengaged plentifully from the soil; springs often issue at a very high temperature, and their waters are very commonly impregnated with the same mineral matters which are discharged by volcanos during eruptions.

* * *†

[*The remainder of Chapter 18 surveys regions of subterranean disturbance, showing just how widely volcanic action occurs throughout the globe. Chapters 19 and 20 focus on the volcanic eruption best described in written evidence from classical sources: Vesuvius, most notorious for destroying Pompeii in* AD *79. Chapter 21 deals briefly with the phenomena associated with Etna in Sicily, Hecla on Iceland, and Jorullo in Mexico.*

Chapter 22 opens by discussing further cases of volcanic forces in action, beginning with the Canary Islands, which had been the subject of detailed research in 1815 by the Prussian geologist Leopold von Buch. Like his contemporary Alexander von Humboldt, Buch had argued that volcanoes were 'craters of elevation' produced by a blister of molten material underneath the surface of the earth. Lyell contests this theory through a detailed study of the Greek island of Santorin:]

* * *

Grecian Archipelago. – We shall next direct our inquiry to the island of Santorin, as it will afford us an opportunity of discussing the merits of a singular theory, which has obtained no small share of popularity in modern times, respecting 'craters of elevation,' (Erhebungs Cratere, Cratères de soulèvement,) as they have been termed. The three islands of Santorin, Therasia, and Aspronisi surround a gulf almost circular, and above six

---

† Omissions within chapters are indicated by asterisks. – *Ed.*

miles in diameter. They are chiefly composed of trachytic conglomerates and tuffs, covered with pumice; but in one part of Santorin clay-slate is seen to be the fundamental rock. The beds in all these isles dip at a slight

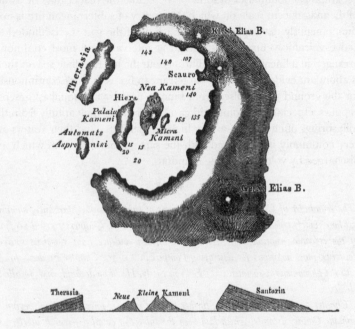

15. Chart and section of Santorin and the contiguous islands in the Grecian Archipelago

angle towards the exterior of the group, and lose themselves in the surrounding sea; whereas, on the contrary, they present a high and steep escarpment towards the centre of the inclosed space. The gulf, therefore, is nearly on all sides environed by precipices; those of Santorin, which form two-thirds of the circumference, being two leagues in extent, and in some parts three hundred feet high. These rocky cliffs plunge at once into the sea, so that close to the shore soundings are only reached at a depth of eight hundred feet, and at a little distance farther at a depth of one thousand feet. In the middle of this gulf, the small isle of Hiera, now called Palaia Kameni, rose up, 144 years before the Christian era. In 1427 this isle received new accessions. In 1573 the Little Kameni was raised in the middle of the basin, its elevation being accompanied by the discharge of large quantities of pumice and a great disengagement of vapour. Lastly, in 1707 and 1709 the New Kameni was formed, which still exhales sulphureous vapours. These isles are formed of rocks of brown trachyte, which has a resinous lustre, and is full of crystals of glassy felspar. Although the birth of New Kameni was attended by an eruption, it is certain that it was upraised from a great depth by earthquakes, and was not a heap of volcanic ejections, nor of lava poured out on the spot. There were shells upon it when it first appeared; and beds of limestone and marine shells are described by several authors as entering, together with igneous rocks, into the structure of other parts of this group. In order, therefore, to explain the formation of such circular gulfs, which are common in other archipelagos, Von Buch supposes, and Humboldt adopts the same opinion, that the different beds of lava, pumice, and whatever else may be inter-stratified, were first horizontally disposed along the floor of the ocean. An expansive force from below then burst an opening through them, and, acting towards a central point, raised symmetrically on every side all which resisted its action, so that the uplifted strata were made to dip away on all sides from the centre outwards, as is usual in volcanic cones, while a deep hollow was left in the middle, resembling in all essential particulars an ordinary volcanic crater.

In the first instance we should inform the reader, that this theory is not founded on actual observations of analogous effects produced by the elevating forces of earthquakes, or the escape of elastic fluids in any part of the globe; for the inflation from below, of the rocks in the plain of Malpais, during the eruption of Jorullo, was, as before stated, an hypothesis proposed, long after that eruption, to account for appearances which admit of a very different explanation. Besides, in the case of Jorullo, there was no great 'crater of elevation' formed in the centre. All our modern analogies, therefore, being in favour of the origin of cones and craters exclusively by *eruptions*, we are entitled to scrutinize with no small severity the new hypothesis; and we have a right to demand demonstrative evidence, that known and ordinary causes are perfectly insufficient to produce the observed phenomena. Had Von Buch and Humboldt, for instance, in the course of those extensive travels which deservedly render their opinions, in regard to all volcanic operations, of high authority, discovered a single cone composed exclusively of marine or lacustrine strata, without a fragment of any igneous rock intermixed; and in the centre a great cavity, encircled by a precipitous escarpment; then we should have been compelled at once to concede, that the cone and crater-like configuration, whatever be its mode of formation, may sometimes have no reference whatever to ordinary volcanic eruptions.

But it is not pretended that, on the whole face of the globe, a single example of this kind can be pointed out. In Europe and North America thousands of square leagues of territory have been examined, composed of marine strata, which have been elevated to various heights, sometimes to more than ten thousand feet above the level of the sea, sometimes in horizontal tabular masses; in other cases with every degree of inclination, from the horizontal to the vertical. Some have been moved without great derangement, others have been rent, contorted, or shattered with the utmost violence. Sometimes large districts, at others small spaces, appear to have changed their position. Yet, amidst the innumerable accidents to which these rocks have been subject, never have they assumed that form, exactly representing a large truncated volcanic cone, with a great cavity in the centre. Are we then called upon to believe that whenever elastic fluids generated in the subterranean regions burst through horizontal strata, so as to upheave them in the peculiar manner before adverted to, they always select, as if from choice, those spots of comparatively insignificant area, where a certain quantity of volcanic matter happens to lie, while they carefully avoid purely lacustrine and marine strata, although they often lie immediately contiguous? Why on the southern borders of the Limagne d'Auvergne, where several eruptions burst through, and elevated the horizontal marls and limestones, did these freshwater beds never acquire in any instance a conical and crateriform disposition?

But let us proceed to examine some of the most celebrated examples adduced of craters of elevation. The most perfect type of this peculiar configuration is said to be afforded by the Isle of Palma; and while we controvert Von Buch's theoretical opinions, we ought not to forget how much geology is indebted to his talents and zeal, and amongst other works for his clear and accurate description of this isle. In the middle of Palma rises a mountain to the height of four thousand feet, presenting the general form of a great cone, the upper part of which had been truncated and replaced by an enormous funnel-shaped cavity, about four thousand feet deep; and the surrounding borders of which attain, at their highest point, an elevation of seven thousand feet above the sea. The external flanks of this cone are gently inclined, and, in part, cultivated; but the bottom and the walls of the central cavity, called by the inhabitants the Caldera, present on all sides rugged and uncultivated rocks, almost completely devoid of vegetation.

16. View of the Isle of Palma, and of the Caldera in its centre

So steep are the sides of the Caldera, that there is no path by which they can be descended; and the only entrance is by a great ravine, which, cutting through the rocks environing the circus, runs down to the sea. The sides of this gorge are jagged, broken, and precipitous. In the mural escarpments surrounding the Caldera are seen nothing but beds of basalt, and conglomerates composed of broken fragments of basalt, which dip away with the greatest regularity, from the centre to the circumference of the cone. Now, according to the theory of 'elevation craters,' we are called upon to suppose that, in the first place, a series of horizontal beds of volcanic matter accumulated over each other, to the enormous depth of more than four thousand feet – a circumstance which alone would imply the proximity, at least, of a vent from which immense quantities of igneous rocks had proceeded. After the aggregation of the mass, the expansive force was directed on a given point with such extraordinary energy, as to lift up bodily the whole mass, so that it should rise to the height of seven thousand feet above the sea, leaving a great gulf or cavity in the middle. Yet, notwithstanding this prodigious effort of gaseous explosions, concentrated on so small a point, the beds, instead of being shattered, contorted, and thrown into the utmost disorder, have acquired that gentle inclination, and that regular and symmetrical arrangement, which characterize the flanks of a large cone of eruption, like Etna! We admit that earthquakes, when they act on extensive tracts of country, may elevate and depress them without deranging, considerably, the relative position of hills, valleys, and ravines. But is it possible to conceive that elastic fluids could break through a mere point as it were of the earth's crust, and that too where the beds were not composed of soft, yielding clay, or incoherent sand, but of solid basalt, thousands of feet thick, and that they could inflate them, as it were, in the manner of a bladder? Would not the rocks, on the contrary, be fractured, fissured, thrown into a vertical, and often into a reversed position; and, ere they attained the height of seven thousand feet, would they not be reduced to a mere confused and chaotic heap?

The Great Canary is an island of a circular form, analogous to that of Palma. Barren Island, also, in the Bay of Bengal, is proposed as a striking illustration of the same phenomenon; and here it is said we have the advantage of being able to contrast the ancient crater of elevation with a cone and crater of eruption in its centre. When seen from the ocean, this island presents, on almost all sides, a surface of bare rocks, which rise up with a moderate declivity towards the interior; but at one point there is a narrow cleft, by which we can penetrate into the centre, and there discover that it is occupied by a great circular basin, filled by the waters

17. Cone and Crater of Barren Island, in the Bay of Bengal

of the sea, bordered all around by steep rocks, in the midst of which rises a volcanic cone, very frequently in eruption. The summit of this cone is 1690 French feet in height, corresponding to that of the circular border which incloses the basin; so that it can only be seen from the sea through the ravine, which precisely resembles the deep gorge by which we penetrate into the Caldera of the Isle of Palma, and of which an equivalent, more or less decided in its characters, is said to occur in all elevation craters.

The cone of the high peak of Teyda, in Teneriffe, is also represented as rising out of the middle of a crater of elevation, standing like a tower surrounded by its foss and bastion; the foss being the remains of the ancient gulf, and the bastion the escarpment of the circular inclosure. So that Teneriffe is an exact counterpart of Barren Island, except that one is raised to an immense height, while the other is still on a level with the sea, and in part concealed beneath its waters.

Now, without enumerating more examples, let us consider what form the products of submarine volcanos may naturally be expected to assume. There is every reason to conclude, from the few accounts which we possess of eruptions at the bottom of the sea, that they take place in the same manner there as on the open surface of a continent. That the volcanic phenomena, if they are ever developed at unfathomable depths, may be extremely different, is very possible; but when they have been witnessed by the crews of vessels casually passing, the explosions of aëriform fluids beneath the waters have closely resembled those of volcanos on the land. Rocky fragments, ignited scoriæ, and comminuted ashes, are thrown up, and in several cases conical islands have been formed, which afterwards disappeared; as when, in 1691 and 1720, small isles were thrown up off St. Michael in the Azores, or as Sabrina rose in 1811 near the same spot, and, in 1783, Nyöe, off the coast of Iceland. Where the cones have disappeared, they probably consisted of loose matters, easily reduced by the waves and currents to a submarine shoal. When islands have remained firm, as in the case of Hiera, and the New and Little Kameni in the Gulf of Santorin (see wood-cut No. 15), they have consisted in part of solid lava. Whatever doubts might have been entertained as to the action of volcanos entirely submarine, yet it must always have been clear, that in those numerous cases where they just raise their peaks above the waves, the ejected sand, scoriæ, and fragments of rock, must accumulate round the vent into a cone with a central crater, while the lighter will be borne to a distance by tides and currents, as by winds during eruptions in open air. The lava which issues from the crater spreads over the subaqueous bottom, seeking the lowest levels, or accumulating upon itself, according to its liquidity, volume, and rapidity of congelation; following, in short, the same laws as when flowing in the atmosphere.

But we may next enquire, what characters may enable a geologist to distinguish between cones formed entirely, or in great part beneath the waters of the sea, and those formed on land. In the first place, large beds of shells and corals often grow on the sloping sides of submarine cones, particularly in the Pacific, and these often become interstratified with lavas. Instead of alluvions containing land-shells, like some of those which cover Herculaneum, great beds of tufaceous sand and conglomerate, mixed with marine remains, might be expected on such parts of the flanks of a volcano like Stromboli as are submerged beneath the waters. The pressure of a column of water exceeding many times that of the atmosphere, must impede the escape of the elastic fluids and of lava, until the resistance is augmented in the same proportion; hence the explosions will be more violent, and when a cone is formed it will be liable to be blown up and truncated at a lower level than in shallower water or in the open air. Add to this, that when a submarine volcano has repaired its cone, it is liable to be destroyed again by the waves, as in several cases before adverted to. The vent will then become choked up with strata of sand and fragments of rock, swept in by the tides and currents. These materials are far more readily consolidated under water than in the air, especially as mineral matter is so copiously introduced by the springs which issue from the ground in all volcanic regions hitherto carefully investigated. Beds of solid travertin, also, and in hot countries coral reefs, must often, during long intervals of quiescence, obstruct the vent, and thus increase the repressive force and augment the violence of eruptions. The probabilities, therefore, in a submarine volcano, of the destruction of a larger part of the cone, and the formation of a more extensive crater, are obvious; nor can the dimensions of 'craters of elevation,' if referred to such operations, surprise us. During an eruption in 1444, accompanied by a tremendous earthquake, the summit of Etna was destroyed, and an enormous crater was left, from which lava flowed. The segment of that crater may still be seen near the Casa Inglese, and, when complete, it must have measured several miles in diameter. The cone was afterwards repaired; but this would not have happened so easily had Etna been placed like Stromboli in a deep sea, with its peak exposed to the fury of the waves. Let us suppose the Etnean crater of 1444 to have been filled up with beds of coral and conglomerate, and that during succeeding eruptions these were thrown out by violent explosions, so that the cone became truncated down to the upper margin of the woody region, a circular basin would then be formed thirty Italian miles in circumference, exceeding by five or six miles the circuit of the Gulf of Santorin. Yet we know by numerous sections that the strata of trachyte, basalt, and trachytic breccia, would, in that part of the great cone of Etna, dip on all sides off from the centre at a gentle angle to every point of the compass, except where irregularities were occasioned, at points where the small buried cones before mentioned occurred. If this gulf were then again choked up, and the vent obstructed, so that new explosions of great violence should truncate the cone once more down to the inferior border of the forest zone of Etna, the circumference of the gulf would then be fifty Italian miles. Yet even then the ruins of the cone of Etna might form a circular island entirely composed of volcanic rocks, sloping gently outwards on all sides at a very slight angle; and this island might be between seventy and eighty English miles in its exterior circuit, while the circular bay within might be between forty and fifty miles round. In fertility it would rival the Island of Palma; and the deep gorge which leads down from the Valley of Calanna to Zafarana, might well serve as an equivalent to the grand defile which leads into the Caldera.

It is most probable, then, that the exterior inclosure of Barren Island, *c d*, in the annexed diagram, is nothing more than the remains of the truncated cone *c a b d*, a great portion of which has been carried away, partly by the action of the waves, and partly by explosions which preceded the formation of the new interior cone, *f e g*. Whether the outer and larger cone has in this particular case, together with the bottom of the ocean on

18. Supposed section of Barren Island, in the Bay of Bengal

which it rests, been upheaved, or whether it originally projected in great part like Stromboli above the level of the sea, may, probably, be determined by geological investigations; for, in the former case, some beds replete with marine remains may be interstratified with volcanic ejections.

Some of the accounts transmitted to us by eye-witnesses, of the gradual manner in which New Kameni first rose covered with living shells in the Gulf of Santorin, appear, certainly, to establish the possibility of the elevation of small masses from a depth of several hundred feet during an eruption, and during the emission of lava. But the protrusion of isolated masses, under such circumstances, affords no analogy to the supposed action of the expansive force in the formation of craters of elevation. It is hardly necessary, after the observations now made, to refer the reader again to our section of Somma and Vesuvius, and to say that we ascribed the formation of the ancient and the modern cone to operations precisely analogous.

M. Necker long ago pointed out the correspondence of their structure, and explained most distinctly the origin of the form of Somma; and his views were afterwards confirmed by Mr. Scrope. But, notwithstanding the juxta-position of the entire and the ruined cone, the identity of the slope and quâquâ-versal dip of the beds, the similarity of their mineral composition, and the intersection of both cones by porphyritic dikes, the defenders of the 'elevation' theory have declared that the lavas and breccias of Somma were once horizontal, and were afterwards raised into a conical mass, while they admit that those in Vesuvius have always been as highly inclined as they are now.

In controverting Von Buch's theory, we might have adduced as the most conclusive argument against it, that it would lead its advocates, if consistent with themselves, to the extravagant conclusion, that the two cones of Vesuvius had derived their form from very distinct causes. But as these geologists are not afraid to follow their system into all its consequences, and have even appealed to Somma as confirmatory of their views, it would be vain to hope, by pointing out the closest analogies between the effects of ordinary volcanic action and 'craters of elevation,' to induce them to abandon their hypothesis.

The marine shelly strata, interstratified with basalt, through which the great cone of Etna rises, are also said to have constituted an ancient crater of elevation; but when we allude more particularly to the geology of Sicily, it will appear that the strata in question do not dip so as to countenance in the least degree such an hypothesis. The nearest approach, perhaps, to the production of a conical mass by elevation from below, is in the Cantal in Central France. The volcanic eruptions which produced at some remote period the volcanic mountain called the Plomb du Cantal, broke up through fresh-water strata, which must have been deposited originally in an horizontal position, on rocks of granitic schist. During the gradual formation of the great cone, beds of lava and tuff, thousands of feet in thickness, were thrown out from one or more central vents, so as to cover great part of the lacustrine strata, and these at the same time were traversed by dikes, and in parts lifted up together with the subjacent granitic rocks; so that if the igneous products could now be removed, and the marls, limestones, and fundamental schists, supported at their present elevation, they would form a kind of dome-shaped protuberance. But the outline of this shattered mass would be very unlike that of a regular cone, and the dip of the beds would be often horizontal, as near Aurillac, often vertical, often reversed, nor would there be in the centre any great cavity or crater of elevation. On the other hand, the *volcanic* beds of the Plomb du Cantal are arranged in a conical form, like those of Etna, not by elevation from below, but because they flowed down during successive eruptions *from above*.

We may observe that the Fossa Grande on Vesuvius, a deep ravine washed out by the winter-torrents which descend from the Atrio del Cavallo, may represent, on a small scale, the Valley of Calanna, and its continuation, the Valley of St. Giacomo on Etna. In the Fossa Grande, a small body of water has cut through tuff, and in some parts solid beds of lava of considerable thickness; and the channel, although repeatedly blocked up by modern lavas, has always been re-excavated. It is natural that on one side of every large hollow, such as the crater of a truncated cone, there should be a channel to drain off the water; and this becoming in the course of ages a deep ravine, may have caused such gorges as exist in Palma and other isles of similar conformation.

Mineral Composition of Volcanic Products. – The mineral called felspar, forms in general more than half of the mass of modern lavas. When it is in great excess, lavas are called trachytic; when augite (or pyroxene) predominates, they are termed basaltic. But lavas of composition, precisely intermediate, occur, and from their colour have been called graystones. A great abundance of quartz characterizes the granitic and other ancient rocks, now generally considered by geologists as of igneous origin, whereas that mineral, which is nothing more than silex crystallized, is rare in recent lavas, although silex enters largely into their composition. Hornblende, which is so common in ancient rocks, is rare in modern lava, nor does it enter largely into rocks of any age in which augite abounds. Mica occurs plentifully in some recent trachytes, but is rarely present where augite is in excess. We must beware, however, not to refer too hastily to a difference of era, characters which may, in truth, belong to the different circumstances under which the products of fire originate.

When we speak of the igneous rocks of our own times, we mean that small portion which happens in violent eruptions to be forced up by elastic fluids to the surface of the earth. We merely allude to the sand, scoriæ, and lava, which cool in the open air; but we cannot obtain access to that which is congealed under the pressure of many hundred, or many thousand atmospheres. We may, indeed, see in the dikes of Vesuvius rocks consolidated from a liquid state, under a pressure of perhaps a thousand feet of lava, and the rock so formed is more crystalline and of greater specific gravity than ordinary lavas. But the column of melted matter raised above the level of the sea during an eruption of Vesuvius must be more than three thousand feet in height, and more than ten thousand feet in Etna; and we know not how many miles deep may be the ducts which communicate between the mountain and those subterranean lakes or seas of burning matter which supply for thousands of years, without being exhausted, the same volcanic vents. The continual escape of hot vapours from many craters during the interval between eruptions, and the chemical changes which are going on for ages in the fumeroles of volcanos, prove that the volcanic foci retain their intense heat constantly, nor can we suppose it to be otherwise; for as lava-currents of moderate thickness require many years to cool down in the open air, we must suppose the great reservoirs of melted matter at vast depths in the nether regions to preserve their high temperature and fluidity for thousands of years.

During the last century, about fifty eruptions are recorded of the five European volcanos, Vesuvius, Etna, Volcano, Santorin, and Iceland, but many beneath the sea in the Grecian Archipelago and near Iceland may doubtless have passed unnoticed. If some of them produced no lava, others on the contrary, like that of Skaptár Jokul in 1783, poured out melted matter for five or six years consecutively, which cases, being reckoned as single eruptions, will compensate for those of inferior strength. Now, if we consider the active volcanos of Europe to constitute about a fortieth part of those already known on the globe, and calculate, that, one with another, they are about equal in activity to the burning mountains in other districts, we may then compute that there happen on the earth about two thousand eruptions in the course of a century, or about twenty every year.

However inconsiderable, therefore, may be the superficial rocks which the operations of fire produce on the surface, we must suppose the subterranean changes now constantly in progress to be on the grandest scale. The loftiest volcanic cones must be as insignificant, when contrasted to the products of fire in the nether regions, as are the deposits formed in shallow estuaries when compared to submarine formations accumulating in the abysses of the ocean. In regard to the characters of these volcanic rocks, formed in our own times in the bowels of the earth, whether in rents and caverns, or by the cooling of lakes of melted lava, we may safely infer that the rocks are heavier and less porous than true lavas, and more crystalline, although composed of the same mineral ingredients. As the hardest crystals produced artificially in the laboratory, require the longest time for their formation, so we must suppose that where the cooling down of melted matter takes place by insensible degrees, in the course of ages a variety of minerals will be produced far harder than any formed by natural processes within the short period of human observation.

These subterranean volcanic rocks, moreover, cannot be stratified in the same manner as sedimentary deposits from water, although it is evident that when great masses consolidate from a state of fusion, they may separate into natural divisions; for this is seen to be the case in many lava-currents. We may also expect that the rocks in question will often be rent by earthquakes, since these are common in volcanic regions, and the fissures will be often injected with similar matter, so that dikes of crystalline rock will traverse masses of similar composition. It is also clear that no organic remains can be included in such masses, unless where sedimentary strata have subsided to great depths, and in this case the fossil substances will probably be so acted upon by heat, that all signs of organization will be obliterated. Lastly, these deep-seated igneous formations must underlie all the strata containing organic remains, because the heat proceeds from below upwards, and the intensity required to reduce the mineral ingredients to a fluid state must destroy all organic bodies in rocks either subjacent or included in the midst of them. If, by a continued series of elevatory movements, such masses shall hereafter be brought up to the surface, in the same manner as sedimentary marine strata have, in the course of ages, been upheaved to the summit of the loftiest mountains, it is not difficult to foresee what perplexing problems may be presented to the geologist. He may then, perhaps, study in some mountain chain the very rocks produced at the depth of several miles beneath the Andes, Iceland, or Java, in the time of Leibnitz, and draw from them the same conclusion which that philosopher derived from certain igneous products of high antiquity; for he conceived our globe to have been, for an indefinite period, in the state of a comet, without an ocean, and uninhabitable alike by aquatic or terrestrial animals.

Earthquakes and their Effects

We have already stated, in our sketch of the geographical boundaries of volcanic regions, that, although the points of eruption are but thinly scattered, and form mere spots on the surface of those vast districts, yet the subterranean movements extend, simultaneously, over immense areas. We shall now proceed to consider the changes which these movements have been observed to produce on the surface, and in the internal structure of the earth's crust.

It is only within the last century and a half, since Hooke first promulgated his views respecting the connexion between geological phenomena and earthquakes, that the permanent changes effected by these convulsions have excited attention. Before that time, the narrative of the historian was almost exclusively confined to the number of human beings who perished, the number of cities laid in ruins, the value of property destroyed, or certain atmospheric appearances which dazzled or terrified the observers. The creation of a new lake, the engulphing of a city, or the raising of a new island, are sometimes, it is true, adverted to, as being too obvious, or of too much geographical interest, to be passed over in silence. But no researches were made expressly with a view of ascertaining the precise amount of depression or elevation of the ground, or the particular alterations in the relative position of sea and land; and very little distinction was made between the raising of soil by volcanic ejections, and the upheaving of it by forces acting from below. The same remark applies to a very large proportion of modern accounts; and how much reason we have to regret this deficiency of information is apparent from the fact, that in every instance where a spirit of scientific inquiry has animated the eye-witnesses of these events, facts calculated to throw much light on former modifications of the earth's structure have been recorded.

As we shall confine ourselves almost entirely, in our notice of certain earthquakes, to the changes brought about by them in the configuration of the earth's crust, we may mention, generally, some accompaniments of these terrible events which are almost uniformly commemorated in history, that it may be unnecessary to advert to them again. Irregularities in the seasons precede or follow the shocks; sudden gusts of wind, interrupted by dead calms; violent rains, in countries or at seasons when such phenomena are unusual or unknown; a reddening of the sun's disk, and a haziness in the air, often continued for months; an evolution of electric matter, or of inflammable gas from the soil, with sulphureous and mephitic vapours; noises underground, like the running of carriages, or the discharge of artillery, or distant thunder; animals utter cries of distress, and evince extraordinary alarm, being more sensitive than men of the slightest movement; a sensation like sea-sickness, and a dizziness in the head, are experienced: – these, and other phenomena which do not immediately bear on our present subject, have recurred again and again at distant ages, and in all parts of the globe.

We shall now begin our enumeration with the latest authentic narratives of earthquakes, and so carry back our survey retrospectively, that we may bring before the reader, in the first place, the minute and circumstantial details of modern times, and enable him, by observing the extraordinary amount of change within the last hundred and fifty years, to perceive how great must be the deficiency in the meagre annals of earlier eras.

* * *

[*Beginning with a tremor in Spain on 21 March 1829 – announced in the newspapers only months before the* Principles *went to press – Chapter 23 moves backwards through the catalogue of disaster produced by fourteen major earthquakes. Chapter 24 uses the best documented example to show just how great their effects can be:*]

Of the numerous earthquakes which have occurred in different parts of the globe, during the last hundred years, that of Calabria, in 1783, is the only one of which the geologist can be said to have such a circumstantial account as to enable him fully to appreciate the changes which this cause is capable of producing in the lapse of ages. The shocks began in February, 1783, and lasted for nearly four years, to the end of 1786. Neither in duration, nor in violence, nor in the extent of territory moved, was this convulsion remarkable, when contrasted with many experienced in other countries, both during the last and present century; nor were the alterations which it occasioned in the relative level of hill and valley, land and sea, so great as those effected by some subterranean movements in South America, in our own times. The importance of the earthquake in question arises from the circumstance, that Calabria is the only spot hitherto visited, both during and after the convulsions, by men possessing sufficient leisure, zeal, and scientific information, to enable them to collect and describe with accuracy the physical facts which throw light on geological questions.

Among the numerous authorities, Vivenzio, physician to the King of Naples, transmitted to the court a regular statement of his observations during the continuance of the shocks; and his narrative is drawn up with care and clearness. Francesco Antonio Grimaldi, then secretary of war, visited the different provinces at the king's command, and published a most detailed description of the permanent changes in the surface. He measured the length, breadth, and depth of the different fissures and gulphs which opened, and ascertained their number in many provinces. His comments, moreover, on the reports of the inhabitants, and his explanations of their relations, are judicious and instructive. Pignataro, a physician residing at Monteleone, a town placed in the very centre of the convulsions, kept a register of the shocks, distinguishing them into four classes, according to their degree of violence. From his work, it appears that, in the year 1783, the number was nine hundred and forty-nine, of which five hundred and one were shocks of the first degree of force; and in the following year there were one hundred and fifty-one, of which ninety-eight were of the first magnitude. Count Ippolito, also, and many others, wrote descriptions of the earthquake; and the Royal Academy of Naples, not satisfied with these and other observations, sent a deputation from their own body into Calabria, before the shocks had ceased, who were accompanied by artists instructed to illustrate by drawings the physical changes of the district, and the state of ruined towns and edifices. Unfortunately these artists were not very successful in their representations of the condition of the country, particularly when they attempted to express, on a large scale, the extraordinary revolutions which many of the great and minor river-courses underwent. But many of the plates published by the Academy are valuable; and we shall frequently avail ourselves of them to illustrate the facts about to be described. In addition to these Neapolitan sources of information, our countryman, Sir William Hamilton, surveyed the district, not without some personal risk, before the shocks had ceased; and his sketch, published in the Philosophical

Transactions, supplies many facts that would otherwise have been lost. He has explained in a rational manner many events which, as related in the language of some eye-witnesses, appeared marvellous and incredible. Dolomieu also examined Calabria, soon after the catastrophe, and wrote an account of the earthquake, correcting a mistake into which Hamilton had fallen, who supposed that a part of the tract shaken had consisted of volcanic tuff. It is, indeed, a circumstance which enhances the geological interest of the commotions which so often modify the surface of Calabria, that they are confined to a country where there are neither ancient nor modern rocks of igneous origin; so that at some future time, when the era of disturbance shall have passed by, the cause of former revolutions will be as latent as in parts of Great Britain now occupied exclusively by ancient marine formations.

The convulsion of the earth, sea, and air, extended over the whole of Calabria Ultra, the south-east part of Calabria Citra, and across the sea to Messina and its environs – a district lying between the 38th and 39th degrees of latitude. The concussion was perceptible over a great part of Sicily, and as far north as Naples; but the surface over which the shocks acted so forcibly as to excite intense alarm, did not generally exceed five hundred square miles in circumference. The soil of that part of Calabria is composed chiefly, like the southern part of Sicily, of calcareo-argillaceous strata of great thickness, containing marine shells. This clay is sometimes associated with beds of sand and limestone. For the most part these formations resemble in appearance and consistency the Subapennine marls, with their accompanying sands and sandstones; and the whole group bears considerable resemblance, in the yielding nature of its materials, to most of our tertiary deposits in France and England. Chronologically considered, however, the Calabrian formations are comparatively of very modern date, and abound in fossil shells referrible to species now living in the Mediterranean.

We learn from Vivenzio, that on the 20th and 26th of March, 1783, earthquakes occurred in the islands of Zante, Cephalonia, and St. Maura; and in the last-mentioned isle several public edifices and private houses were overthrown, and many people destroyed. We have already shown that the Ionian Isles fall within the line of the same great volcanic region as Calabria; so that both earthquakes were probably derived from a common source, and it is not improbable that the bed of the whole intermediate sea was convulsed.

If the city of Oppido, in Calabria, be taken as a centre, and round that centre a circle be described with a radius of twenty-two miles, this space will comprehend the surface of the country which suffered the greatest alteration, and where all the towns and villages were destroyed. But if we describe the circle with a radius of seventy-two miles, this will then comprehend the whole country that had any permanent marks of having been affected by the earthquake. The first shock, of February 5th, 1783, threw down, in two minutes, the greater part of the houses in all the cities, towns, and villages, from the western flanks of the Apennines in Calabria Ultra, to Messina in Sicily, and convulsed the whole surface of the country. Another occurred on the 28th of March, with almost equal violence. The granitic chain which passes through Calabria from north to south, and attains the height of many thousand feet, was shaken but slightly; but it is said that a great part of the shocks which were propagated with a wave-like motion through the recent strata from west to east, became very violent when they reached the point of junction with the granite, as if a reaction was produced where the undulatory movement of the soft strata was suddenly arrested by the more solid rocks. The surface of the country often heaved like the billows of a swelling sea, which produced a swimming in the head like sea-sickness. It is particularly stated, in almost all the accounts, that just before each shock the clouds appeared motionless; and although no explanation is offered of this phenomenon, it is obviously the same as that observed in a ship at sea when it pitches violently. The clouds seem arrested in their career as often as the vessel rises in a direction contrary to their course; so that the Calabrians must have experienced precisely the same motion on the land.

We shall first consider that class of physical changes produced by the earthquake, which are connected with changes in the relative level of the different parts of the land; and afterwards describe those which are more immediately connected with the derangement of the regular drainage of the country, and where the force of running water co-operated with that of the earthquake.

In regard to alterations of relative level, none of the accounts establish that they were on a considerable scale; but it must always be remembered, that in proportion to the area moved is the difficulty of proving that the general level has undergone any change, unless the sea-coast happens to have participated in the principal movement. Even then it is often impossible to determine whether an elevation or depression even of several feet has occurred, because there is nothing novel in a band of sand and shingle of unequal breadth above the level of the sea, marking the point reached by the waves during spring-tides or the most violent tempests. The scientific investigator has not sufficient topographical knowledge to discover whether the extent of beach has diminished or increased; and he who has the necessary local information feels no interest in ascertaining the amount of the rise or fall of the ground. Add to this the great difficulty of making correct observations, in consequence of the enormous waves which roll in upon a coast during an earthquake, and efface every landmark near the shore.

It is evidently in sea-ports alone that we can look for very accurate indications of slight changes of level; and when we find them, we may presume that they would not be rare at other points, if equal facilities of comparing relative altitudes were afforded. Grimaldi states (and his account is confirmed by Hamilton and others) that at Messina in Sicily the shore was rent; and the soil along the port, which before the shock was perfectly level, was found afterwards to be inclined towards the sea, the sea itself near the 'Banchina' becoming deeper, and its bottom in several places disordered. The quay also sank down about fourteen inches below the level of the sea, and the houses in its vicinity were much fissured. Among various proofs of partial elevation and depression in the interior, the Academicians mention, in their Survey, that the ground was sometimes

19. Deep fissure near Polistena in Calabria, caused by the earthquake of 1783

*20. Shift or 'fault' in the round tower of Terranuova in Calabria,
occasioned by the earthquake of 1783*

on the same level on both sides of new ravines and fissures, but sometimes
there had been a considerable shifting, either by the upheaving of one
side or the subsidence of the other. Thus, on the sides of long rents in the
territory of Soriano, the stratified masses had altered their relative position
to the extent of from eight to fourteen palms (six to ten and a half feet).
Similar shifts in the strata are alluded to in the territory of Polistena,
where there appeared innumerable fissures in the earth. One of these was
of great length and depth; and in parts, the level of the corresponding
sides was greatly changed. In the town of Terranuova, some houses were
seen uplifted above the common level, and others adjoining sunk down
into the earth. In several streets, the soil appeared thrust up, and abutted
against the walls of houses; a large circular tower of solid masonry, which
had withstood the general destruction, was divided by a vertical rent, and
one side was upraised, and the foundations heaved out of the ground. It
was compared by the Academicians to a great tooth half extracted from
the alveolus, with the upper part of the fangs exposed. (See cut No. 20.)

Along the line of this shift, or 'fault' as it would be termed technically
by miners, the walls were found to adhere firmly to each other, and to fit

so well, that the only signs of their having been disunited was the want of correspondence in the courses of stone on either side of the rent.

In some walls which had been thrown down, or violently shaken, in Monteleone, the separate stones were parted from the mortar so as to leave an exact mould where they had rested, whereas in other cases the mortar was ground to dust between the stones.

It appears that the wave-like motions, and those which are called vorticose or whirling in a vortex, often produced effects of the most capricious kind. Thus, in some streets of Monteleone, every house was thrown down but one; in others, all but two; and the buildings which were spared were often scarcely in the least degree injured.

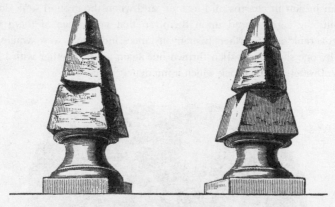

21. Shift in the stones of two obelisks in the Convent of S. Bruno

In many cities of Calabria, all the most solid buildings were thrown down, while those which were slightly built, escaped; but at Rosarno, as also at Messina, in Sicily, it was precisely the reverse, the massive edifices being the only ones that stood. Two obelisks (No. 21) placed at the extremities of a magnificent façade in the convent of S. Bruno, in a small town called Stefano del Bosco, were observed to have undergone a movement of a singular kind. The shock which agitated the building is described as having been horizontal and vorticose. The pedestal of each obelisk remained in its original place; but the separate stones above were turned partially round, and removed sometimes nine inches from their position, without falling.

It appears evident that a great part of the rending and fissuring of the ground was the effect of a violent motion from below upwards; and in a multitude of cases where the rents and chasms opened and closed alternately, we must suppose that the earth was by turns heaved up, and then let fall again. We may conceive the same effect to be produced on a small scale, if, by some mechanical force, a pavement composed of large flags of stone should be raised up and then allowed to fall suddenly, so as to resume its original position. If any small pebbles happened to be lying on the line of contact of two flags, they would fall into the opening when the pavement rose, and be swallowed up, so that no trace of them would appear after the subsidence of the stones. In the same manner, when the earth was upheaved, large houses, trees, cattle, and men were engulphed in an instant in chasms and fissures; and when the ground sank down again, the earth closed upon them, so that no vestige of them was discoverable on the surface. In many instances, individuals were swallowed up by one shock, and then thrown out again alive, together with large jets of water, by the shock which immediately succeeded.

22. *Fissures near Jerocarne in Calabria, caused by the earthquake of 1783*

At Jerocarne, a country which, according to the Academicians, was *lacerated* in a most extraordinary manner, the fissures ran in every direction like cracks on a broken pane of glass (see cut No. 22); and, as a great portion of them remained open after the shocks, it is very possible that this country was permanently upraised.

In the vicinity of Oppido, the central point from which the earthquake diffused its violent movements, many houses were swallowed up by the yawning earth, which closed immediately over them. In the adjacent district also of Cannamaria, four farm-houses, several oil-stores, and some spacious dwelling-houses were so completely engulphed in one chasm, that not a vestige of them was afterwards discernible. The same phenomenon occurred at Terranuova, S. Christina, and Sinopoli. The Academicians state particularly that when deep abysses had opened in the argillaceous strata of Terranuova, and houses had sunk into them, the sides of the chasms closed with such violence, that, on excavating afterwards to recover articles of value, the workmen found the contents and detached parts of the buildings jammed together so as to become one compact mass. It is unnecessary to accumulate examples of similar occurrences; but so many are well authenticated during this earthquake in Calabria, that we may, without hesitation, yield assent to the accounts of catastrophes of the same kind repeated again and again in history, where whole towns are declared to have been engulphed, and nothing but a pool of water or tract of sand left in their place.

* * *

Sir W. Hamilton was shown several deep fissures in the vicinity of Mileto, which, although not one of them was above a foot in breadth, had opened so wide during the earthquake as to swallow up an ox and near one hundred goats. The Academicians also found, on their return through districts which they had passed at the commencement of their tour, that many rents had in that short interval gradually closed in, so that their width had diminished several feet, and the opposite walls had sometimes nearly met. It is natural that this should happen in argillaceous strata, while in more solid rocks we may expect that fissures will remain open for ages. Should this be ascertained to be a general fact in countries convulsed by earthquakes, it would afford a satisfactory explanation of a common phenomenon in mineral veins. Such veins often retain their full size so long as the rocks consist of limestone, granite, or other indurated materials; but they contract their dimensions, become mere threads, or are even entirely cut off, where masses of an argillaceous nature are interposed. If we suppose the filling up of fissures with metallic and other ingredients to be a process requiring ages for its completion, it is obvious that the opposite walls of rents, where strata consist of yielding materials, must collapse or approach very near to each other before sufficient time is allowed for the accretion of a large quantity of veinstone.

* * *

The next class of effects to be considered, are those more immediately connected with the formation of valleys, in which the action of water was often combined with that of the earthquake. The country agitated was composed, as we before stated, chiefly of argillaceous strata, intersected by deep narrow valleys, sometimes from five to six hundred feet deep. As the boundary cliffs were in great part vertical, it will readily be conceived that, amidst the various movements of the earth, the precipices over-hanging the rivers, being without support on one side, were often thrown down. We find, indeed, that inundations produced by obstructions in river-courses are among the most disastrous consequences of great earthquakes in all parts of the world; for the alluvial plains in the bottoms of valleys are usually the most fertile and well peopled parts of the whole country, and whether the site of a town is above or below a temporary barrier in the channel of a river, it is exposed to injury by the waters either of a lake or a flood.

From each side of the deep valley or ravine of Terranuova, enormous masses of the adjoining flat country were detached and cast down into the course of the river, so as to give rise to great lakes. Oaks, olive-trees, vineyards, and corn, were often seen growing at the bottom of the ravine, as little injured as their companions from which they were separated in the plain above at least five hundred feet higher, and at the distance of about three-quarters of a mile. In one part of this ravine was an enormous mass, two hundred feet high, and about four hundred feet in diameter at its basis, which had been detached by some former earthquake. It is well attested that this mass travelled down the ravine near four miles, having been put in motion by the earthquake of the 5th of February. Hamilton, after examining the locality, declared that this phenomenon might be accounted for by the declivity of the valley, the great abundance of rain which fell, and the great weight of the alluvial matter which pressed behind it. The momentum of the 'terre movitine,' or lavas, as the flowing mud is called in the country, is no doubt very great; but the transportation of masses that might be compared to small hills, for a distance of several miles at a time, is an effect which could never have been anticipated: and the fact should serve as a hint to those geologists who are fond of appealing to alluvial phenomena as proofs of the superior violence of aqueous causes in former ages.

* * *

It would be tedious, and our space would not permit us, to follow the different authors through their local details of landslips produced in numerous minor valleys; but they are highly interesting, as showing to how great an extent the power of rivers to widen valleys, and to carry away large portions of soil towards the sea, is increased where earthquakes are of periodical occurrence. Among other territories, that of Cinquefrondi was greatly convulsed, various portions of soil being raised or sunk, and innumerable fissures traversing the country in all directions (see cut No. 27). Along the flanks of a small valley in this district there appears to have been an almost uninterrupted line of landslips.

27. Landslips near Cinquefrondi, caused by the earthquake of 1783

Vivenzio states, that near Sitizzano a valley was very nearly filled up to a level with the high grounds on each side, by the enormous masses detached from the boundary hills, and cast down into the course of two streams. By this barrier a lake was formed of great depth, about two miles long and a mile broad. The same author mentions that upon the whole, there were fifty lakes occasioned during the convulsions, and he assigns localities to all of these. The government surveyors enumerated two hundred and fifteen lakes, but they included in this number many small and insignificant ponds.

Near S. Lucido, among other places, the soil is described as having been 'dissolved,' so that large torrents of mud inundated all the low grounds, like lava. Just emerging from this mud, the tops only of trees and of the ruins of farm-houses, were seen. Two miles from Laureana the swampy soil in two ravines became filled with calcareous matter, which oozed out from the ground immediately before the first great shock. This mud, rapidly accumulating, began, ere long, to roll onward like a flood of lava into the valley, where the two streams uniting, moved forward with increased impetus from east to west. It now presented a breadth of three hundred palms by twenty in depth, and before it ceased to move, covered a surface equal in length to an Italian mile. In its progress it overwhelmed a flock of thirty goats, and tore up by the roots many olive and mulberry-trees, which floated like ships upon its surface. When this calcareous lava had ceased to move, it gradually became dry and hard, during which process the mass was lowered ten palms. It contained fragments of earth of a ferruginous colour, and emitting a sulphureous smell.

Many of the appearances exhibited in the alluvial plains indicate clearly the alternate rising and sinking of the ground. The first effect of the more violent shocks was usually to dry up the rivers, but they immediately afterwards overflowed their banks. Along the alluvial plains, and in marshy places, an immense number of cones of sand were thrown up. These appearances Hamilton explains, by supposing that the first movement raised the fissured plain from below upwards, so that the rivers and stagnant waters in bogs sank down, or at least were not upraised with the soil. But when the ground returned with violence to its former position, the water was thrown up in jets through fissures.

* * *

Along the sea coast of the straits of Messina, near the celebrated rock of Scilla, the fall of huge masses detached from the bold and lofty cliffs overwhelmed many villas and gardens. At Gian Greco a continuous line of cliff, for a mile in length, was thrown down. Great agitation was frequently observed in the bed of the sea during the shocks, and, on those parts of the coast where the movement was most violent, all kinds of fish were taken in greater abundance, and with much greater facility. Some rare species, as that called Cicirelli, which usually lie buried in the sand, were taken on the surface of the waters in great quantity. The sea is said to have boiled up near Messina, and to have been agitated as if by a copious discharge of vapours from its bottom. The Prince of Scilla had persuaded a great part of his vassals to betake themselves to their fishing-boats for safety, and he himself had gone on board. On the night of the 5th of February, when some of the people were sleeping in the boats, and others on a level plain slightly elevated above the sea, the earth rocked, and suddenly a great mass was torn from the contiguous Mount Jaci, and thrown down with a dreadful crash upon the plain. Immediately afterwards, the sea rising thirty palms above the level of this low tract, rolled foaming over it, and swept away the multitude. It then retreated, but soon rushed back again with greater violence, bringing with it some of the people and animals it had carried away. At the same time every boat was sunk or dashed against the beach, and some of them were swept far inland. The aged Prince, with one thousand four hundred and thirty of his people, was destroyed. The number of persons who perished during the earthquake in the two Calabrias and Sicily is estimated by Hamilton at about forty thousand, and about twenty thousand more died by epidemics which were caused by insufficient nourishment, exposure to the atmosphere, and malaria, arising from the new stagnant lakes and pools. By far the greater number were buried under the ruins of their houses; while some were burnt to death in the conflagrations which almost invariably followed the shocks, and consumed immense magazines of oil and other provisions. A small number were engulphed in chasms and fissures, and their skeletons are perhaps buried in the earth to this day, at the depth of several hundred feet, for such was the profundity of some of the openings which did not close in again.

The inhabitants of Pizzo remarked, that on the 5th of February, 1783, when the first great shock afflicted Calabria, the volcano of Stromboli, which is in full view of that town, and at the distance of about fifty miles, smoked less, and threw up a less quantity of inflamed matter, than it had done for some years previously. On the other hand, the great crater of Etna is said to have given out a considerable quantity of vapour towards the beginning, and Stromboli towards the close of the commotions. But as no eruption happened from either of these great vents during the whole earthquake, the sources of the Calabrian convulsions, and of the volcanic fires of Etna and Stromboli, appear to be very independent of each other;

unless, indeed, they have the same mutual relation as Vesuvius and the volcanos of the Phlegræan Fields and Ischia, a violent disturbance in one district serving as a safety-valve to the other, and both never being in full activity at once.

It is impossible for the geologist to consider attentively the effect of this single earthquake of 1783, and to look forward to the alterations in the physical condition of the country to which a continued series of such movements will hereafter give rise, without perceiving that the formation of valleys by running water can never be understood, if we consider the question independently of the agency of earthquakes. Rivers do not begin to act, as some seem to imagine, when a country is already elevated far above the level of the sea, but while it is *rising* or *sinking* by successive movements. Whether Calabria is now undergoing any considerable change of relative level, in regard to the sea, or is, upon the whole, nearly stationary, is a question which our observations, confined almost entirely to the last half century, cannot possibly enable us to determine. But we know that strata, containing species of shells identical with those now living in the contiguous parts of the Mediterranean, have been raised in this country, as they have in Sicily, to the height of several thousand feet. Now those geologists who merely grant that the present course of Nature, in the inanimate world, has been unchanged since the existing species of animals were in being, will not feel surprise that the Calabrian streams and rivers have cut out of such comparatively modern strata a great system of valleys varying in depth from fifty to six hundred feet, and often several miles wide, when they consider how numerous must have been the earthquakes which lifted those recent marine strata to so prodigious a height. Some speculators, indeed, who disregard the analogy of existing Nature, and who are as prodigal of violence as they are thrifty of time, may suppose that Calabria 'rose like an exhalation'[31] from the deep, after the manner of Milton's Pandemonium. But such an hypothesis will deprive them of that peculiar removing force required to form a regular system of deep and wide valleys, for *time* is essential to the operation. Landslips must be cleared away in the intervals between subterranean movements, otherwise fallen masses will serve as buttresses to the precipitous cliffs bordering a valley, so that the succeeding earthquake will be unable to exert its full power. Barriers must be worn through and swept away, and steep or overhanging cliffs again left without support, before another shock can take effect in the same manner.

If a single convulsion be too violent, and agitate at once an entire hydrographical basin, or if the shocks follow each other too rapidly, the previously-existing valleys will be annihilated, instead of being modified and enlarged. Every stream will be compelled to begin its operations anew, and to open for itself a passage through strata before undisturbed, instead of continuing to deepen and widen channels already in great part excavated. On the other hand, if, consistently with all that is known from observation of the laws which regulate subterranean movements, we consider their action to have been intermittent – if sufficient periods have always intervened between the severer shocks to allow the drainage of the country to be nearly restored to its original state, then are both the kind and degree of force supplied which may enable running water to hollow out a valley of any depth and size consistent with the degree of elevation above the sea which the district in question may happen at any time to have attained during a succession of physical revolutions.

Nothwithstanding the great derangement caused by violent earthquakes, there is an evident tendency in running water to remain constant to the same connected series of valleys. The softening of the soil is invariably greatest in the channels of rivers and in alluvial plains. The water is absorbed in an infinite number of rents, and when the ground is swelled with water it is reduced almost to a state of mud by the vehement agitation of the ground in every direction, and often for several years consecutively. The erosive and transporting action of running water is, therefore, facilitated in the tracts already excavated.

When we read of the drying up and desertion of the channels of rivers, the accounts most frequently refer to their deflection into some other part of the same alluvial plain, perhaps several miles distant. Under certain circumstances a change of level may undoubtedly force the water to flow over into some distinct hydrographical basin; but even then it will fall immediately into valleys already formed. Provided, therefore, we suppose the elevation and subsidence of mountain-chains to be a gradual process, there is no difficulty in explaining how the rivers draining our continents have converted ravines into valleys, and enlarged and deepened valleys to an enormous extent. On the contrary, the signs of slow and gradual action so manifest in the sinuosities and other characters of valleys are admirably reconcileable with the great width and depth of the excavations, if we are content not only to suppose a great succession of ordinary earthquakes, but also the usual intervals of time between the shocks.

We may observe that earthquakes alone could never give rise to a regular system of valleys ramifying from a main trunk like the veins from the great arteries of the human body. On the contrary, they would, in the course of time, destroy every system of valleys on the globe, were it not for the agency of aqueous causes. We learn from history that ever since the first Greek colonists, the Bruttii, settled in Calabria, that region has been subject to devastation by earthquakes, and, for the last century and a half, ten years have seldom elapsed without a shock; but the severer convulsions have not only been separated by intervals of twenty, fifty, or one hundred years, but have not affected precisely the same points when they recurred. Thus the earthquake of 1783, although confined within the same geographical limits as that of 1638, and not very inferior in violence, visited, according to Grimaldi, very different localities. The points where the local intensity of the force is developed, being thus perpetually varied, more time is allowed for the removal of separate mountain masses thrown into river channels by each shock.

When chasms and deep hollows open at the bottom of valleys, they must often be filled with those 'mud lavas' before described; and these must be extremely analogous to the enormous ancient deposits of mud which are seen in many countries, as in the basin of the Tay, Isla, and North Esk rivers, for example, in Scotland – alluvions hundreds of feet thick, which are neither stratified nor laminated like the sediment which subsides from water. Whenever a landslip blocks up a river, these currents of mud will be arrested, and accumulate to an enormous depth.

The transportation for several miles at a time, of masses as large as great edifices by the momentum of these floods of mud combined with the motion of the earthquake, and the enveloping of land animals, together with many other facts mentioned in the Calabrian account, cannot but excite in the mind of every geologist a strong desire to become more acquainted with the changes now in progress in those vast regions of the globe which are habitually devastated by earthquakes. To our extreme ignorance of this important class of phenomena we may probably refer the obscurity of many of the appearances of superficial alluvions throughout the greater part of Europe, as well as the diversity of opinion relating to them, and the extravagant theories which have passed current.

The portion of the Calabrian valleys formed within the last three thousand years, must, undoubtedly, be inconsiderable in amount, compared to that previously formed, just as the lavas which have flowed from

Etna since the historical era constitute but a small proportion of the whole cone. But as a continued series of such eruptions as man has witnessed would reproduce another cone like Etna, so a sufficient number of earthquakes like that of 1783 would enable torrents and rivers to re-excavate all the Calabrian valleys if they were now to be entirely obliterated. It must be evident that more change is effected in two centuries in the width and depth of the valleys of that region, than in many thousand years in a country as undisturbed by earthquakes as Great Britain. For the same reason, therefore, that he who desires to comprehend the volcanic phenomena of Central France will repair to Vesuvius, Etna, or Hecla, so they who aspire to explain the mode in which valleys are formed must visit countries where earthquakes are of frequent occurrence. For we may be assured, that the power which uplifted our more ancient tertiary strata of marine origin to more than a thousand feet above the level of the sea, co-operated at some former epoch with the force of rivers in the removal of large portions of rock and soil, just as the elevatory power which has upraised newer strata to the height of several thousand feet in the south of Italy has caused those formations to be already intersected by deep valleys and ravines.

He who studies the hydrographical basin of the Thames, and compares its present state with its condition when it was a Roman province, may have good reason to declare that if that river and its tributaries had since their origin been always as inactive, and as impotent as they are now, they could never, not even in millions of years, have excavated the valleys through which they flow: but, if he concludes from these premises, that the valleys in this basin were not formed by ordinary causes,[32] he reasons like one, who having found a solfatara which for many centuries has thrown out nothing more than vapour and a few handfuls of sand and scoriæ, infers that a lofty cone, composed of successive streams of lava and ejections, can no longer be produced by volcanic agency.

CHAPTER 25

Earthquakes, continued – *Temple of Serapis*

In the preceding chapters we have considered a small part of those earthquakes only which have occurred during the last fifty years, of which accurate and authentic descriptions happen to have been recorded. We shall next proceed to examine some of earlier date, respecting which information of geological interest has been obtained.

[*The chronological account of earthquakes continues with nineteen further examples, from Java in 1772, through great disasters at Lisbon in 1755 and Conception in 1750, to Jamaica in 1692.*]

* * *

We have now only enumerated the earthquakes of the last hundred and forty years, respecting which, facts illustrative of geological inquiries are on record. Even if our limits permitted, it would be a tedious and unprofitable task to examine all the obscure and ambiguous narratives of similar events of earlier epochs, although, if the localities were now examined by geologists well practised in the art of interpreting the monuments of physical changes, many events which have happened within the historical era might still be determined with precision. The reader must not imagine, that in our sketch of the occurrences in the short period above alluded to, we have given an account of all, or even the greater part of the mutations which the earth has undergone, by the agency of subterranean movements. Thus, for example, the earthquake of Aleppo, in the present century, and of Syria in the middle of the eighteenth, would doubtless have afforded numerous phenomena of great geological importance, had those catastrophes been described by scientific observers. The shocks in Syria in 1759, were protracted for three months, throughout a space of ten thousand square leagues, an area compared to which that of the Calabrian earthquake, of 1783, was insignificant. Accon, Saphat, Balbeck, Damascus, Sidon, Tripoli, and many other places, were almost entirely levelled to the ground. Many thousands of the inhabitants perished in each, and in the valley of Balbeck alone twenty thousand men are said to have been victims to the convulsion. It would be as irrelevant to our present purpose to enter into a detailed account of such calamities, as to follow the track of an invading army, to enumerate the cities burnt or rased to the ground, and reckon the number of individuals who perished by famine or the sword. If such then be the amount of ascertained changes in the last one hundred and forty years, notwithstanding the extreme deficiency of our records during that brief period, how important must we presume the physical revolutions to have been in the course of thirty or forty centuries, during which, some countries habitually convulsed by earthquakes have been peopled by civilized nations! Towns engulphed during one earthquake may, by repeated shocks, have sunk to enormous depths beneath the surface, while their ruins remain as imperishable as the hardest rocks in which they are inclosed. Buildings and cities submerged for a time beneath seas or lakes, and covered with sedimentary deposits, must, in some places, have been re-elevated to considerable heights above the level of the ocean. The signs of these events have probably been rendered visible by subsequent mutations, as by the encroachments of the sea upon the coast, by deep excavations made by torrents and rivers, by the opening of new ravines and chasms, and other effects of natural agents, so active in districts agitated by subterranean movements. If it be asked why if such wonderful monuments exist, so few have hitherto been brought to light – we reply – because they have not been searched for. In order to rescue from oblivion the memorials of former occurrences, we must know what we may reasonably expect to discover; and under what peculiar local circumstances. The inquirer, moreover, must be acquainted with the action and effects of physical causes, in order to recognise, explain, and describe, correctly, the phenomena when they present themselves.

The best known of the great volcanic regions of which we sketched the boundaries, in the eighteenth chapter, is that which includes Southern Europe, Northern Africa, and Central Asia, yet nearly the whole even of this region must be laid down in a geological map as 'Terra Incognita.' Even Calabria may be regarded as unexplored, as also Spain, Portugal, the Barbary states, the Ionian Isles, the Morea, Asia Minor, Cyprus, Syria, and the countries between the Caspian and Black Seas. We are, in truth, beginning to obtain some insight into one small spot of that great zone of volcanic disturbance, the district around Naples, a tract by no means remarkable for the violence of the earthquakes which have convulsed it.

If, in this part of Campania, we are enabled to establish, that considerable changes in the relative level of land and sea have taken place since the Christian era, it is all that we could have expected, and it is to recent antiquarian and geological research, not to history, that we are principally indebted for the information. We shall proceed to lay before the reader some of the results of modern investigations in the Bay of Baiæ and the adjoining coast.

Temple of Jupiter Serapis. – This celebrated monument of antiquity affords, in itself alone, unequivocal evidence, that the relative level of land and sea has changed twice at Puzzuoli, since the Christian era, and each movement both of elevation and subsidence has exceeded twenty feet. Before examining these proofs we may observe, that a geological examination of the coast of the Bay of Baiæ, both on the north and south of Puzzuoli, establishes in the most satisfactory manner an elevation at no remote period, of more than twenty feet, and the evidence of this change would have been complete even if the temple had to this day remained undiscovered. If we coast along the shore from Naples to Puzzuoli we find, on approaching the latter place, that the lofty and precipitous cliffs of indurated tuff, resembling that of which Naples is built, retire slightly from the sea, and that a low level tract of fertile land, of a very different aspect, intervenes between the present sea-beach, and what was evidently the ancient line of coast. The inland cliff is in many parts eighty feet high near Puzzuoli, and as perpendicular as if it was still undermined by the waves. At its base, the new deposit attains a height of about twenty feet above the sea, and as it consists of regular sedimentary deposits, containing marine shells, its position proves that since its formation there has been a change of more than twenty feet in the relative level of land and sea.

The sea encroaches on these new incoherent strata, and as the soil is valuable, a wall has been built for its protection; but when I visited the spot in 1828, the waves had swept away part of this rampart, and exposed to view a regular series of strata of tuff, more or less argillaceous, alternating with beds of pumice and lapilli, and containing great abundance of marine shells, of species now common on this coast, and amongst them

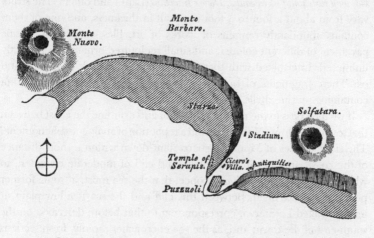

30. *Ground plan of the coast of the Bay of Baiæ in the environs of Puzzuoli.*

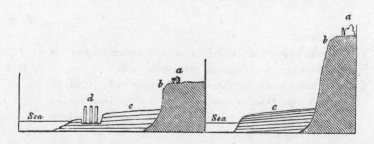

a. Remains of Cicero's villa, N. side of Puzzuoli.	*a.* Antiquities on hill S.E. of Puzzuoli.
b. Ancient cliff now inland.	*b.* Ancient cliff now inland.
c. Terrace composed of recent submarine deposits.	*c.* Terrace composed of recent submarine deposits.
d. Temple of Serapis.	

31. *Two sections, the one exhibiting the relation of the recent marine deposits to the more ancient in the Bay of Baiæ to the north of Puzzuoli, and the other exhibiting the same relation to the south-east*

Cardium rusticum, Ostrea edulis, Donax trunculus (Lam.) and others. The strata vary from about a foot to a foot and half in thickness, and one of them contains abundantly remains of works of art, tiles, squares of mosaic pavement of different colours, and small sculptured ornaments, perfectly uninjured. Intermixed with these I collected some teeth of the pig and ox. These fragments of building occur below as well as above strata containing marine shells.

If we then pass to the north of Puzzuoli and examine the coast between that town and Monte Nuovo, we find a repetition of analogous phenomena. The sloping sides of Monte Barbaro slant down within a short distance of the coast, and terminate in an inland cliff of moderate elevation, to which the geologist perceives at once, that the sea must, at some former period, have extended. Between this cliff and the sea is a low plain or terrace, called La Starza, corresponding to that before described on the south-east of the town; and, as the sea encroaches rapidly, fresh sections of the strata may readily be obtained, of which the annexed is an example.

Section on the shore north of the town of Puzzuoli.

	Ft.	In.
1. Vegetable soil	I	0
2. Horizontal beds of pumice and scoriæ, with broken fragments of unrolled bricks, bones of animals, and marine shells	I	6
3. Beds of lapilli, containing abundance of marine shells, principally *Cardium rusticum, Donax trunculus* Lam., *Ostrea edulis, Triton cutaceum,* Lam. and *Buccinum serratum,* Brocchi, the beds varying in thickness from one to eighteen inches	10	0
4. Argillaceous tuff containing bricks and fragments of buildings not rounded by attrition	I	6

The thickness of many of these beds varies greatly as we trace them along the shore, and sometimes the whole group rises to a greater height than at the point above described. The surface of the tract which they compose appears to slope gently upwards towards the base of the old cliffs. Puzzuoli itself stands chiefly on a promontory of the older tufaceous formation, which cuts off the new deposit, although I detected a small patch of the latter in a garden under the town.

Now if these appearances presented themselves on the eastern or southern coast of England, a geologist would naturally endeavour to seek an explanation in some local depression of high water-mark, in consequence of a change in the set of the tides and currents: for towns have been built, like ancient Brighton, on sandy tracts intervening between the old cliff and the sea, and in some cases they have been finally swept away by the return of the ocean. On the other hand, the inland cliff at Lowestoff, in Suffolk, remains, as we stated in the fifteenth chapter, at some distance from the shore, and the low green tract called the Ness may be compared to the low flat called La Starza, near Puzzuoli. But there are no tides in the Mediterranean; and to suppose that sea to have sunk generally from twenty to twenty-five feet since the shores of Campania were covered with sumptuous buildings, is an hypothesis obviously untenable. The observations, indeed, made during modern surveys on the moles and cothons (docks) constructed by the ancients in various ports of the Mediterranean, have proved that there has been no sensible variation of level in that sea during the last two thousand years. A very slight change would have been perceptible; and had any been ascertained to have taken place, and had it amounted only to a difference of a few feet, it would not have appeared very extraordinary, since the equilibrium of the Mediterranean is only restored by a powerful current from the Atlantic.

Thus we arrive, without the aid of the celebrated temple, at the conclusion that the recent marine deposit at Puzzuoli was upraised in modern times above the level of the sea, and that not only this change of position, but the accumulation of the modern strata, was posterior to the destruction of many edifices, of which they contain the imbedded remains. If we now examine the evidence afforded by the temple itself, it appears, from the most authentic accounts, that the three pillars now standing erect, continued, down to the middle of the last century, half buried in the new marine strata before described. The upper part of the columns, being concealed by bushes, had not attracted the notice of antiquaries; but, when the soil was removed in 1750, they were seen to form part of the remains of a splendid edifice, the pavement of which was still preserved, and upon it lay a number of columns of African breccia and of granite. The original plan of the building could be traced distinctly; it was of a quadrangular form, seventy feet in diameter, and the roof had been supported by forty-six noble columns, twenty-four of granite, and the rest of marble. The large court was surrounded by apartments, supposed to have been used as bathing-rooms; for a thermal spring, still used for medicinal purposes, issues now just behind the building, and the water, it is said, of this spring, was conveyed by marble ducts into the chambers.

Many antiquaries have entered into elaborate discussions as to the deity to which this edifice was consecrated; but Signor Carelli, who has written the last able treatise on the subject, endeavours to show that all the religious edifices of Greece were of a form essentially different – that the building, therefore, could never have been a temple – that it corresponded to the public bathing-rooms at many of our watering-places, and, lastly, that if it had been a temple, it could not have been dedicated to Serapis, – the worship of the Egyptian god being strictly prohibited at the time when this edifice was in use, by the senate of Rome.

It is not for the geologist to offer an opinion on these topics, and we shall, therefore, designate this valuable relic of antiquity by its generally received name, and proceed to consider the memorials of physical changes, inscribed on the three standing columns in most legible characters by the hand of nature. (See Frontispiece.) The pillars are forty-two feet in height; their surface is smooth and uninjured to the height of about twelve feet above their pedestals. Above this, is a zone, twelve feet in height, where the marble has been pierced by a species of marine perforating bivalve – *Lithodomus,* Cuv. The holes of these animals are pear-shaped, the external opening being minute, and gradually increasing downwards. At the bottom of the cavities, many shells are still found, notwithstanding the great numbers that have been taken out by visitors. The perforations are so considerable in depth and size, that they manifest a long continued abode of the Lithodomi in the columns; for, as the inhabitant grows older and increases in size, it bores a larger cavity, to correspond with the increasing magnitude of its shell. We must, consequently, infer a long continued immersion of the pillars in sea-water, at a time when the lower part was covered up and protected by strata of tuff and the rubbish of buildings, the highest part at the same time projecting above the waters, and being consequently weathered, but not materially injured. On the pavement of the temple, lie some columns of marble, which are perforated in the same manner in certain parts, one, for example, to the length of eight feet, while, for the length of four feet, it is uninjured. Several of these broken columns are eaten into, not only on the exterior, but on the cross fracture, and, on some of them, other marine animals have fixed themselves. All the granite pillars are untouched by Lithodomi. The platform of the Temple is at present about one foot below high-water mark, (for there are small tides in the Bay of Naples,) and the sea, which is only one hundred feet distant, soaks through the intervening soil. The upper part of the perforations then are at least twenty-three feet above high-water mark, and it is clear, that the columns must have continued for a long time in an erect position, immersed in salt-water. After remaining for many years submerged, they must have been upraised to the height of about twenty-three feet above the level of the sea.

So far the information derived from the Temple corroborates that before obtained from the new strata in the plain of La Starza, and proves nothing more. But as the temple could not have been built originally at the bottom of the sea, it must have first sunk down below the waves, and afterwards have been elevated. Of such subsidences there are numerous independent proofs in the Bay of Baiæ. Not far from the shore, to the north-west of the Temple of Serapis, are the ruins of a Temple of Neptune, and a Temple of the Nymphs, now under water. These buildings probably participated in the movement which raised the Starza, but, either they were deeper under water than the Temple of Serapis, or they were not raised up again to so great a height. There are also two Roman roads under water in the Bay, one reaching from Puzzuoli towards the Lucrine Lake, which may still be seen, and the other near the Castle of Baiæ. The ancient mole too, which exists at the Port of Puzzuoli, and which is commonly called that of Caligula, has the water up to a considerable height of the arches; whereas Brieslak justly observes, it is next to certain, that the piers must formerly have reached the surface before the springing of the arches. A modern writer also reminds us, that these effects are not so local as some would have us believe; for on the opposite side of the Bay of Naples, on the Sorrentine coast, which, as well as Puzzuoli, is subject to earthquakes, a road, with some fragments of Roman buildings, is covered to some depth by the sea. In the island of Capri, also, which is situated some way at sea, in the opening of the Bay of Naples, one of the palaces of Tiberius is now covered with water. They who have attentively considered the effects of earthquakes before enumerated by us during the last one hundred and forty`years, will not feel astonished at these signs of alternate elevation and depression of the bed of the sea and the adjoining coast during the course of eighteen centuries, but, on the contrary, they will be very much astonished if future researches fail to bring to light similar indications of change in all regions of volcanic disturbances. That buildings should have been submerged, and afterwards upheaved, without being entirely reduced to a heap of ruins, will appear no anomaly, when we recollect that in the year 1819, when the delta of the Indus sank down, the houses within the fort of Sindree subsided beneath the waves without being overthrown. In like manner, in the year 1692, the buildings around the harbour of Port Royal, in Jamaica, descended suddenly to the depth of between thirty and fifty feet under the sea without falling. Even on small portions of land, transported to a distance of a mile, down a declivity, tenements like those near Mileto, in Calabria, were carried entire. At Valparaiso, buildings were left standing when their foundations, together with a long tract of the Chilian coast, were permanently upraised to the height of several feet in 1822. It is true that, in the year 1750, when the bottom of the sea in the harbour of Penco was suddenly uplifted to the extraordinary elevation of twenty-four feet above its former level, the buildings of that town were thrown down; but we might still suppose that a great portion of them would have escaped, had the walls been supported on the exterior and interior with a deposit, like that which surrounded and filled to the height of ten or twelve feet the Temple of Serapis at Puzzuoli.

The next subject of inquiry, is the era when these remarkable changes took place in the Bay of Baiæ. It appears, that in the Atrium of the Temple of Serapis, inscriptions were found in which Septimus Severus and Marcus Aurelius record their labours in adorning it with precious marbles. We may, therefore, conclude, that it existed at least down to the third century of our era in its original position. On the other hand, we have evidence that the marine deposit forming the flat land called La Starza was still covered by the sea in the year 1530, or just eight years anterior to the tremendous explosion of Monte Nuovo. Mr. Forbes has lately pointed out the distinct testimony of an old Italian writer Loffredo, in confirmation of this important point. Writing in 1580, Loffredo declares that fifty years previously, the sea washed the base of the hills which rise from the flat land before alluded to, and at that time he expressly tells us that a person *might have fished* from the site of those ruins which are now called the Stadium. (See wood cut, No. 30.) Hence it follows, that the subsidence of the ground on which the Temple stood, happened at some period between the third century and the beginning of the sixteenth century. Now in this interval the only two events which are recorded in the imperfect annals of the dark ages, are the eruption of the Solfatara in 1198, and an earthquake in 1488 by which Puzzuoli was ruined. It is at least highly probable, that earthquakes, which preceded the eruption of the Solfatara, which is very near the Temple, (see wood cut, No. 30) caused a subsidence, and the pumice and other matters ejected from that volcano might have fallen in heavy showers into the sea, and would thus immediately have covered up the lower part of the columns. The action of the waves might afterwards have thrown down many pillars, and formed strata of broken fragments of the building intermixed with volcanic ejections, before the Lithodomi had time to perforate the lower part of the columns. In like manner, the sea acting on other submerged buildings, would naturally have caused a similar stratum, containing works of art and shells for several miles along the coast.

Now it is perfectly evident from Loffredo's statement, that the re-elevation of the low tract called La Starza took place after the year 1530, and long before the year 1580; and from this alone we might confidently conclude that the change happened in the year 1538 when Monte Nuovo was formed. But fortunately we are not left in the slightest doubt that such was the date of this remarkable event. Sir William Hamilton has given us two original letters describing the eruption of 1538, the first of which by Falconi, dated 1538, contains the following passages. 'It is now two years since there have been frequent earthquakes at Puzzuoli, Naples, and the neighbouring parts. On the day and in the night before the eruption (of Monte Nuovo), above twenty shocks great and small were felt. – The next morning (after the formation of Monte Nuovo) the poor inhabitants of Puzzuoli quitted their habitations, &c., some with their children in their arms, some with sacks full of their goods, others carrying quantities of birds of various sorts that had fallen dead at the beginning of the eruption, others again with fish which they had found, and which were to be met with in plenty on the shore, the sea having *left them dry for a considerable time.* – I accompanied Signor Moramaldo to behold the wonderful effects of the eruption. The sea had retired on the side of Baiæ, *abandoning a considerable tract,* and the shore appeared almost entirely dry from the quantity of ashes and broken pumice-stones thrown up by the eruption. I saw two springs *in the newly discovered ruins,* one before the house that was the Queen's, of hot and salt-water, &c.'[33] So far Falconi – the other account is by Pietro Giacomo di Toledo, which begins thus: 'It is now two years since this province of Campagna has been afflicted with earthquakes, the country about Puzzuoli much more so than any other parts: but the 27th and the 28th of the month of September last, the earthquakes did not cease day or night in the town of Puzzuoli; that plain which lies between lake Avernus, the Monte Barbaro and the sea was *raised a little*, and many cracks were made in it, from some of which issued water; at the same time the sea immediately adjoining the plain *dried up about two hundred paces*, so that the fish were left on the sand a prey to the inhabitants of Puzzuoli. At last, on the 29th of the same month, about two o'clock in the night, the earth opened, &c.'[34] Now both these accounts, written immediately after the birth of Monte Nuovo, agree in expressly stating, that the sea retired, and one mentions that its bottom was upraised. To this elevation we have already seen that Hooke, writing at the close of the seventeenth century, alludes as to a well known fact. The preposterous theories, therefore, that have been advanced in order to dispense with the elevation of the land, in the face of all this historical and physical evidence, are not entitled to a serious refutation. The flat land, when first upraised, must have been more extensive than now, for the sea encroaches somewhat rapidly, both to the north and south-east of Puzzuoli. The coast has of late years given way more than a foot in a twelve-month, and I was assured by fishermen in the bay, that it has lost ground near Puzzuoli, to the extent of thirty feet, within their memory. It is, probably, this gradual encroachment which has led many authors to imagine that the level of the sea is slowly rising in the Bay of Baiæ, an opinion by no means warranted by such circumstances. In the course of time the whole of the low land will, perhaps, be carried away, unless some earthquake shall remodify the surface of the country, before the waves reach the ancient coast-line; but the removal of this narrow tract will by no means restore the country to its former state, for the old tufaceous hills and the interstratified current of trachytic lava which has flowed from the Solfatara, must have participated in the movement of 1538; and these will remain upraised even though the sea may regain its ancient limits.

In 1828 excavations were made below the marble pavement of the Temple of Serapis, and another costly pavement of mosaic was found, at the depth of five feet or more below the other. The existence of these two pavements at different levels seems clearly to imply some subsidence previously to all the changes already alluded to, which had rendered it necessary to construct a new floor at a higher level. But to these and other circumstances bearing on the history of the Temple antecedently to the revolutions already explained, we shall not refer at present, trusting that future investigations will set them in a clearer light.

In concluding this subject, we may observe, that the interminable controversies to which the phenomena of the Bay of Baiæ gave rise, have sprung from an extreme reluctance to admit that the land rather than the sea is subject alternately to rise and fall. Had it been assumed that the level of the ocean was invariable, on the ground that no fluctuations have as yet been clearly established, and that, on the other hand, the continents are inconstant in their level, as has been demonstrated by the most unequivocal proofs again and again, from the time of Strabo to our own times, the appearances of the temple at Puzzuoli could never have been regarded as enigmatical. Even if contemporary accounts had not distinctly attested the upraising of the coast, this explanation should have been proposed in the first instance as the most natural, instead of being now adopted unwillingly when all others have failed. To the strong prejudices still existing in regard to the mobility of the land, we may attribute the rarity of such discoveries as have been recently brought to light in the Bay of Baiæ and the Bay of Conception. A false theory it is well known may render us blind to facts, which are opposed to our prepossessions, or may conceal from us their true import when we behold them. But it is time that the geologist should in some degree overcome those first and natural impressions which induced the poets of old to select the rock as the emblem of firmness – the sea as the image of inconstancy. Our modern poet, in a more philosophical spirit, saw in the latter 'The image of Eternity,' and has finely contrasted the fleeting existence of the successive empires which have flourished and fallen, on the borders of the ocean, with its own unchanged stability.

> ——Their decay
> Has dried up realms to deserts: – not so thou,
> Unchangeable, save to thy wild waves' play:
> Time writes no wrinkle on thine azure brow;
> Such as creation's dawn beheld, thou rollest now.
>
> CHILDE HAROLD, Canto iv.[35]

CHAPTER 26

Causes of Earthquakes and Volcanos

When we consider attentively the changes brought about by earthquakes during the last century, and reflect on the light which they already throw on the ancient history of the globe, we cannot but regret that investigations into the effects of this powerful cause have hitherto been prosecuted with so little zeal. The disregard of this important subject may be attributed to the general persuasion, that former revolutions of the earth were not brought about by causes now in operation, – a theory which, if true, would fully justify a geologist in neglecting the study of such phenomena. We may say of the superficial alterations arising from subterranean movements, as we have already declared of the visible effects of active volcanos, that, important as they are in themselves, they are still more so as indicative of far greater changes in the interior of the earth's crust. That both the chemical and mechanical changes in the subterranean regions must often be of a kind to which no counterpart can possibly be found in progress within the reach of our observation, may be confidently inferred; and speculations on these subjects ought not to be discouraged, since a great step is gained if they render us more conscious of the extent of our inability to define the amount and kind of results to which ordinary subterranean operations are now giving rise. It is no longer disputed that a great series of convulsions have carried up deposits once formed on the bottom of the ocean to the height of several miles above its level; and it is not difficult to perceive that the same movements must in numerous places have raised rocks to elevations above the level of the sea, which were once formed at the depth of several miles in the bowels of the earth. If, then, there were no spots discoverable which exhibited signs of extraordinary mechanical and chemical changes, the effects at some former period of immense pressure, intense heat, and other conditions far different from those developed on the surface, it might be urged as a triumphant argument against those who are dissatisfied with the proofs hitherto adduced in favour of the mutability of the course of Nature.

In order to set this in a clear light, let the reader suppose himself acquainted with just one-tenth part of the words of some living language, and that he is presented with several books purporting to be written in the same tongue ten centuries ago. If he should find that he comprehends a tenth part of the terms in the ancient volumes, and that he cannot divine the meaning of the other nine-tenths, would he not be strongly disposed to believe that, for a thousand years, the language has remained *unaltered*? Could he, without great labour and study, interpret the greater part of what is written in the antique documents, he must feel at once convinced that, in the interval of ten centuries, a great revolution in the language had taken place. He might, undoubtedly, by comparing the conventional signs already known to him, with those not previously acquired, and by observing the analogies and associations of terms in many of the old books, come at length to discover the true import of much of the ancient writings, and guess at the meaning of nearly all the rest; but if he is entirely shut out from all communication with those who now use the same language, he will never fully understand the value of some terms. So if a student of Nature, who, when he first examines the monuments of former changes upon our globe, is acquainted only with one-tenth part of the processes now going on upon or far below the surface, or in the depths of the sea, should still find that he comprehends at once the import of the signs of all, or even half the changes that went on in the same regions some hundred or thousand centuries ago, he might declare without hesitation that the ancient laws of nature have been subverted. Even after toiling for centuries, and learning more both of the present and former state of things, he must never expect to gain a perfect insight into all that formerly happened, so long as his acquaintance is very limited in regard to much that is now going on. So completely has the force of this line of argument been overlooked, that when any one has ventured to presume that all former changes were simply the result of causes now in operation, they have invariably been called upon to explain every obscure phenomenon in geology, and if they failed, it was considered as conclusive against their assumption. Whereas, in truth, there is no part of the evidence in favour of the uniformity of the system, more cogent than the fact, that with much that is intelligible, there is still more which is yet novel, mysterious, and inexplicable in the monuments of ancient mutations in the earth's crust.

Before the immense depth of the sources of volcanic fire was generally admitted, the causes of subterranean movements were sought in peculiar states of the atmosphere. These were imagined to afford not only prognostics of the convulsions, but to have considerable influence in their production. But the supposed signs of approaching earthquakes were of a most uncertain and contradictory character. Aristotle, Pliny, and Seneca, taught that earthquakes were preceded by a serene state of the air; whereas several modern writers have been of opinion that a cloudy sky and sudden storms are the forerunners of these commotions. That there is an intimate connexion between subterranean convulsions and particular states of the weather is unquestionable; but as Michell truly remarked, 'it is more probable that the air should be affected by the causes of earthquakes, than that the earth should be affected in so extraordinary a manner, and to so great a depth, by a cause residing in the air.'[36]

After violent earthquakes the regular drainage of a country is obstructed; lakes and pools are caused by local subsidences or landslips, and the evaporation of an extensive surface of shallow water produces unseasonable rains. Fogs proceed from the damp soil which is traversed by numerous rents and crevices filled with water. In addition to these circumstances, the electrical effect produced by the movement and friction of great masses of rock against each other may cause lightning, gusts of wind, luminous exhalations, and other atmospheric phenomena. Rains, moreover, are sometimes derived from volcanic eruptions accompanying earthquakes; for eruptions, as we before stated, are attended with a copious discharge of aqueous vapour.

Before we attempt to enquire farther into the true causes of earthquakes, we shall briefly recapitulate our reasons for considering them as originating from the same sources as volcanic phenomena. In the first place, the regions convulsed by violent earthquakes include within them the site of all the active volcanos. Earthquakes, sometimes local, sometimes extending over vast areas, precede volcanic eruptions. Both the subterranean movement and the eruption return again and again, at unequal intervals of time, and with unequal degrees of force, to the same places. The duration of both may continue for a few hours, or for several consecutive years. Paroxysmal convulsions of both kinds are usually followed by long periods of tranquillity. Thermal springs, and those containing abundance of mineral matter in solution, are characteristic of countries where active volcanos or earthquakes are frequent. In districts considerably distant from volcanic vents, the temperature of hot springs has been sometimes raised by subterranean movements. In addition to these signs of relation and analogy, we may observe, that it is not very easy to conceive how columns of melted matter can be raised to such great heights, as we know them to attain in volcanos, without exerting an hydrostatic pressure capable of moving enormous masses of land; nor can we be surprised that elastic fluids capable of forcing up so great a weight of rock in fusion, and of projecting large stones to immense heights in the air, should also cause tremors, vibrations, and violent movements in the solid crust of the earth. The volcano of Cotopaxi has thrown a mass of rock, about one hundred cubic yards in volume, to the distance of eight or nine miles, and we may well conceive that the slightest obstruction to the escape of such an expansive force may convulse a considerable tract in South America. 'If these vapours,' says Michell, 'when they find a vent are capable of shaking a country to the distance of ten or twenty miles round the volcano, what may we not expect from them when they are confined?'[37] As there is no doubt that aqueous vapour constitutes the most abundant of the aëriform products of volcanic eruptions, it may be well to consider attentively a case in which steam is exclusively the moving power – the Geysers of Iceland. These intermittent hot springs rise from a large tract, covered to a considerable depth by a stream of lava; and where thermal waters, and apertures evolving steam, are very common. The great Geyser rises out of a spacious basin at the summit of a circular mound, composed of siliceous incrustations deposited from the spray of its waters. The diameter of the basin or crater, in one direction, is fifty-six feet, and forty-six in another.

In the centre is a pipe seventy-eight feet in perpendicular depth, and from eight to ten feet in diameter, but gradually widening as it opens into the basin. The inside of the basin is whitish, consisting of a siliceous incrustation, and perfectly smooth, as are two small channels on the sides of the mound, down which the water makes its escape when filled to the margin. The circular basin is sometimes empty, as represented in the above sketch [p. 166–Ed.], but is usually filled with beautifully transparent water in a state of ebullition. During the rise of the boiling water up the pipe, especially when the ebullition is most violent, and when the water flows over or is thrown up in jets, subterranean noises are heard, like the

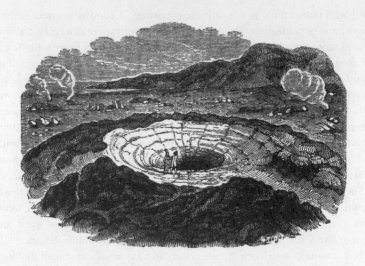

32. View of the crater of the great Geyser in Iceland

distant firing of cannon, and the earth is slightly shaken. The sound then increases and the motion becomes more violent, until at length a column of water is thrown up perpendicularly with loud explosions, to the height of one or two hundred feet. After playing for a time like an artificial fountain, and giving off great clouds of vapour, the pipe is evacuated, and a column of steam then rushes up with amazing force and a thundering noise, after which the eruption terminates. If stones are thrown into the crater they are instantly ejected, and such is the explosive force, that very hard rocks are sometimes shivered into small pieces. Henderson found that by throwing a great quantity of large stones into the pipe of Strockr, one of the Geysers, he could bring on an eruption in a few minutes. The fragments of stone as well as the boiling water were thrown in that case to a much greater height than usual. After the water had been ejected, a column of steam continued to rush up with a deafening roar for nearly an hour; but the Geyser, as if exhausted by this effort, did not give symptoms of a fresh eruption when its usual interval of rest had elapsed.

In the different explanations offered of this singular phenomenon, all writers agree in supposing a subterranean cavity where water and steam collect, and where the free escape of the steam is intercepted at intervals, until it acquires sufficient force to discharge the water. Suppose water

33. Supposed section of the subterranean reservoir and pipe of a Geyser in Iceland

percolating from the surface of the earth to penetrate into the subterranean cavity A D by the fissures F F, while at the same time, steam, at an extremely high temperature, such as is commonly given out from the rents of lava-currents during congelation, emanates from the fissures C C. A portion of the steam is at first condensed into water, and the temperature of the water is raised by the latent heat evolved, until, at last, the lower part of the cavity is filled with boiling water and the upper with steam under high pressure. The expansive force of the steam becomes, at length, so great, that the boiling water is forced up the fissure or pipe E B, and a considerable quantity runs over the rim of the basin. When the pressure is thus diminished, the steam in the upper part of the cavity A expands until all the water D is driven to E; when this happens, the steam, being the lighter of the two fluids, rushes up with great velocity, as on the opening of the valve of a steam-boiler. If the pipe be choked up artificially with stones, even for a few minutes, a great increase of heat must take place, for it is prevented from escaping in a latent form in steam, so that the water is made to boil up in a few minutes, and this brings on an eruption.

Now if we suppose a great number of large subterranean cavities at the depth of several miles below the surface of the earth, wherein melted lava accumulates, and that water penetrating to these is converted into steam, this steam, together with other gases generated by the decomposition of melted rocks, may press upon the lava and force it up the duct of a volcano, in the same manner as it drives a column of water up the pipe of a Geyser. But the weight of the lava being immense, the hydrostatic pressure exerted on the sides and roofs of such large cavities and fissures may well be supposed to occasion not merely slight tremors, such as agitate the ground before an eruption of the Geyser, but violent earthquakes. Sometimes the lateral pressure of the lower extremity of the high column of lava may cause the more yielding strata to give way, and to fold themselves in numerous convolutions, so as to occupy less space, and thereby give relief, for a time, to the fused and dilated matter. Sometimes, on the contrary, a weight equal to that of the vertical column of lava, pressing on every part of the roof, may heave up the superincumbent mass, and force lava into every fissure which, on consolidation, may support the arch, and cause the land above to be permanently elevated. On the other hand, subsidences may follow the condensation of vapour when cold water descends through fissures, or when heat is lost by the cooling down of lava.

That lava should often break out from the side or base, rather than from the summit of a lofty cone like Etna, has always been attributed to the immense hydrostatic pressure which the sides of the mountain undergo, before the lava can rise to the crater. This conclusion is too obvious not to have met with a general reception; yet how trifling must this pressure be when compared to that which the same column imparts to the reservoirs of aëriform fluids and melted rock, at the depth of many miles or leagues below the surface!

If earthquakes be derived from the expansion by heat of elastic fluids and melted rock, it is perfectly natural that they should terminate, either when a volcanic vent permits a portion of the pent up vapours or lava to escape, or when the earth has been so fissured that the vapour is condensed by its admission into cooler regions, or by its coming in contact with water. Or relief may be obtained when lava and gaseous fluids have, by distending the strata, made more room for themselves, so that the weight of the superincumbent mass is sufficient to repress them. If we regard earthquakes as abortive volcanic eruptions at a great depth, we must expect them to succeed each other for an indefinite number of times in the same place, for the same reason that eruptions do; and it is easy to conceive that, if the matter has failed several times to reach the surface, the consolidation of the lava first raised and congealed will strengthen the earth's crust, and become an additional obstacle to the protrusion of other fused matter during subsequent convulsions.

As most volcanos are in islands or maritime tracts, the neighbourhood of the sea seems one of the conditions necessary for the ascent of lava to great heights. Even those volcanos which lie inland form part of a chain of volcanic hills, and may be supposed to have a subterranean communication with the extremities of the chain which are in the neighbourhood of large masses of salt-water. Thus Jorullo, in Mexico, though itself no less than forty leagues from the nearest ocean, seems, nevertheless, connected with the volcano of Tuxtla on the one hand, and that of Colima on the other, the one bordering on the Atlantic, the other on the Pacific ocean. This communication is rendered the more probable by the parallelism that exists between these and several volcanic hills intermediate. Perhaps the quantity of water which percolates from the surface of the land is sufficient to contribute to the violence of earthquakes, without producing so much steam as is required to bring on a volcanic eruption. But when the sea overlies a mass of incandescent lava, and the intermediate crust of the globe is shaken and fissured by earthquakes, it may well be supposed that a convulsion of a different kind will ensue. If an open fissure be caused like that which traversed the plain of S. Lio, on Etna, in 1669, so that the water descends at once upon a mass of melted lava, eruptions will probably burst forth along the line of this aperture, the steam rushing up, together with gaseous emanations from the lava, and carrying up scoriæ with it. But from what we know of the wave-like motion of the ground during earthquakes, there is good reason to conclude that a continuous communication will rarely be formed between the sea and a bed of lava at great depth below, because the alternate rising and falling of the earth causes chasms to open and again to close in violently. In the same manner, therefore, as yawning fissures shut again after engulfing trees and houses, so great masses of water may be swallowed up, and the sea may immediately afterwards be excluded. Suppose then a volcanic vent to be once formed by a submarine eruption, all the water engulphed will, on penetrating to subterranean reservoirs of heated lava, be converted into steam, and this steam making its way through the same channels by which elastic fluids escape in the intervals between eruptions, will drive melted lava before it. Successive eruptions will have a tendency to seek the same vent, especially if the peak of a cone is raised above the water; for then there will probably be no more than the pressure of the atmosphere in a great part of the duct leading to the crater.

Volcanos exhale, during eruptions, besides aqueous vapour, the following gases: muriatic acid, sulphur combined with hydrogen or oxygen, carbonic acid and nitrogen, the greater part of which would result from the decomposition of salt-water, a fact which, when taken in conjunction with the proximity of nearly two hundred active vents to the sea, and their absence in the interior of large continents, is almost conclusive as to the co-operation of water and fire in the raising of lava to the surface.

We have before suggested the great probability that, in existing volcanic regions, there are enormous masses of matter in a constant state of fusion far below the surface: this opinion is confirmed by numerous phenomena. Perennial supplies of hot vapour and aëriform fluids rise to certain craters, as in Stromboli for example, and Nicaragua, which are in a state of ceaseless eruption. Sangay in Quito, Popocatepetl in Mexico, and the volcano of the isle of Bourbon, have continued in incessant activity for periods of sixty or one hundred and fifty years. Numerous solfataras, evolving the same gases as volcanos, serve as permanent vents of heat generated in the subterranean regions. The plentiful evolution, also, of carbonic acid, from springs and fissures throughout hundreds of square leagues, is another regular source of communication between the interior and the surface. Steam, often above the boiling temperature, is emitted for ages without intermission from 'stufas,' as the Italians term them. Hot springs in great numbers, especially in tracts where earthquakes are frequent, serve also as regular conductors of heat from the interior upwards. Silex, carbonate of lime, muriate of soda, and many earths, alkalies and metals are poured out in a state of solution by springs, and the solid matter which is tranquilly removed in this manner may, perhaps, exceed that which issues in the shape of lava.

It is to the efficacy of this ceaseless discharge of heat, and of solid as well as gaseous matter, that we probably owe the general tranquillity of our globe; for were it not that some kind of equilibrium is established between fresh accessions of heat and its discharge, we might expect perpetual convulsions, if we conceive the land and the ocean itself to be incumbent in many extensive districts on subterranean reservoirs of lava. If there be reason for wonder, it is, as Pliny observed, that a single day should pass without some dreadful explosion. 'It really exceeds all other wonders, that one single day should pass in which everything is not consumed.'[38] But the circulation of heat from the interior to the surface, is probably regulated like that of water from the continents to the sea, in such a manner that it is only when some obstruction occurs to the regular discharge, that the usual repose of Nature is broken. Any interruption to the regular drainage of a country causes a flood, and, if there be any obstruction in the passages by which volcanic matter continually rises, an earthquake or a paroxysmal eruption is the consequence.

Michell has observed, that the wave-like motion of the ground during earthquakes, appears less extraordinary if we call to mind the extreme elasticity of the earth, and that even the most solid materials are easily compressible. If we suppose large districts to rest upon the surface of subterranean lakes of melted matter, through which violent motions are propagated, it is easy to conceive that superincumbent solid masses may be made to vibrate or undulate. The following ingenious speculations are suggested by the above mentioned writer. 'As a small quantity of vapour almost instantly generated at some considerable depth below the surface of the earth will produce a vibratory motion, so a very large quantity (whether it be generated almost instantly, or in any small portion of time) will produce a wave-like motion. The manner in which this wave-like motion will be propagated may in some measure be represented by the following experiment. Suppose a large cloth, or carpet (spread upon a floor) to be raised at one edge, and then suddenly brought down again to the floor, the air under it being by this means propelled, will pass along, till it escapes at the opposite side, raising the cloth in a wave all the way as it goes. In like manner, a large quantity of vapour may be conceived to raise the earth in a wave, as it passes along between the strata which it may easily separate in an horizontal direction, there being little or no cohesion between one stratum and another. The part of the earth that is first raised, being bent from its natural form, will endeavour to restore itself by its elasticity, and the parts next to it being to have their weight supported by the vapour, which will insinuate itself under them, will be raised in their turn, till it either finds some vent, or is again condensed by the cold into water, and by that means prevented from proceeding any farther.'[39]

In order to account for the retreat of the ocean from the shores before or during an earthquake, the same author imagines a subsidence at the bottom of the sea, from the giving way of the roof of some cavity in consequence of a vacuum produced by the condensation of steam. For such condensation, he observes, might be the first effect of the introduction of a large body of water into fissures and cavities already filled with steam, before there has been sufficient time for the heat of the incandescent lava to turn so large a supply of water into steam, which being soon accomplished causes a greater explosion. Sometimes the rising of the coast must give rise to the retreat of the sea, and the subsequent wave may be occasioned by the subsiding of the shore to its former level; but this will not always account for the phenomena. During the Lisbon earthquake, for example, the retreat preceded the wave not only on the coast of Portugal, but also at the island of Madeira and several other places. If the upheaving of the coast of Portugal had caused the retreat, the motion of the waters when propagated to Madeira would have produced a wave previous to the retreat. Nor could the motion of the waters at Madeira have been caused by a different local earthquake, for the shock travelled from Lisbon to Madeira in two hours, which agrees with the time which it required to reach other places equally distant.

We shall not indulge at present in further speculations on the mode whereby subterranean heat may give rise to the phenomena of earthquakes and volcanos. No one, however, can fail to be convinced, if he turns his thoughts to the subject, that a great part of the reasoning of the most profound natural philosophers and chemists can be regarded as little more than mere conjecture on matters where the circumstances are so far removed from those which fall under actual observation. Many processes must be carried on in situations where the pressure exceeds as much that produced by the weight of the loftiest mountains, as the weight of the unfathomed ocean surpasses that of the atmosphere. The mechanical effects, therefore, of earthquakes at vast depths, may be such as can never be paralleled on the surface. The intensity of heat must often be so far removed from that which we can imitate by experiments, that the elements of solid rocks or fluids may enter into combinations such as can never take place within the limited range of our observations. Water at a certain depth may, as Michell boldly suggested, become incandescent without expanding, and remain at rest without any tendency to produce an earthquake. Air, if it ever penetrate to such depths, may become a fluid. Sir James Hall's experiments prove, that, under a pressure of about one thousand seven hundred feet of sea, corresponding to that of only six hundred feet of liquid lava, limestone melts without giving off its carbonic acid, so that it is only when calcareous lavas are forced up to within a slight distance of the surface, or into a sea of moderate depth, that the carbonic acid begins to assume a gaseous form, and to assist in bringing on a volcanic eruption.

But let us now turn our attention to those superficial changes brought about by so many of the earthquakes within the last century and a half, before described. Besides the undulatory movements, and the opening of fissures, it was shewn that certain parts of the earth's crust often of considerable area, both above and below the level of the sea, have been permanently elevated or depressed; examples of elevation by single earthquakes having occurred, to the amount of from one to about twenty-five feet, and of subsidence from a few inches to about fifty feet, exclusively of those limited tracts, as the forest of Aripao, where a sinking down to the amount of three hundred feet took place. It is evident, that the force of subterranean movement does not operate at random, but the same continuous tracts are agitated again and again; and however inconsiderable may be the alterations produced during a period sufficient only for the production of ten or fifteen eruptions of an active volcano, it is obvious that, in the time required for the formation of a lofty cone, composed of thousands of lava-currents, shallow seas may be converted into lofty mountains, and low lands into deep seas. We need, therefore, cherish none of the apprehensions entertained by Buffon, that the inequalities of the earth's surface, or the height and area of our continents, will be reduced by the action of running water; nor need we participate in the wonder of Ray, that the dry land should not lose ground more rapidly. Neither need we anticipate with Hutton the waste of successive continents followed by the creation of others by paroxysmal convulsions. The renovating as well as the destroying causes are unceasingly at work, the repair of land being as constant as its decay, and the deepening of seas keeping pace with the formation of shoals. If, in the course of a century, the Ganges and other great rivers have carried down to the sea a mass of matter equal to many lofty mountains, we also find that a district in Chili, one hundred thousand square miles in area, has been uplifted to the average height of a foot or more, and the cubic contents of the granitic mass thus added in a few hours to the land, may have counterbalanced the loss effected by the aqueous action of many rivers in a century. On the other hand, if the water displaced by fluviatile sediment cause the mean level of the ocean to rise in a slight degree, such subsidences of its bed, as that of Cutch in 1819, or St. Domingo in 1751, or Jamaica in 1692, may have compensated by increasing the capacity of the great oceanic basin. No river can push forward its delta without raising the level of the whole ocean, although in an infinitesimal degree; and no lowering can take place in the bed of any part of the ocean, without a general sinking of the water, even to the antipodes.

If the separate effects of different agents, whether aqueous or igneous, are insensible, it is because they are continually counteracted by each other, and a perfect adjustment takes place before any appreciable disturbance is occasioned. How many considerable earthquakes there may be upon an average in the course of one year, throughout the whole globe, is a question that we cannot decide at present; but as we have calculated that there are about twenty volcanic eruptions annually, we shall, perhaps, not overrate the earthquakes, if we estimate their number to be equal. A large number of eruptions are attended by local earthquakes of sufficient violence to modify the surface in some slight degree, and there are many earthquakes, on the other hand, not followed by eruptions. Even if we do not assume, as many have done, that the submarine convulsions exceed in number and violence those on the land, in spaces of equal area, we must, nevertheless, reckon about three shocks exclusively submarine, for one exclusively confined to the continents.

We have said in a former chapter that the aqueous and igneous agents may be regarded as antagonist forces, the aqueous labouring incessantly to reduce the inequalities of the earth's surface to a level, while the igneous are equally active in restoring the unevenness of the crust of the globe. But an erroneous theory appears to have been entertained by many geologists, and is indeed as old as the time of Lazzoro Moro, that the levelling power of running water was opposed rather to the *elevating* force of earthquakes than to their action generally. To such an opinion the numerous well-attested facts of subsidences must always have appeared a serious objection, but the same hypothesis would lead to other assumptions of a very arbitrary and improbable kind, inasmuch as it would be necessary to imagine the magnitude of our planet to be always on the increase if the elevation of the earth's surface by subterranean movements exceeded the depression. The sediment carried into the depths of the sea by rivers, tides, and currents, tends to diminish the height of the land; but, on the other hand, it tends, in a degree, to augment the height of the ocean, since water, equal in volume to the matter carried in, is displaced. The mean distance, therefore, of the surface, whether occupied by land or water from the centre of the earth, remains unchanged by the action of rivers, tides, and currents. Now suppose that while these agents are destroying islands and continents, the restoration of land should take place solely by the forcing out of the earth's envelope – it will be seen that this would imply a continual distension of the whole mass of the earth. For the greater number of earthquakes would be submarine, and they would cause the sea to rise and submerge the low lands even in a greater degree than would the influx of sediment. Two causes would, therefore, tend to destroy the land; submarine earthquakes, and the destroying and transporting power of water; and in order to counterbalance these effects, shallow seas must be upraised into continents, and low lands into mountains.

If we first consider the question simply, in regard to the manner whereby earthquakes may prevent running water from altering the relative proportion of land and sea, or the height of the land and depth of the ocean, we shall find that if the rising and sinking be equal, things would remain upon the whole in the same state: because rivers, tides, and currents, add as much to the height of lands which are rising, as they take from those which have risen.

Suppose a large river to carry down sediment into a certain part of the ocean where there is a depth of two thousand feet, and that the whole space is reduced by the fluviatile depositions to a shoal only covered by water at high tide: then let a series of two hundred earthquakes strike the shoal, each raising the ground ten feet; the result will be a mountain two thousand feet high. But suppose the same earthquakes had visited the same hollow in the bottom of the sea before the sediment of the river had filled it up, their whole force would then have been expended in converting a deep sea into a shoal, instead of changing a shoal into a mountain two thousand feet high. The superior altitude, then, of a district may often be due to the transportation of matter at a former period *to lower levels*. It would probably be more consistent with the natural course of events, if, instead of a succession of elevatory movements, we were to suppose considerable oscillations before the district attained its full height. Let there be, for example, three hundred instead of two hundred shocks, each separated from the other by intervals of about fifty years. Let the mean alteration of level produced by each earthquake be ten feet, two hundred and fifty shocks causing a rise, and the other fifty a sinking in of the ground; although more time will have been consumed by this operation than by the former, we shall still have the same result, for a tract will be raised to the height of about two thousand feet. The chief difference will consist in the superior breadth and depth of the valleys, which will be greater nearly in the proportion of one-third, in consequence of the number of landslips, floods, opening of chasms, and other effects produced by one hundred additional earthquakes. It should be borne in mind, moreover, that some of the lowering movements, happening towards the close of the period of disturbance, may have given rise to strange anomalies, should an attempt be made to reconcile the whole excavation in various hydrographical basins to the levels finally retained. Perhaps, for example, the middle portion of a valley may have sunk down, so that a deep lake may intervene between mountains and certain low plains, to which their debris had been previously carried.

But to return to the consideration of the proportion between the elevation and depression of the earth's crust, which may be necessary to preserve the uniformity of the general relations of land and sea, on the surface. The circumstances are in truth more complicated than those before stated, for, independently of the transfer of matter by running water from the continents to the ocean, there is a constant transportation of mineral ingredients from below upwards, by mineral springs and volcanic vents. As mountain masses are in the course of ages created by the pouring forth of successive streams of lava, so others originate from the carbonate of lime and other mineral ingredients with which springs are impregnated. The surface of the land, and parts of the bottom of the sea are thus raised, and if we conceive the dimensions of the planet to remain uniform, we must suppose these external accessions to be counteracted by some action of an opposite kind. A considerable quantity of earthy matter may sink down into fissures caused by earthquakes, but this cannot be deemed sufficient to counterbalance the addition of mountain masses by the causes before adverted to, and we must therefore suppose, that the subsidences of the earth's crust exceed the elevations caused by subterranean movements. It is to be expected, on mechanical principles, that the constant subtraction of matter from the interior will cause vacuities, so that the surface undermined will fall in during convulsions which shake the earth's crust even to great depths, and the sinking down will be occasioned partly by the hollows left when portions of the solid crust are heaved up, and partly when they are undermined by the subtraction of lava and the ingredients of decomposed rocks. The geological consequences which will follow if we embrace the theory now proposed are very important, for if there be upon the whole more subsidence than elevation, then we must consider the depth to which former surfaces have sunk down beneath their original level, to exceed the height which ancient marine strata have attained above the sea. If, for example, marine strata about the age of our chalk and green-sand have been lifted up in Europe to an extreme elevation of more than eleven thousand feet, and to a mean height of some hundreds above the level of the sea, we may conclude that certain parts of the earth's surface, which existed whether above or below the waters when those strata were deposited, have subsequently sunk down to an extreme depth of *more than* eleven thousand feet below their original level, and to a mean depth of *more than* a few hundreds.

In regard to faults, also, we must, according to the hypothesis now proposed, infer that a greater number have arisen from the sinking down than from the elevation of rocks. If we find, therefore, ancient deposits full of fresh-water remains which evidently originated in a delta or shallow estuary, covered subsequently by purely marine formations of vast thickness, we shall not be surprised; for we must expect that a greater number of existing deltas and estuary formations will sink below, than those which will rise above their present level. Although it would be rash to attempt to confirm these speculations by reference to the scanty observations hitherto made on the effects of earthquakes, yet we cannot but remark, that the instances of subsidence on record are far more numerous than are those of elevation.

Those writers who have most strenuously contended for the analogy of the effects of earthquakes in ancient and modern times, have nevertheless declared that the energy of the force has considerably abated. But they do not appear to have been aware that, in order to adduce plausible grounds for such an hypothesis, they must possess a most extensive knowledge of the economy of the whole terrestrial system. We can only estimate the relative amount of change produced at two distinct periods, by a particular cause in a given lapse of time, when we have obtained some common standard for the measurement of equal portions of time at both periods. We have shown that, within the last one hundred and forty years, some hundred thousand square miles of territory have been upheaved to the height of several feet, and that an area of equal, if not greater extent, has been depressed. Now, they who contend, that formerly more movement was accomplished by earthquakes in the space of one hundred and forty years, must first explain the measure of time referred to, for it is obvious that they cannot in geology avail themselves of the annual revolution of our planet round the sun. Suppose they assume that the power of volcanos to emit lava, and of running water to transport sediment from one part of the globe to the other, has remained uniform from the earliest periods, they might then attempt to compare the effects of subterranean movements in ancient and modern times by reference to one common standard, and to show that, while a certain number of lava-currents were produced, or so many cubic yards of sediment accumulated, the elevation and depression of the earth's crust were once much greater than they are now. Or, if they should declare that the progressive rate of change of species in the animal and vegetable kingdoms had always been uniform, they might then endeavour to disparage the degree of energy now exerted by earthquakes, by showing that, in relation to the mutations of assemblages of organic species, earthquakes had become comparatively feeble. But our present scanty acquaintance, both with the animate and inanimate world, can by no means warrant such generalizations; nor have they who contend for the gradual decline of the activity of natural agents, attempted to support such a line of argument. That it would be most premature, in the present state of natural history, to reason on the comparative rate of fluctuation in the species of organic beings in ancient and modern times, will be more fully demonstrated when we proceed, in the next division of our subject, to consider the intimate connexion between geology, and the study of the present condition of the animal and vegetable kingdoms.

To conclude: it appears, from the views above explained, respecting the agency of subterranean movements, that the constant repair of the dry land, and the subserviency of our planet to the support of terrestrial as well as aquatic species, are secured by the elevating and depressing power of earthquakes. This cause, so often the source of death and terror to the inhabitants of the globe, which visits, in succession, every zone, and fills the earth with monuments of ruin and disorder, is, nevertheless, a conservative principle in the highest degree, and, above all others, essential to the stability of the system.

View of the Valle del Bove Etna

VOLUME II

'The inhabitants of the globe, like all the other parts of it, are subject to change. It is not only the individual that perishes, but whole species.'

'A change in the animal kingdom seems to be part of the order of nature, and is visible in instances to which human power cannot have extended.'

Playfair, *Illustrations of the Huttonian Theory*, § 413[1]

CHAPTER I

Changes of the Organic World –
Reality of Species

In our first volume we treated of the changes which have taken place in the inorganic world within the historical era, and we must next turn our attention to those now in progress in the animate creation. In examining this class of phenomena, we shall treat first of the vicissitudes to which *species* are subject, and afterwards consider the influence of the powers of vitality in modifying the surface of the earth and the material constituents of its crust.

The first of these divisions will lead us, among other topics, to inquire, first, whether species have a real and permanent existence in nature; or whether they are capable, as some naturalists pretend, of being indefinitely modified in the course of a long series of generations? Secondly, whether, if species have a real existence, the individuals composing them have been derived originally from many similar stocks, or each from one only, the descendants of which have spread themselves gradually from a particular point over the habitable lands and waters? Thirdly, how far the duration of each species of animal and plant is limited by its dependance on certain fluctuating and temporary conditions in the state of the animate and inanimate world? Fourthly, whether there be proofs of the successive extermination of species in the ordinary course of nature, and whether there be any reason for conjecturing that new animals and plants are created from time to time, to supply their place?

Before we can advance a step in our proposed inquiry, we must be able to define precisely the meaning which we attach to the term species. This is even more necessary in geology than in the ordinary studies of the naturalist; for they who deny that such a thing as a species exists, concede nevertheless that a botanist or zoologist may reason as if the specific character were constant, because they confine their observations to a brief period of time. Just as the geographer, in constructing his maps from century to century, may proceed as if the apparent places of the fixed stars remained absolutely the same, and as if no alteration was brought about by the precession of the equinoxes, so it is said in the organic world, the stability of a species may be taken as absolute, if we do not extend our views beyond the narrow period of human history; but let a sufficient number of centuries elapse, to allow of important revolutions in climate, physical geography, and other circumstances, and the characters, say they, of the descendants of common parents may deviate indefinitely from their original type.

Now, if these doctrines be tenable, we are at once presented with a principle of incessant change in the organic world, and no degree of dissimilarity in the plants and animals which may formerly have existed, and are found fossil, would entitle us to conclude that they may not have been the prototypes and progenitors of the species now living. Accordingly, M. Geoffroy St. Hilaire has declared his opinion, that there has been an uninterrupted succession in the animal kingdom effected by means of generation, from the earliest ages of the world up to the present day; and that the ancient animals whose remains have been preserved in the strata, however different, may nevertheless have been the ancestors of those now in being. Although this notion is not generally received, we feel that we are not warranted in assuming the contrary, without fully explaining the data and reasoning by which we conceive it may be refuted.

We shall begin by stating as concisely as possible all the facts and ingenious arguments by which the theory has been supported, and for this purpose we cannot do better than offer the reader a rapid sketch of Lamarck's statement of the proofs which he regards as confirmatory of the doctrine, and which he has derived partly from the works of his predecessors, and in part from original investigations.

We shall consider his proofs and inferences in the order in which they appear to have influenced his mind, and point out some of the results to which he was led while boldly following out his principles to their legitimate consequences.

The name of species, observes Lamarck, has been usually applied to 'every collection of similar individuals, produced by other individuals like themselves.'[2] This definition, he admits, is correct, because every living individual bears a very close resemblance to those from which it springs. But this is not all which is usually implied by the term species, for the majority of naturalists agree with Linnæus in supposing that all the individuals propagated from one stock have certain distinguishing characters in common which will never vary, and which have remained the same since the creation of each species.

In order to shake this opinion, Lamarck enters upon the following line of argument. The more we advance in the knowledge of the different organized bodies which cover the surface of the globe, the more our embarrassment increases, to determine what ought to be regarded as a species, and still more how to limit and distinguish genera. In proportion as our collections are enriched, we see almost every void filled up, and all our lines of separation effaced; we are reduced to arbitrary determinations, and are sometimes fain to seize upon the slight differences of mere varieties, in order to form characters for what we choose to call a species, and sometimes we are induced to pronounce individuals but slightly differing, and which others regard as true species, to be varieties.

The greater the abundance of natural objects assembled together, the more do we discover proofs that everything passes by insensible shades into something else; that even the more remarkable differences are evanescent, and that nature has, for the most part, left us nothing at our disposal for establishing distinctions, save trifling and, in some respects, puerile particularities.

We find that many genera amongst animals and plants are of such an extent, in consequence of the number of species referred to them, that the study and determination of these last has become almost impracticable. When the species are arranged in a series, and placed near to each other, with due regard to their natural affinities, they each differ in so minute a degree from those next adjoining, that they almost melt into each other, and are in a manner confounded together. If we see isolated species, we may presume the absence of some more closely connected, and which have not yet been discovered. Already are there genera, and even entire orders, – nay, whole classes, which present an approximation to the state of things here indicated.

If, when species have been thus placed in a regular series, we select one, and then, making a leap over several intermediate ones, we take a second, at some distance from the first, these two will, on comparison, be seen to be very dissimilar; and it is in this manner that every naturalist begins to study the objects which are at his own door. He then finds it an easy task to establish generic and specific distinctions; and it is only when his experience is enlarged, and when he has made himself master of the intermediate links, that his difficulties and ambiguities begin. But while we are thus compelled to resort to trifling and minute characters in our attempt to separate species, we find a striking disparity between individuals which we know to have descended from a common stock, and these newly-acquired peculiarities are regularly transmitted from one generation to another, constituting what are called *races*.

From a great number of facts, continues the author, we learn that, in proportion as the individuals of one of our species change their situation, climate, and manner of living, they change also, by little and little, the consistence and proportions of their parts, their form, their faculties, and even their organization, in such a manner, that everything in them comes at last to participate in the mutations to which they have been exposed. Even in the same climate a great difference of situation and exposure causes individuals to vary; but if these individuals continue to live and to be reproduced under the same difference of circumstances, distinctions are brought about in them which become in some degree essential to their existence. In a word, at the end of many successive generations, these individuals, which originally belonged to another species, are transformed into a new and distinct species.

Thus, for example, if the seeds of a grass, or any other plant which grows naturally in a moist meadow, be accidentally transported, first to the slope of some neighbouring hill, where the soil, although at a greater elevation, is damp enough to allow the plant to live; and if, after having lived there, and having been several times regenerated, it reaches by degrees the drier and almost arid soil of a mountain declivity, it will then, if it succeeds in growing and perpetuates itself for a series of generations, be so changed that botanists who meet with it will regard it as a particular species. The unfavourable climate in this case, deficiency of nourishment, exposure to the winds, and other causes, give rise to a stunted and dwarfish race, with some organs more developed than others, and having proportions often quite peculiar.

What nature brings about in a great lapse of time we occasion suddenly by changing the circumstances in which a species has been accustomed to live. All are aware that vegetables taken from their birth-place and cultivated in gardens, undergo changes which render them no longer recognizable as the same plants. Many which were naturally hairy become smooth or nearly so; a great number of such as were creepers and trailed along the ground, rear their stalks and grow erect. Others lose their thorns or asperities; others again, from the ligneous state which their stem possessed in hot climates, where they were indigenous, pass to the herb-aceous, and, among them, some which were perennials become mere annuals. So well do botanists know the effects of such changes of circum-stances, that they are averse to describe species from garden specimens, unless they are sure that they have been cultivated for a very short period.

'Is not the cultivated wheat,' (*Triticum sativum*) asks Lamarck, 'a vegetable brought by man into the state in which we now see it? Let any one tell me in what country a similar plant grows wild, unless where it has escaped from cultivated fields? Where do we find in nature our cabbages, lettuces, and other culinary vegetables, in the state in which they appear in our gardens? Is it not the same in regard to a great quantity of animals which domesticity has changed or considerably modified?'[3] Our domestic fowls and pigeons are unlike any wild birds. Our domestic ducks and geese have lost the faculty of raising themselves into the higher regions of the air, and crossing extensive countries in their flight, like the wild ducks and wild geese from which they were originally derived. A bird which we breed in a cage cannot, when restored to liberty, fly like others of the same species which have been always free. This small alteration of circumstances, however, has only diminished the power of flight, without modifying the form of any part of the wings. But when individuals of the same race are retained in captivity during a considerable length of time, the form even of their parts is gradually made to differ, especially if climate, nourishment, and other circumstances, be also altered.

The numerous races of dogs which we have produced by domesticity are nowhere to be found in a wild state. In nature we should seek in vain for mastiffs, harriers, spaniels, greyhounds, and other races, between which the differences are sometimes so great, that they would be readily admitted as specific between wild animals; 'yet all these have sprung originally from a single race, at first approaching very near to a wolf, if, indeed, the wolf be not the true type which at some period or other was domesticated by man.'[4]

Although important changes in the nature of the places which they inhabit modify the organization of animals as well as vegetables, yet the former, says Lamarck, require more time to complete a considerable degree of transmutation, and, consequently, we are less sensible of such occurrences. Next to a diversity of the medium in which animals or plants may live, the circumstances which have most influence in modifying their organs are differences in exposure, climate, the nature of the soil, and other local particulars. These *circumstances* are as varied as are the characters of species, and, like them, pass by insensible shades into each other, there being every intermediate gradation between the opposite extremes. But each locality remains for a very long time the same, and is altered so slowly that we can only become conscious of the reality of the change, by consulting geological monuments, by which we learn that the order of things which now reigns in each place has not always prevailed, and by inference anticipate that it will not always continue the same.

Every considerable alteration in the local circumstances in which each race of animals exists, causes a change in their wants, and these new wants excite them to new actions and habits. These actions require the more frequent employment of some parts before but slightly exercised, and then greater development follows as a consequence of their more frequent use. Other organs no longer in use are impoverished and diminished in size, nay, are sometimes entirely annihilated, while in their place new parts are insensibly produced for the discharge of new functions.

We must here interrupt the author's argument, by observing that no positive fact is cited to exemplify the substitution of some *entirely new* sense, faculty, or organ, in the room of some other suppressed as useless. All the instances adduced go only to prove that the dimensions and strength of members and the perfection of certain attributes may, in a long succession of generations, be lessened and enfeebled by disuse; or, on the contrary, be matured and augmented by active exertion, just as we know that the power of scent is feeble in the greyhound, while its swiftness of pace and its acuteness of sight are remarkable – that the harrier and stag-hound, on the contrary, are comparatively slow in their movements, but excel in the sense of smelling.

We point out to the reader this important chasm in the chain of the evidence, because he might otherwise imagine that we had merely omitted the illustrations for the sake of brevity, but the plain truth is, that there were no examples to be found; and when Lamarck talks 'of the efforts of internal sentiment,' 'the influence of subtle fluids,' and the 'acts of organization,'[5] as causes whereby animals and plants may acquire *new organs*, he gives us names for things, and with a disregard to the strict rules of induction, resorts to fictions, as ideal as the 'plastic virtue,' and other phantoms of the middle ages.

It is evident, that if some well authenticated facts could have been adduced to establish one complete step in the process of transformation, such as the appearance, in individuals descending from a common stock, of a sense or organ entirely new, and a complete disappearance of some other enjoyed by their progenitors, that time alone might then be supposed sufficient to bring about any amount of metamorphosis. The gratuitous assumption, therefore, of a point so vital to the theory of transmutation, was unpardonable on the part of its advocate.

But to proceed with the system; it being assumed as an undoubted fact, that a change of external circumstances may cause one organ to become entirely obsolete, and a new one to be developed such as never before belonged to the species, the following proposition is announced, which, however staggering and absurd it may seem, is logically deduced from the assumed premises. 'It is not the organs, or, in other words, the nature and form of the parts of the body of an animal which have given rise to its habits, and its particular faculties, but on the contrary, its habits, its manner of living, and those of its progenitors have in the course of time determined the form of its body, the number and condition of its organs, in short, the faculties which it enjoys. Thus otters, beavers, water-fowl, turtles, and frogs, were not made web-footed in order that they might swim; but their wants having attracted them to the water in search of prey, they stretched out the toes of their feet to strike the water and move rapidly along its surface. By the repeated stretching of their toes, the skin which united them at the base acquired a habit of extension, until in the course of time the broad membranes which now connect their extremities were formed.'

'In like manner the antelope and the gazelle were not endowed with light agile forms, in order that they might escape by flight from carnivorous animals; but having been exposed to the danger of being devoured by lions, tigers, and other beasts of prey, they were compelled to exert themselves in running with great celerity, a habit which, in the course of many generations, gave rise to the peculiar slenderness of their legs, and the agility and elegance of their forms.'

'The cameleopard was not gifted with a long flexible neck because it was destined to live in the interior of Africa, where the soil was arid and devoid of herbage, but being reduced by the nature of that country to support itself on the foliage of lofty trees, it contracted a habit of stretching itself up to reach the high boughs, until its fore-legs became longer than the hinder, and its neck so elongated, that it could raise its head to the height of twenty feet above the ground.'[6]

Another line of argument is then entered upon, in farther corroboration of the instability of species. In order it is said that individuals should perpetuate themselves unaltered by generation, those belonging to one species ought never to ally themselves to those of another: but such sexual unions do take place, both among plants and animals; and although the offspring of such irregular connexions are usually sterile, yet such is not always the case. Hybrids have sometimes proved prolific where the disparity between the species was not too great; and by this means alone, says Lamarck, varieties may gradually be created by near alliances, which would become races, and in the course of time would constitute what we term species.

But if the soundness of all these arguments and inferences be admitted, we are next to inquire, what were the original types of form, organization, and instinct, from which the diversities of character, as now exhibited by animals and plants, have been derived? We know that individuals which are mere varieties of the same species, would, if their pedigree could be traced back far enough, terminate in a single stock; so according to the train of reasoning before described, the species of a genus, and even the genera of a great family, must have had a common point of departure. What then was the single stem from which so many varieties of form have ramified? Were there many of these, or are we to refer the origin of the whole animate creation, as the Egyptian priests did that of the universe, to a single egg?

In the absence of any positive data for framing a theory on so obscure a subject, the following considerations were deemed of importance to guide conjecture.

In the first place, if we examine the whole series of known animals, from one extremity to the other, when they are arranged in the order of their natural relations, we find that we may pass progressively, or at least with very few interruptions, from beings of more simple to those of a more compound structure; and in proportion as the complexity of their organization increases, the number and dignity of their faculties increase also. Among plants a similar approximation to a graduated scale of being is apparent. Secondly, it appears from geological observations, that plants and animals of more simple organization existed on the globe before the appearance of those of more compound structure, and the latter were successively formed at later periods: each new race being more fully developed than the most perfect of the preceding era.

Of the truth of the last-mentioned geological theory, Lamarck seems to have been fully persuaded; and he also shews that he was deeply impressed with a belief prevalent amongst the older naturalists, that the primeval ocean invested the whole planet long after it became the habitation of living beings, and thus he was inclined to assert the priority of the types of marine animals to those of the terrestrial, and to fancy, for example, that the testacea of the ocean existed first, until some of them, by gradual evolution, were *improved* into those inhabiting the land.

These speculative views had already been, in a great degree, anticipated by Delamétherie in his Teliamed, and by several modern writers, so that the tables were completely turned on the philosophers of antiquity, with whom it was a received maxim, that created things were always most perfect when they came first from the hands of their Maker, and that there was a tendency to progressive deterioration in sublunary things when left to themselves –

> Thus all below, whether by Nature's curse,
> Or Fates Decree, degen'rate still to worse.[7]

So deeply was the faith of the ancient schools of philosophy imbued with this doctrine, that to check this universal proneness to degeneracy, nothing less than the re-intervention of the Deity was thought adequate; and it was held, that thereby the order, excellence, and pristine energy of the moral and physical world had been repeatedly restored.

But when the possibility of the indefinite modification of individuals descending from common parents was once assumed, as also the geological generalization respecting the progressive development of organic life, it was natural that the ancient dogma should be rejected, or rather reversed; and that the most simple and imperfect forms and faculties should be conceived to have been the originals whence all others were developed. Accordingly, in conformity to these views, inert matter was supposed to have been first endowed with life; until in the course of ages, sensation was superadded to mere vitality: sight, hearing, and the other senses, were afterwards acquired; and then instinct and the mental faculties; until, finally, by virtue of the tendency of things to *progressive improvement*, the irrational was developed into the rational.

The reader, however, will immediately perceive, that if all the higher orders of plants and animals were thus supposed to be comparatively modern, and to have been derived in a long series of generations from those of more simple conformation, some further hypothesis became indispensable, in order to explain why, after an indefinite lapse of ages, there were still so many beings of the simplest structure. Why have the majority of existing creatures remained stationary throughout this long succession of epochs, while others have made such prodigious advances? Why are there still such multitudes of infusoria and polypes, or of confervæ and other cryptogamic plants? Why, moreover, has the process of development acted with such unequal and irregular force on those classes of beings which have been greatly perfected, so that there are wide chasms in the series; gaps so enormous, that Lamarck fairly admits we can never expect to fill them up by future discoveries?

The following hypothesis was provided to meet these objections. Nature, we are told, is not an intelligence, nor the Deity, but a delegated power – a mere instrument – a piece of mechanism acting by necessity – an order of things constituted by the Supreme Being, and subject to laws which are the expressions of his will. This nature is *obliged* to proceed gradually in all her operations; she cannot produce animals and plants of all classes at once, but must always begin by the formation of the most simple kinds; and out of them elaborate the more compound, adding to them successively, different systems of organs, and multiplying more and more their number and energy.

This Nature is daily engaged in the formation of the elementary rudiments of animal and vegetable existence, which correspond to what the ancients termed *spontaneous generations*. She is always beginning anew, day by day, the work of creation, by forming monads, or 'rough draughts' (ébauches), which are the only living things she ever gives birth to *directly*.

There are distinct primary rudiments of plants and animals, and *probably* of each of the great divisions of the animal and vegetable kingdoms. These are gradually developed into the higher and more perfect classes by the slow, but unceasing agency of two influential principles: first, *the tendency to progressive advancement* in organization, accompanied by greater dignity in instinct, intelligence, &c.; secondly, *the force of external circumstances*, or of variations in the physical condition of the earth, or the mutual relations of plants and animals. For as species spread themselves gradually over the globe, they are exposed from time to time to variations in climate, and to changes in the quantity and quality of their food; they meet with new plants and animals which assist or retard their development, by supplying them with nutriment, or destroying their foes. The nature also of each locality is in itself fluctuating, so that even if the relation of other animals and plants were invariable, the habits and organization of species would be modified by the influence of local revolutions.

Now, if the first of these principles, *the tendency to progressive development*, were left to exert itself with perfect freedom, it would give rise, says Lamarck, in the course of ages, to a graduated scale of being, where the most insensible transition might be traced from the simplest to the most compound structure, from the humblest to the most exalted degree of intelligence. But in consequence of the perpetual interference of the *external causes* before mentioned, this regular order is greatly interfered with, and an approximation only to such a state of things is exhibited by the animate creation, the progress of some races being retarded by unfavourable, and that of others accelerated by favourable, combinations of circumstances. Hence, all kinds of anomalies interrupt the continuity of the plan, and chasms, into which whole genera or families might be inserted, are seen to separate the nearest existing portions of the series.

Such is the machinery of the Lamarckian system; but our readers will hardly, perhaps, be able to form a perfect conception of so complicated a piece of mechanism, unless we exhibit it in motion, and shew in what manner it can work out, under the author's guidance, all the extraordinary effects which we behold in the present state of the animate creation. We have only space for exhibiting a small part of the entire process by which a complete metamorphosis is achieved, and shall, therefore, omit the mode whereby, after a countless succession of generations, a small gelatinous body is transformed into an oak or an ape. We pass on at once to the last grand step in the progressive scheme, whereby the orang-outang, having been already evolved out of a monad, is made slowly to attain the attributes and dignity of man.

One of the races of quadrumanous animals which had reached the highest state of perfection, lost, by constraint of circumstances, (concerning the exact nature of which tradition is unfortunately silent,) the habit of climbing trees, and of hanging on by grasping the boughs with their feet as with hands. The individuals of this race being obliged for a long series of generations to use their feet exclusively for walking, and ceasing to employ their hands as feet, were transformed into bimanous animals, and what before were thumbs became mere toes, no separation being required when their feet were used solely for walking. Having acquired a habit of holding themselves upright, their legs and feet assumed insensibly a conformation fitted to support them in an erect attitude, till at last these animals could no longer go on all fours without much inconvenience.

The Angola orang, *Simia troglodytes*, Linn., is the most perfect of animals, much more so than the Indian orang, *Simia Satyrus*, which has been called the orang-outang, although *both* are *very inferior* to man in corporeal powers and intelligence. These animals frequently hold themselves upright, but their organization has *not yet* been sufficiently modified to sustain them habitually in this attitude, so that the standing posture is very uneasy to them. When the Indian orang is compelled to take flight from pressing danger, he immediately falls down upon all fours, shewing clearly that this was the original position of the animal. Even in man, whose organization, in the course of a long series of generations, has advanced so much farther, the upright posture is fatiguing and can only be supported for a limited time, and by aid of the contraction of many muscles. If the vertebral column formed the axis of the human body, and supported the head and all the other parts in equilibrium, then might the upright position be a state of repose; but as the human head does not articulate in the centre of gravity; as the chest, belly, and other parts, press almost entirely forward with their whole weight, and as the vertebral column reposes upon an oblique base, a watchful activity is required to prevent the body from falling. Children which have large heads and prominent bellies can hardly walk at the end even of two years, and their frequent tumbles indicate the natural tendency in man to resume the quadrupedal state.

Now, when so much progress had been made by the quadrumanous animals before mentioned, that they could hold themselves habitually in an erect attitude, and were accustomed to a wide range of vision, and ceased to use their jaws for fighting, and tearing, or for clipping herbs for food, their snout became gradually shorter, their incisor teeth became vertical, and the facial angle grew more open.

Among other ideas which the natural *tendency to perfection* engendered, the desire of ruling suggested itself, and this race succeeded at length in getting the better of the other animals, and made themselves masters of all those spots on the surface of the globe which best suited them. They drove out the animals which approached nearest to them in organization and intelligence, and which were in a condition to dispute with them the good things of this world, forcing them to take refuge in deserts, woods and wildernesses, where their multiplication was checked, and the progressive development of their faculties retarded, while in the mean time the dominant race spread itself in every direction, and lived in large companies where new wants were successively created, exciting them to industry, and gradually perfecting their means and faculties.

In the supremacy and increased intelligence acquired by the ruling race, we see an illustration of the natural tendency of the organic world to grow more perfect, and in their influence in repressing the advance of others, an example of one of those disturbing causes before enumerated, that *force of external circumstances*, which causes such wide chasms in the regular series of animated beings.

When the individuals of the dominant race became very numerous, their ideas greatly increased in number, and they felt the necessity of communicating them to each other, and of augmenting and varying the signs proper for the communication of ideas. Meanwhile the inferior quadrumanous animals, although most of them were gregarious, acquired no new ideas, being persecuted and restless in the deserts, and obliged to fly and conceal themselves, so that they conceived no new wants. Such ideas as they already had remained unaltered, and they could dispense with the communication of the greater part of these. To make themselves, therefore, understood by their fellows, required merely a few movements of the body or limbs – whistling, and the uttering of certain cries varied by the inflexions of the voice.

On the contrary, the individuals of the ascendant race, animated with a desire of interchanging their ideas, which became more and more numerous, were prompted to multiply the means of communication, and were no longer satisfied with mere pantomimic signs, nor even with all the possible inflexions of the voice, but made continual efforts to acquire the power of uttering articulate sounds, employing a few at first, but afterwards varying and perfecting them according to the increase of their wants. The habitual exercise of their throat, tongue and lips, insensibly modified the conformation of these organs, until they became fitted for the faculty of speech.

In effecting this mighty change, 'the exigencies of the individuals were the sole agents, they gave rise to efforts, and the organs proper for articulating sounds were developed by their habitual employment.'[8] Hence, in this peculiar race, the origin of the admirable faculty of speech; hence also the diversity of languages, since the distance of places where the individuals composing the race established themselves, soon favoured the corruption of conventional signs.

CHAPTER 2

Theory of the Transmutation
of Species Untenable

The theory of the transmutation of species, considered in the last chapter, has met with some degree of favour from many naturalists, from their desire to dispense, as far as possible, with the repeated intervention of a First Cause, as often as geological monuments attest the successive appearance of new races of animals and plants, and the extinction of those pre-existing. But, independently of a predisposition to account, if possible, for a series of changes in the organic world, by the regular action of secondary causes, we have seen that many perplexing difficulties present themselves to one who attempts to establish the nature and the reality of the specific character. And if once there appears ground of reasonable doubt, in regard to the constancy of species, the amount of transformation which they are capable of undergoing, may seem to resolve itself into a mere question of the quantity of time assigned to the past duration of animate existence.

Before we enter upon our reasons for rejecting Lamarck's hypothesis, we shall recapitulate, in a few words, the phenomena, and the whole train of thought, by which we conceive it to have been suggested, and which have gained for this and analogous theories, both in ancient and modern times, a considerable number of votaries.

In the first place, the various groups into which plants and animals may be thrown, seem almost invariably, to a beginner, to be so natural, that he is usually convinced at first, as was Linnæus to the last, 'that genera are as much founded in nature as the species which compose them.'[9] When, by examining the numerous intermediate gradations, the student finds all lines of demarcation to be in most instances obliterated, even where they at first appeared most distinct, he grows more and more sceptical as to the real existence of genera, and finally regards them as mere arbitrary and artificial signs, invented like those which serve to distinguish the heavenly constellations for the convenience of classification, and having as little pretensions to reality.

Doubts are then engendered in his mind as to whether species may not also be equally unreal. The student is probably first struck with the phenomenon, that some individuals are made to deviate widely from the ordinary type by the force of peculiar circumstances, and with the still more extraordinary fact, that the newly-acquired peculiarities are faithfully transmitted to the offspring. How far, he asks, may such variations extend in the course of indefinite periods of time, and during great vicissitudes in the physical condition of the globe? His growing incertitude is at first checked by the reflection, that nature has forbidden the intermixture of the descendants of distinct original stocks, or has, at least, entailed sterility on their offspring, thereby preventing their being confounded together, and pointing out that a multitude of distinct types must have been created in the beginning, and must have remained pure and uncorrupted to this day.

Relying on this general law, he endeavours to solve each difficult problem by direct experiment, until he is again astounded by the phenomenon of a prolific hybrid, and still more by an example of a hybrid perpetuating itself throughout several generations in the vegetable world. He then feels himself reduced to the dilemma of choosing between two alternatives, either to reject the test, or to declare that the two species, from the union of which the fruitful progeny has sprung, were mere varieties. If he prefer the latter, he is compelled to question the reality of the distinctness of all other supposed species which differ no more than the parents of such prolific hybrids; for although he may not be enabled immediately to procure, in all such instances, a fruitful offspring, yet experiments show, that after repeated failures the union of two recognized species may at last, under very favourable circumstances, give birth to a fertile progeny. Such circumstances, therefore, the naturalist may conceive to have occurred again and again, in the course of a great lapse of ages.

His first opinions are now fairly unsettled, and every stay at which he has caught has given way one after another; he is in danger of falling into any new and visionary doctrine which may be presented to him; for he now regards every part of the animate creation as void of stability, and in a state of continual flux. In this mood he encounters the Geologist, who relates to him how there have been endless vicissitudes in the shape and structure of organic beings in former ages – how the approach to the present system of things has been gradual – that there has been a progressive development of organization subservient to the purposes of life, from the most simple to the most complex state – that the appearance of man is the last phenomenon in a long succession of events – and, finally, that a series of physical revolutions can be traced in the inorganic world, coeval and coextensive with those of organic nature.

These views seem immediately to confirm all his preconceived doubts as to the stability of the specific character, and he thinks he can discern an inseparable connexion between a series of changes in the inanimate world, and the capability of species to be indefinitely modified by the influence of external circumstances. Henceforth his speculations know no definite bounds; he gives the rein to conjecture, and fancies that the outward form, internal structure, instinctive faculties, nay, that reason itself, may have been gradually developed from some of the simplest states of existence, – that all animals, that man himself, and the irrational beings, may have had one common origin; that all may be parts of one continuous and progressive scheme of development from the most imperfect to the more complex; in fine, he renounces his belief in the high genealogy of his species, and looks forward, as if in compensation, to the future perfectibility of man in his physical, intellectual, and moral attributes.

Let us now proceed to consider what is defective in evidence, and what fallacious in reasoning, in the grounds of these strange conclusions. Blumenbach judiciously observes, 'that no general rule can be laid down for determining the distinctness of species, as there is no particular class of characters which can serve as a criterion. In each case we must be guided by *analogy* and *probability*.'[10] The multitude, in fact, and complexity of the proofs to be weighed, is so great, that we can only hope to obtain presumptive evidence, and we must, therefore, be the more careful to derive our general views as much as possible from those observations where the chances of deception are least. We must be on our guard not to tread in the footsteps of the naturalists of the middle ages, who believed the doctrine of spontaneous generation to be applicable to all those parts of the animal and vegetable kingdoms which they least understood, in direct contradiction to the analogy of all the parts best known to them; and who, when at length they found that insects and cryptogamous plants were also propagated from eggs and seeds, still persisted in retaining their old prejudices respecting the infusory animalcules and other minute beings, the generation of which had not then been demonstrated by the microscope to be governed by the same laws.

Lamarck has indeed attempted to raise an argument in favour of his system, out of the very confusion which has arisen in the study of some orders of animals and plants, in consequence of the slight shades of difference which separate the new species discovered within the last half century. That the embarrassment of those who attempt to classify and distinguish the new acquisitions poured in such multitudes into our museums, should increase with the augmentation of their number is quite natural; for to obviate this it is not enough that our powers of discrimination should keep pace with the increase of the objects, but we ought to possess greater opportunities of studying each animal and plant in all stages of its growth, and to know profoundly their history, their habits and physiological characters, throughout several generations. For, in proportion as the series of known animals grows more complete, none can doubt that there is a nearer approximation to a graduated scale of being; and thus the most closely allied species will possess a greater number of characters in common.

But, in point of fact, our new acquisitions consist, more and more as we advance, of specimens brought from foreign and often very distant and barbarous countries. A large proportion have never even been seen alive by scientific inquirers. Instead of having specimens of the young, the adult, and the aged individuals of each sex, and possessing means of investigating the anatomical structure, the peculiar habits and instincts of each, what is usually the state of our information? A single specimen, perhaps, of a dried plant, or a stuffed bird or quadruped; a shell without the soft parts of the animal; an insect in one stage of its numerous transformations; these are the scanty and imperfect data, which the naturalist possesses. Such information may enable us to separate species which stand at a considerable distance from each other; but we have no right to expect anything but difficulty and ambiguity, if we attempt, from such imperfect opportunities, to obtain distinctive marks for defining the characters of species, which are closely related.

If Lamarck could introduce so much certainty and precision into the classification of several thousand species of recent and fossil shells, notwithstanding the extreme remoteness of the organization of these animals from the type of those vertebrated species which are best known, and in the absence of so many of the living inhabitants of shells, we are led to form an exalted conception of the degree of exactness to which specific distinctions are capable of being carried, rather than to call in question their reality.

When our data are so defective, the most acute naturalist must expect to be sometimes at fault, and, like the novice, to overlook essential points of difference, passing unconsciously from one species to another, until, like one who is borne along in a current, he is astonished, on looking back, at observing that he has reached a point so remote from that whence he set out.

It is by no means improbable that when the series of species of certain genera is very full, they may be found to differ less widely from each other, than do the mere varieties or races of certain species. If such a fact could be established, it would by no means overthrow our confidence in the reality of species, although it would certainly diminish the chance of our obtaining certainty in our results.

It is almost necessary, indeed, to suppose, that varieties will differ in some cases, more decidedly than some species, if we admit that there is a graduated scale of being, and assume that the following laws prevail in the economy of the animate creation: – first, that the organization of individuals is capable of being modified to a limited extent by the force of external causes; secondly, that these modifications are, to a certain extent, transmissible to their offspring; thirdly, that there are fixed limits beyond which the descendants from common parents can never deviate from a certain type; fourthly, that each species springs from one original stock, and can never be permanently confounded, by intermixing with the progeny of any other stock; fifthly, that each species shall endure for a considerable period of time. Now if we assume, for the present, these rules hypothetically, let us see what consequences may naturally be expected to result.

We must suppose, that when the Author of Nature creates an animal or plant, all the possible circumstances in which its descendants are destined to live are foreseen, and that an organization is conferred upon it which will enable the species to perpetuate itself and survive under all the varying circumstances to which it must be inevitably exposed. Now the range of variation of circumstances will differ essentially in almost every case. Let us take for example any one of the most influential conditions of existence, such as temperature. In some extensive districts near the equator, the thermometer might never vary throughout several thousand centuries for more than 20° Fahrenheit; so that if a plant or animal be provided with an organization fitting it to endure such a range, it may continue on the globe for that immense period, although every individual might be liable at once to be cut off by the least possible excess of heat or cold beyond the determinate quantity. But if a species be placed in one of the temperate zones, and have a constitution conferred on it capable of supporting a similar range of temperature only, it will inevitably perish before a single year has passed away.

The same remark might be applied to any other condition, as food for example; it may be foreseen that the supply will be regular throughout indefinite periods in one part of the world, and in another very precarious and fluctuating both in kind and quantity. Different qualifications may be required for enabling species to live for a considerable time under circumstances so changeable. If, then, temperature and food be among those external causes, which according to certain laws of animal and vegetable physiology modify the organization, form, or faculties of individuals, we instantly perceive that the degrees of variability from a common standard must differ widely in the two cases above supposed, since there is a necessity of accommodating a species in one case to a much greater latitude of circumstances than in the other.

If it be a law, for instance, that scanty sustenance should check those individuals in their growth which are enabled to accommodate themselves to privations of this kind, and that a parent prevented in this manner from attaining the size proper to its species should produce a dwarfish offspring, a stunted race will arise, as is remarkably exemplified in some varieties of the horse and dog. The difference of stature in some races of dogs in comparison to others, is as one to five in linear dimensions, making a difference of a hundred-fold in volume. Now there is good reason to believe that species in general are by no means susceptible of existing under a diversity of circumstances, which may give rise to such a disparity in size, and consequently, there will be a multitude of distinct species, of which no two adult individuals can ever depart so widely from a certain standard of dimensions as the mere varieties of certain other species, – the dog for instance. Now we have only to suppose that what is true of size, may also hold in regard to colour and many other attributes, and it will at once follow that the degree of possible discordance between varieties of the same species, may in certain cases exceed the utmost disparity which can even arise between two individuals of many distinct species.

The same remarks may hold true in regard to instincts; for if it be foreseen that one species will have to encounter a great variety of foes, it may be necessary to arm it with great cunning and circumspection, or with courage or other qualities capable of developing themselves on certain occasions; such for example as those migratory instincts which are so remarkably exhibited at particular periods, after they have remained dormant for many generations. The history and habits of one variety of such a species, may often differ more considerably from some other than those of many distinct species which have no such latitude of accommodation to circumstances.

Lamarck has somewhat misstated the idea commonly entertained of a species, for it is not true that naturalists in general assume that the organization of an animal or plant remains absolutely constant, and that it can never vary in any of its parts. All must be aware that circumstances influence the habits, and that the habits may alter the state of the parts and organs. But the difference of opinion relates to the extent to which these modifications of the habits and organs of a particular species may be carried.

Now let us first inquire what positive facts can be adduced in the history of known species, to establish a great and permanent amount of change in the form, structure, or instinct of individuals descending from some common stock. The best authenticated examples of the extent to which species can be made to vary, may be looked for in the history of domesticated animals and cultivated plants. It usually happens that those species, both of the animal and vegetable kingdom, which have the greatest pliability of organization, those which are most capable of accommodating themselves to a great variety of new circumstances, are most serviceable to man. These only can be carried by him into different climates, and can have their properties or instincts variously diversified by differences of nourishment and habits. If the resources of a species be so limited, and its habits and faculties be of such a confined and local character, that it can only flourish in a few particular spots, it can rarely be of great utility.

We may consider, therefore, that in perfecting the arts of domesticating animals and cultivating plants, mankind have first selected those species which have the most flexible frames and constitutions, and have then been engaged for ages in conducting a series of experiments, with much patience and at great cost, to ascertain what may be the greatest possible deviation from a common type which can be elicited in these extreme cases.

The modifications produced in the different races of dogs, exhibit the influence of man in the most striking point of view. These animals have been transported into every climate, and placed in every variety of circumstances; they have been made, as a modern naturalist observes, the servant, the companion, the guardian, and the intimate friend of man, and the power of a superior genius has had a wonderful influence, not only on their forms, but on their manners and intelligence. Different races have undergone remarkable changes in the quantity and colour of their clothing: the dogs of Guinea are almost naked, while those of the Arctic circle are covered with a warm coat both of hair and wool, which enables them to bear the most intense cold without inconvenience. There are differences also of another kind no less remarkable, as in size, the length of their muzzles, and the convexity of their foreheads.

But if we look for some of those essential changes which would be required to lend even the semblance of a foundation for the theory of Lamarck, respecting the growth of new organs and the gradual obliteration of others, we find nothing of the kind. For in all these varieties of the dog, says Cuvier, the relation of the bones with each other remain essentially the same; the form of the teeth never changes in any perceptible degree, except that in some individuals, one additional false grinder occasionally appears, sometimes on the one side, and sometimes on the other. The greatest departure from a common type, and it constitutes the maximum of variation as yet known in the animal kingdom, is exemplified in those races of dogs which have a supernumerary toe on the hind foot with the corresponding tarsal bones, a variety analogous to one presented by six-fingered families of the human race.

Lamarck has thrown out as a conjecture, that the wolf may have been the original of the dog, but he has adduced no data to bear out such an hypothesis. 'The wolf,' observes Dr. Prichard, 'and the dog differ, not only with respect to their habits and instincts, which in the brute creation are very uniform within the limits of one species; but some differences have also been pointed out in their internal organization, particularly in the structure of a part of the intestinal canal.'[11]

It is well known that the horse, the ox, the boar and other domestic animals, which have been introduced into South America, and have run wild in many parts, have entirely lost all marks of domesticity, and have reverted to the original characters of their species. But the dog has also become wild in Cuba, Hayti, and in all the Caribbean islands. In the course of the seventeenth century, they hunted in packs from twelve to fifty, or more in number, and fearlessly attacked herds of wild-boars and other animals. It is natural, therefore, to enquire to what form they reverted? Now they are said by many travellers to have resembled very nearly the shepherd's dog; but it is certain that they were never turned into wolves. They were extremely savage, and their ravages appear to have been as much dreaded as those of wolves, but when any of their whelps were caught, and brought from the woods to the towns, they grew up in the most perfect submission to man.

As the advocates of the theory of transmutation trust much to the slow and insensible changes which time may work, they are accustomed to lament the absence of accurate descriptions, and figures of particular animals and plants, handed down from the earliest periods of history, such as might have afforded data for comparing the condition of species, at two periods considerably remote. But fortunately, we are in some measure independent of such evidence, for by a singular accident, the priests of Egypt have bequeathed to us, in their cemeteries, that information, which the museums and works of the Greek philosophers have failed to transmit.

For the careful investigation of these documents, we are greatly indebted to the skill and diligence of those naturalists who accompanied the French armies during their brief occupation of Egypt: that conquest of four years, from which we may date the improvement of the modern Egyptians in the arts and sciences, and the rapid progress which has been made of late in our knowledge of the arts and sciences of their remote predecessors. Instead of wasting their whole time as so many preceding travellers had done, in exclusively collecting human mummies, M. Geoffroy and his associates examined diligently, and sent home great numbers of embalmed bodies of consecrated animals, such as the bull, the dog, the cat, the ape, the ichneumon, the crocodile, and the ibis.

To those who have never been accustomed to connect the facts of Natural History with philosophical speculations, who have never raised their conceptions of the end and import of such studies beyond the mere admiration of isolated and beautiful objects, or the exertion of skill in detecting specific differences, it will seem incredible that amidst the din of arms, and the stirring excitement of political movements, so much enthusiasm could have been felt in regard to these precious remains.

In the official report drawn up by the Professors of the Museum at

Paris, on the value of these objects, there are some eloquent passages which may appear extravagant, unless we reflect how fully these naturalists could appreciate the bearing of the facts thus brought to light on the past history of the globe.

'It seems,' say they, 'as if the superstition of the ancient Egyptians had been inspired by Nature, with a view of transmitting to after ages a monument of her history. That extraordinary and whimsical people, by embalming with so much care the brutes which were the objects of their stupid adoration, have left us, in their sacred grottoes, cabinets of zoology almost complete. The climate has conspired with the art of embalming to preserve the bodies from corruption, and we can now assure ourselves by our own eyes what was the state of a great number of species three thousand years ago. We can scarcely restrain the transports of our imagination, on beholding thus preserved with their minutest bones, with the smallest portions of their skin, and in every particular most perfectly recognizable, many an animal, which at Thebes or Memphis, two or three thousand years ago, had its own priests and altars.'[12]

Among the Egyptian mummies thus procured were not only those of numerous wild quadrupeds, birds, and reptiles, but, what was perhaps of still greater importance in deciding the great question under discussion, there were the mummies of domestic animals, among which those above mentioned, the bull, the dog, and the cat, were frequent. Now such was the conformity of the whole of these species to those now living, that there was no more difference, says Cuvier, between them than between the human mummies and the embalmed bodies of men of the present day. Yet some of these animals have since that period been transported by man to almost every variety of climate, and forced to accommodate their habits to new circumstances, as far as their nature would permit. The cat, for example, has been carried over the whole earth, and, within the last three centuries, has been naturalized in every part of the new world, from the cold regions of Canada to the tropical plains of Guiana; yet it has scarcely undergone any perceptible mutation, and is still the same animal which was held sacred by the Egyptians.

Of the ox, undoubtedly there are many very distinct races; but the bull Apis, which was led in solemn processions by the Egyptian priests, did not differ from some of those now living. The black cattle that have run wild in America, where there were many peculiarities in the climate not to be found, perhaps, in any part of the old world, and where scarcely a single plant on which they fed was of precisely the same species, instead of altering their form and habits, have actually reverted to the exact likeness of the aboriginal wild cattle of Europe.

In answer to the arguments drawn from the Egyptian mummies, Lamarck said that they were identical with their living descendants in the same country, because the climate and physical geography of the banks of the Nile have remained unaltered for the last thirty centuries. But why, we ask, have other individuals of these species retained the same characters in so many different quarters of the globe, where the climate and many other conditions are so varied?

The evidence derived from the Egyptian monuments was not confined to the animal kingdom; the fruits, seeds, and other portions of twenty different plants, were faithfully preserved in the same manner; and among these the common wheat was procured by Delille, from closed vessels in the sepulchres of the kings, the grains of which retained not only their form, but even their colour, so effectual has proved the process of embalming with bitumen in a dry and equable climate. No difference could be detected between this wheat and that which now grows in the East and elsewhere, and similar identifications were made in regard to all the other plants.

And here we may observe, that there is an obvious answer to Lamarck's objection, that the botanist cannot point out a country where the common wheat grows wild, unless in places where it may have been derived from neighbouring cultivation. All naturalists are well aware that the geographical distribution of a great number of species is extremely limited, and that it was to be expected that every useful plant should first be cultivated successfully in the country where it was indigenous, and that, probably, every station which it partially occupied, when growing wild, would be selected by the agriculturist as best suited to it when artificially increased. Palestine has been conjectured, by a late writer on the Cerealia, to have been the original habitation of wheat and barley, a supposition which appears confirmed by Hebrew and Egyptian traditions, and by tracing the migrations of the worship of Ceres, as indicative of the migrations of the plant.

If we are to infer that some one of the wild grasses has been transformed into the common wheat, and that some animal of the genus *canis*, still unreclaimed, has been metamorphosed into the dog, merely because we cannot find the domestic dog, or the cultivated wheat, in a state of nature, we may be next called upon to make similar admissions in regard to the camel; for it seems very doubtful whether any race of this species of quadruped is now wild.

But if agriculture, it will be said, does not supply examples of extraordinary changes of form and organization, the horticulturist can, at least, appeal to facts which may confound the preceding train of reasoning. The crab has been transformed into the apple; the sloe into the plum: flowers have changed their colour and become double; and these new characters can be perpetuated by seed, – a bitter plant with wavy sea-green leaves has been taken from the sea-side where it grew like wild charlock, has been transplanted into the garden, lost its saltness, and has been metamorphosed into two distinct vegetables as unlike each other as is each to the parent plant – the red cabbage and the cauliflower. These, and a multitude of analogous facts, are undoubtedly among the wonders of nature, and attest more strongly, perhaps, the extent to which species may be modified, than any examples derived from the animal kingdom. But in these cases we find, that we soon reach certain limits, beyond which we are unable to cause the individuals, descending from the same stock, to vary; while, on the other hand, it is easy to show that these extraordinary varieties could seldom arise, and could never be perpetuated in a wild state for many generations, under any imaginable combination of accidents. They may be regarded as extreme cases brought about by human interference, and not as phenomena which indicate a capability of indefinite modification in the natural world.

The propagation of a plant by buds or grafts, and by cuttings, is obviously a mode which nature does not employ; and this multiplication, as well as that produced by roots and layers, seems merely to operate as an extension of the life of an individual, and not as a reproduction of the species, as happens by seed. All plants increased by the former means retain precisely the peculiar qualities of the individual to which they owe their origin, and, like an individual, they have only a determinate existence; in some cases longer and in others shorter. It seems now admitted by horticulturists, that none of our garden varieties of fruit are entitled to be considered strictly permanent, but that they wear out after a time; and we are thus compelled to resort again to seeds; in which case, there is so decided a tendency in the seedlings to revert to the original type, that our utmost skill is sometimes baffled in attempting to recover the desired variety.

The different races of cabbages afford, as we have admitted, an astonishing example of deviation from a common type; but we can scarcely conceive them to have originated, much less to have lasted for several generations, without the intervention of man. It is only by strong manures that these varieties have been obtained, and in poorer soils they instantly degenerate. If, therefore, we suppose in a state of nature the seed of the wild *Brassica oleracea* to have been wafted from the sea-side to some spot enriched by the dung of animals, and to have there become a cauliflower, it would soon diffuse its seed to some comparatively steril soils around, and the offspring would relapse to the likeness of the parent stock, like some individuals which may now be seen growing on the cornice of old London bridge.

But if we go so far as to imagine the soil, in the spot first occupied, to be constantly manured by herds of wild animals, so as to continue as rich as that of a garden, still the variety could not be maintained, because we know that each of these races is prone to fecundate others, and gardeners are compelled to exert the utmost diligence to prevent cross-breeds. The intermixture of the pollen of varieties growing in the poorer soil around, would soon destroy the peculiar characters of the race which occupied the highly-manured tract; for, if these accidents so continually happen in spite of us, among the culinary varieties, it is easy to see how soon this cause might obliterate every marked singularity in a wild state.

Besides, it is well-known that although the pampered races which we rear in our gardens for use or ornament, may often be perpetuated by seed, yet they rarely produce seed in such abundance, or so prolific in quality, as wild individuals; so that, if the care of man were withdrawn, the most fertile variety would always, in the end, prevail over the more steril.

Similar remarks may be applied to the double flowers which present such strange anomalies to the botanist. The ovarium, in such cases, is frequently abortive, and the seeds, when prolific, are generally much fewer than where the flowers are single.

Some curious experiments recently made on the production of blue instead of red flowers in the *Hydrangea hortensis*, illustrate the immediate effect of certain soils on the colours of the petals. In garden-mould or compost, the flowers are invariably red; in some kinds of bog-earth they are blue; and the same change is always produced by a particular sort of yellow loam.

Linnæus was of opinion that the primrose, oxlip, cowslip, and polyanthus, were only varieties of the same species. The majority of modern botanists, on the contrary, consider them to be distinct, although some conceived that the oxlip might be a cross between the cowslip and the primrose. Mr. Herbert has lately recorded the following experiment: – 'I raised from the natural seed of one umbel of a highly-manured red cowslip, a primrose, a cowslip, oxlips of the usual and other colours, a black polyanthus, a hose-in-hose cowslip, and a natural primrose bearing its flower on a polyanthus stalk. From the seed of that very hose-in-hose cowslip I have since raised a hose-in-hose primrose. I therefore consider all these to be only local varieties depending upon soil and situation.'[13] Professor Henslow, of Cambridge, has since confirmed this experiment of Mr. Herbert, so that we have an example, not only of the remarkable varieties which the florist can obtain from a common stock, but of the distinctness of analogous races found in a wild state.

On what particular ingredient, or quality in the earth, these changes depend, has not yet been ascertained. But gardeners are well aware that particular plants, when placed under the influence of certain circumstances, are changed in various ways according to the species; and as often as the experiments are repeated similar results are obtained. The nature of these results, however, depends upon the species, and they are, therefore, part of the specific character; they exhibit the same phenomena again and again, and indicate certain fixed and invariable relations between the physiological peculiarities of the plant, and the influence of certain external agents. They afford no ground for questioning the instability of species, but rather the contrary; they present us with a class of phenomena which, when they are more thoroughly understood, may afford some of the best tests for identifying species, and proving that the attributes originally conferred, endure so long as any issue of the original stock remains upon the earth.

CHAPTER 3

Limits of the Variability of Species

We endeavoured in the last chapter to show, that a belief in the reality of species is not inconsistent with the idea of a considerable degree of variability in the specific character. This opinion, indeed, is little more than an extension of the idea which we must entertain of the identity of an individual, throughout the changes which it is capable of undergoing.

If a quadruped, inhabiting a cold northern latitude, and covered with a warm coat of hair or wool, be transported to a southern climate, it will often, in the course of a few years, shed a considerable portion of its coat, which it gradually recovers on being again restored to its native country. Even there the same changes are, perhaps, superinduced to a certain extent by the returns of winter and summer. We know that the Alpine hare and the ermine become white during winter, and again obtain their full colour during the warmer season; that the plumage of the ptarmigan undergoes a like metamorphosis in colour and quantity, and that the change is equally temporary. We are aware that, if we reclaim some wild animal, and modify its habits and instincts by domestication, it may, if it escapes, become in a few years nearly as wild and untractable as ever; and if the same individual be again retaken, it may be reduced to its former tame state. A plant is placed in a prepared soil in order that the petals of its flowers may multiply, and their colour be heightened or changed; if we then withhold our care, the flowers of this same individual become again single. In these, and innumerable other instances, we must suppose that the individual was produced with a certain number of qualities; and, in the case of animals, with a variety of instincts, some of which may or may not be developed according to circumstances, or which, after having been called forth, may again become latent when the exciting causes are removed.

Now the formation of races seems the necessary consequence of such a capability in individuals to vary, if it be a general law that the offspring should very closely resemble the parent. But, before we can infer that there are no limits to the deviation from an original type which may be brought about in the course of an indefinite number of generations, we ought to have some proof that, in each successive generation, individuals may go on acquiring an equal amount of new peculiarities, under the influence of equal changes of circumstances. The balance of evidence, however, inclines most decidedly on the opposite side, for in all cases we find that the quantity of divergence diminishes from the first in a very rapid ratio.

It cannot be objected, that it is out of our power to go on varying the circumstances in the same manner as might happen in the natural course of events during some great geological cycle. For in the first place, where a capacity is given to individuals to adapt themselves to new circumstances, it does not generally require a very long period for its development; if, indeed, such were the case, it is not easy to see how the modification would answer the ends proposed, for all the individuals would die before new qualities, habits, or instincts, were conferred.

When we have succeeded in naturalizing some tropical plant in a temperate climate, nothing prevents us from attempting gradually to extend its distribution to higher latitudes, or to greater elevations above the level of the sea, allowing equal quantities of time, or an equal number of generations for habituating the species to successive increments of cold. But every husbandman and gardener is aware that such experiments will fail; and we are more likely to succeed in making some plants, in the course of the first two generations, support a considerable degree of difference of temperature than a very small difference afterwards, though we persevere for many centuries.

It is the same if we take any other cause instead of temperature; such as the quality of the food, or the kind of dangers to which an animal is exposed, or the soil in which a plant lives. The alteration in habits, form, or organization, is often rapid during a short period; but when the circumstances are made to vary further, though in ever so slight a degree, all modification ceases, and the individual perishes. Thus some herbivorous quadrupeds may be made to feed partially on fish or flesh, but even these can never be taught to live on some herbs which they reject, and which would even poison them, although the same may be very nutritious to other species of the same natural order. So when man uses force or stratagem against wild animals, the persecuted race soon becomes more cautious, watchful, and cunning; new instincts seem often to be developed, and to become hereditary in the first two or three generations; but let the skill and address of man increase, however gradually, no further variation can take place, no new qualities are elicited by the increasing dangers. The alteration of the habits of the species has reached a point beyond which no ulterior modification is possible, however indefinite the lapse of ages during which the new circumstances operate. Extirpation then follows, rather than such a transformation as could alone enable the species to perpetuate itself under the new state of things.

It has been well observed by M. F. Cuvier and M. Dureau de la Malle, that unless some animals had manifested in a wild state an aptitude to second the efforts of man, their domestication would never have been attempted. If they had all resembled the wolf, the fox, and the hyæna, the patience of the experimentalist would have been exhausted by innumerable failures before he at last succeeded in obtaining some imperfect results; so, if the first advantages derived from the cultivation of plants had been elicited by as tedious and costly a process as that by which we now make some slight additional improvement in certain races, we should have remained to this day in ignorance of the greater number of their useful qualities.

It is undoubtedly true, that many new habits and qualities have not only been acquired in recent times by certain races of dogs, but have been transmitted to their offspring. But in these cases it will be observed, that the new peculiarities have an intimate relation to the habits of the animal in a wild state, and therefore do not attest any tendency to departure to an indefinite extent from the original type of the species. A race of dogs employed for hunting deer in the platform of Santa Fé in Mexico, affords a beautiful illustration of a new hereditary instinct. The mode of attack, observes M. Roulin, which they employ, consists in seizing the animal by the belly and overturning it by a sudden effort, taking advantage of the moment when the body of the deer rests only upon the forelegs. The weight of the animal thus thrown over, is often six times that of its antagonist. The dog of pure breed inherits a disposition to this kind of chase, and never attacks a deer from before while running. Even should the latter, not perceiving him, come directly upon him, the dog steps aside and makes his assault on the flank, whereas other hunting dogs, though of superior strength and general sagacity, which are brought from Europe, are destitute of this instinct. For want of similar precautions, they are often killed by the deer on the spot, the vertebræ of their neck being dislocated by the violence of the shock.

A new instinct also has become hereditary in a mongrel race of dogs employed by the inhabitants of the banks of the Magdalena, almost exclusively in hunting the white-lipped pecari. The address of these dogs consists in restraining their ardour, and attaching themselves to no animal in particular, but keeping the whole herd in check. Now, among these dogs some are found, which, the very first time they are taken to the woods, are acquainted with this mode of attack; whereas, a dog of another breed starts forward at once, is surrounded by the pecari, and whatever may be his strength is destroyed in a moment.

Some of our countrymen, engaged of late in conducting the principal mining association in Mexico, carried out with them some English grey-hounds of the best breed, to hunt the hares which abound in that country. The great platform which is the scene of sport is at an elevation of about nine thousand feet above the level of the sea, and the mercury in the barometer stands habitually at the height of about nineteen inches. It was found that the greyhounds could not support the fatigues of a long chase in this attenuated atmosphere, and before they could come up with their prey, they lay down gasping for breath; but these same animals have produced whelps which have grown up, and are not in the least degree incommoded by the want of density in the air, but run down the hares with as much ease as the fleetest of their race in this country.

The fixed and deliberate stand of the pointer has with propriety been regarded as a mere modification of a habit, which may have been useful to a wild race accustomed to wind game, and steal upon it by surprise, first pausing for an instant in order to spring with unerring aim. The faculty of the Retriever, however, may justly be regarded as more inexplicable and less easily referrible to the instinctive passions of the species. M. Majendie, says a French writer in a recently-published memoir, having learnt that there was a race of dogs in England, which stopped and brought back game of their own accord, procured a pair, and having obtained a whelp from them kept it constantly under his eyes, until he had an opportunity of assuring himself that, without having received any instruction and on the very first day that it was carried to the chase, it brought back game with as much steadiness as dogs which had been schooled into the same manœuvre by means of the whip and collar.

Such attainments, as well as the habits and dispositions which the shepherd's dog and many others inherit, seem to be of a nature and extent which we can hardly explain by supposing them to be modifications of instincts necessary for the preservation of the species in a wild state. When such remarkable habits appear in races of this species, we may reasonably conjecture that they were given with no other view than for the use of man and the preservation of the dog which thus obtains protection.

As a general rule, we fully agree with M. F. Cuvier that, in studying the habits of animals, we must attempt, as far as possible, to refer their domestic qualities to modifications of instincts which are implanted in them in a state of nature; and that writer has successfully pointed out, in an admirable essay on the domestication of the mammalia, the true origin of many dispositions which are vulgarly attributed to the influence of education alone. But we should go too far if we did not admit that some of the qualities of particular animals and plants may have been given solely with a view to the connexion which it was foreseen would exist between them and man – especially when we see that connexion to be in many cases so intimate, that the greater number, and sometimes all the individuals of the species which exist on the earth are in subjection to the human race.

We can perceive in a multitude of animals, especially in some of the parasitic tribes, that certain instincts and organs are conferred for the purpose of defence or attack against some other species. Now if we are reluctant to suppose the existence of similar relations between man and the instincts of many of the inferior animals, we adopt an hypothesis no less violent, though in the opposite extreme to that which has led some to imagine the whole animate and inanimate creation to have been made solely for the support, gratification, and instruction of mankind.

Many species most hostile to our persons or property multiply in spite of our efforts to repress them; others, on the contrary, are intentionally augmented many hundred-fold in number by our exertions. In such instances we must imagine the relative resources of man and of species, friendly or inimical to him, to have been prospectively calculated and adjusted. To withhold assent to this supposition would be to refuse what we must grant in respect to the economy of Nature in every other part of the organic creation; for the various species of contemporary plants and animals have obviously their relative forces nicely balanced, and their respective tastes, passions, and instincts, so contrived, that they are all in perfect harmony with each other. In no other manner could it happen, that each species surrounded as it is by countless dangers should be enabled to maintain its ground for periods of considerable duration.

The docility of the individuals of some of our domestic species extending, as it does, to attainments foreign to their natural habits and faculties, may perhaps have been conferred with a view to their association with man. But lest species should be thereby made to vary indefinitely, we find that such habits are never transmissible by generation.

A pig has been trained to hunt and point game with great activity and steadiness; and other learned individuals, of the same species, have been taught to spell; but such fortuitous acquirements never become hereditary, for they have no relation whatever to the exigencies of the animal in a wild state, and cannot therefore be developments of any instinctive propensities.

An animal in domesticity, says M. F. Cuvier, is not essentially in a different situation in regard to the feeling of restraint from one left to itself. It lives in society without constraint, because without doubt it was a social animal, and it conforms itself to the will of man, because it had a chief to which in a wild state it would have yielded obedience. There is nothing in its new situation that is not conformable to its propensities; it is satisfying its wants by submission to a master, and makes no sacrifice of its natural inclinations. All the social animals when left to themselves form herds more or less numerous, and all the individuals of the same herd know each other, are mutually attached, and will not allow a strange individual to join them. In a wild state, moreover, they obey some individual, which by its superiority has become the chief of the herd. Our domestic species had originally this sociability of disposition, and no solitary species, however easy it may be *to tame it*, has yet afforded true domestic races. We merely, therefore, develope to our own advantage, propensities which propel the individuals of certain species to draw near to their fellows.

The sheep which we have reared is induced to follow us, as it would be led to follow the flock among which it was brought up; and when individuals of gregarious species have been accustomed to one master, it is he alone whom they acknowledge as their chief, he only whom they obey. – 'The elephant only allows himself to be led by the carnac whom he has adopted; the dog itself, reared in solitude with its master, manifests a hostile disposition towards all others; and everybody knows how danger-ous it is to be in the midst of a herd of cows, in pasturages that are little frequented, when they have not at their head the keeper who takes care of them.'

'Everything, therefore, tends to convince us, that formerly men were only, with regard to the domestic animals, what those who are particularly charged with the care of them still are, namely, members of the society, which these animals form among themselves, and that they are only distinguished in the general mass by the authority which they have been enabled to assume from their superiority of intellect. Thus, every social animal which recognizes man as a member, and as the chief of its herd, is a domestic animal. It might even be said that from the moment when such an animal admits man as a member of its society, it is domesticated, as man could not enter into such a society without becoming the chief of it.'[14]

But the ingenious author whose observations we have here cited, admits that the obedience which the individuals of many domestic species yield indifferently to every person is without analogy in any state of things which could exist previously to their subjugation by man. Each troop of wild horses, it is true, has some stallion for its chief, who draws after him all the individuals of which the herd is composed; but when a domesticated horse has passed from hand to hand, and has served several masters, he becomes equally docile towards *any person*, and is subjected to the whole human race. It seems fair to presume, that the capability in the instinct of the horse to be thus modified, was given to enable the species to render greater services to man; and, perhaps, the facility with which many other acquired characters become hereditary in various races of the horse, may be explicable only on a like supposition. The amble, for example, a pace to which the domestic races in Spanish America are exclusively trained, has, in the course of several generations, become hereditary, and is assumed by all the young colts before they are broken in.

It seems also reasonable to conclude, that the power bestowed on the horse, the dog, the ox, the sheep, the cat, and many species of domestic fowls, of supporting almost every climate, was given expressly to enable them to follow man throughout all parts of the globe – in order that we might obtain their services, and they our protection. If it be objected that the elephant, which, by the union of strength, intelligence, and docility, can render the greatest services to mankind, is incapable of living in any but the warmest latitudes, we may observe, that the quantity of vegetable food required by this quadruped would render its maintenance, in the temperate zone, too costly, and in the arctic impossible.

Among the changes superinduced by man, none appear, at first sight, more remarkable than the perfect tameness of certain domestic races. It is well known, that at however early an age we obtain possession of the young of many unreclaimed races, they will retain, throughout life, a considerable timidity and apprehensiveness of danger; whereas, after one or two generations, the descendants of the same will habitually place the most implicit confidence in man. There is good reason, however, to suspect that such changes are not without analogy in a state of nature, or, to speak more correctly, in situations where man has not interfered.

Thus Dr. Richardson informs us, in his able history of the habits of North American animals, that 'in the retired parts of the mountains, where the hunters had seldom penetrated, there is no difficulty in approaching the Rocky Mountain sheep, which there exhibit *the simplicity of character so remarkable in the domestic species*; but where they have been often fired at, they are exceedingly wild, alarm their companions, on the approach of danger, by a hissing noise, and scale the rocks with a speed and agility that baffles pursuit.'[15]

It is probable, therefore, that as man, in diffusing himself over the globe, has tamed many wild races, so also he has made many tame races wild. Had some of the larger carnivorous beasts, capable of scaling the rocks, made their way into the North American mountains before our hunters, a similar alteration in the instincts of the sheep would doubtless have been brought about.

No animal affords a more striking illustration of the principal points we have been endeavouring to establish than the elephant. For in the first place, the wonderful sagacity with which he accommodates himself to the society of man, and the new habits which he contracts are not the result of time nor of modifications produced in the course of many generations. These animals will breed in captivity, as is now ascertained in opposition to the vulgar opinion of many modern naturalists, and in conformity to that of the ancients Ælian and Columella. Yet it has always been the custom, as the least expensive mode of obtaining them, to capture wild individuals in the forests, usually when full grown, and in a few years after they are taken, sometimes, it is said, in the space of a few months, their education is completed.

Had the whole species been domesticated from an early period in the history of man, like the camel, their superior intelligence would doubtless have been attributed to their long and familiar intercourse with the lord of the creation: but we know that a few years is sufficient to bring about this wonderful change of habits; and, although the same individual may continue to receive tuition for a century afterwards, yet it makes no further progress in the general development of its faculties. Were it otherwise, indeed, the animal would soon deserve more than the poet's epithet of 'half-reasoning.'[16]

From the authority of our countrymen employed in the late Burmese war, it appears, in corroboration of older accounts, that when elephants are required to execute extraordinary tasks, they may be made to understand that they will receive unusual rewards. Some favourite dainty is shown to them, in the hope of acquiring which, the work is done. And so perfectly does the nature of the contract appear to be understood, that the breach of it, on the part of the master, is often attended with danger. In this case, a power has been given to the species to adapt their social instincts to new circumstances with surprising rapidity; but the extent of this change is defined by strict and arbitrary limits. There is no indication of a tendency to continued divergence from certain attributes with which the elephant was originally endued, no ground whatever for anticipating, that in thousands of centuries any material alteration could ever be effected. All that we can infer from analogy is, that some useful and peculiar races might probably be formed, if the experiment were fairly tried, and that some individual characteristic, now only casual and temporary, might be perpetuated by generation.

In all cases, therefore, where the domestic qualities exist in animals, they seem to require no lengthened process for their development, and they appear to have been wholly denied to some classes, which from their strength and social nature might have rendered great services to man; as, for example, the greater part of the quadrumana. The orang-outang, indeed, which for its resemblance in form to man, and apparently for no other good reason, has been assumed, by Lamarck, to be the most perfect of the inferior animals, has been tamed by the savages of Borneo, and made to climb lofty trees, and to bring down the fruit. But he is said to yield to his masters an unwilling obedience, and to be held in subjection only by severe discipline. We know nothing of the faculties of this animal which can suggest the idea that it rivals the elephant in intelligence, much less anything which can countenance the dreams of those who have fancied that it might have been transmuted into 'the dominant race.' One of the baboons of Sumatra (*Simia carpolegus*) appears to be more docile, and is frequently trained by the inhabitants to ascend trees for the purpose of gathering cocoa-nuts, a service in which the animal is very expert. He selects, says Sir Stamford Raffles, the ripe nuts with great judgment, and pulls no more than he is ordered. The capuchin and cacajao monkeys are, according to Humboldt, taught to ascend trees in the same manner, and to throw down fruit on the banks of the lower Orinoco.

We leave it to the Lamarckians to explain, how it happens that those same savages of Borneo have not themselves acquired, by dint of longing for many generations for the power of climbing trees, the elongated arms of the orang, or even the prehensile tails of some American monkeys. Instead of being reduced to the necessity of subjugating stubborn and untractable brutes, we should naturally have anticipated 'that their wants would have excited them to efforts, and that continued efforts would have given rise to new organs;' or, rather, to the re-acquisition of organs which, in a manner irreconcileable with the principle of the *progressive* system, have grown obsolete in tribes of men which have such constant need of them.

It follows, then, from the different facts which we have considered in this chapter, that a short period of time is generally sufficient to effect nearly the whole change which an alteration of external circumstances can bring about in the habits of a species, and that such capacity of accommodation to new circumstances is enjoyed in very different degrees by different species.

Certain qualities appear to be bestowed exclusively with a view to the relations which are destined to exist between different species, and, among others, between certain species and man; but these latter are always so nearly connected with the original habits and propensities of each species in a wild state, that they imply no indefinite capacity of varying from the original type. The acquired habits, derived from human tuition, are rarely transmitted to the offspring; and when this happens, it is almost universally the case with those merely which have some obvious connexion with the attributes of the species when in a state of independence.

CHAPTER 4
Hybrids

We have yet to consider another class of phenomena, those relating to the production of hybrids, which have been regarded in a very different light with reference to their bearing on the question of the permanent distinctness of species; some naturalists considering them as affording the strongest of all proofs in favour of the reality of species; others, on the contrary, appealing to them as countenancing the opposite doctrine, that all the varieties of organization and instinct now exhibited in the animal and vegetable kingdoms, may have been propagated from a small number of original types.

In regard to the mammifers and birds, it is found that no sexual union will take place between races which are remote from each other in their habits and organization; and it is only in species that are very nearly allied that such unions produce offspring. It may be laid down as a general rule, admitting of very few exceptions among quadrupeds, that the hybrid progeny is steril, and there seem to be no well-authenticated examples of the continuance of the mule race beyond one generation. The principal number of observations and experiments relate to the mixed offspring of the horse and the ass; and in this case it is well established, that the male-mule can generate and the female-mule produce. Such cases occur in Spain and Italy, and much more frequently in the West Indies and New Holland; but these mules have never bred in cold climates, seldom in warm regions, and still more rarely in temperate countries.

The hybrid offspring of the female-ass and the stallion, the γιννος of Aristotle, and the hinnus of Pliny, differs from the mule, or the offspring of the ass and mare. In both cases, says Buffon, these animals retain more of the mother than of the father, not only in the magnitude but in the figure of the body; whereas, in the form of the head, limbs, and tail, they bear a greater resemblance to the father. The same naturalist infers, from various experiments respecting cross-breeds between the he-goat and ewe, the dog and she-wolf, the goldfinch and canary-bird, that the male transmits his sex to the greatest number, and that the preponderance of males over females exceeds that which prevails where the parents are of the same species.

The celebrated John Hunter has observed, that the true distinction of species must ultimately be gathered from their incapacity of propagating with each other, and producing offspring capable of again continuing itself. He was unwilling, however, to admit, that the horse and the ass were of the same species, because some rare instances had been adduced of the breeding of mules, which he attributed to a degree of monstrosity in the organs of the mule, for these he suggested might not have been those of a mixed animal, but those of the mare or female-ass. 'This, he argues, is not a far-fetched idea, for true species produce monsters, and many animals of distinct sex are incapable of breeding at all; and as we find nature, in its greatest perfection, deviating from general principles, why may it not happen likewise in the production of mules, so that sometimes a mule shall breed from the circumstance of its being a monster respecting mules?'[17]

Yet, in the same memoir, this great anatomist inferred that the wolf, the dog, and the jackal, were all of one species, because he had found, by two experiments, that the dog would breed, both with the wolf and the jackal; and that the mule, in each case, would breed again with the dog. In these cases, however, we may observe, that there was always one parent at least of pure breed, and no proof was obtained that a true hybrid race could be perpetuated; a fact of which we believe no examples are yet recorded, either in regard to mixtures of the horse and ass, or any other of the mammalia.

Should the fact be hereafter ascertained, that two mules can propagate their kind, we must still inquire whether the offspring may not be regarded in the light of a monstrous birth, proceeding from some accidental cause, or rather, to speak more philosophically, from some general law not yet understood, but which may not be permitted permanently to interfere with those laws of generation, whereby species may, in general, be prevented from becoming blended. If, for example, we discovered that the progeny of a mule race degenerated greatly in the first generation, in force, sagacity, or any attribute necessary for its preservation in a state of nature, we might infer that, like a monster, it is a mere temporary and fortuitous variety. Nor does it seem probable that the greater number of such monsters could ever occur unless obtained by art; for in Hunter's experiments, stratagem or force was, in most instances, employed to bring about the irregular connexion.

It seems rarely to happen that the mule offspring is truly intermediate in character between the two parents. Thus Hunter mentions, that, in his experiments, one of the hybrid pups resembled the wolf much more than the rest of the litter; and we are informed by Wiegmann, that in a litter lately obtained in the Royal Menagerie at Berlin, from a white pointer and a she-wolf, two of the cubs resembled the common wolf-dog, but the third was like a pointer with hanging ears.

There is, undoubtedly, a very close analogy between these phenomena and those presented by the intermixture of distinct races of the same species, both in the inferior animals and in man. Dr. Prichard, in his 'Physical History of Mankind,' cites examples where the peculiarities of the parents have been transmitted very unequally to the offspring; as where children, entirely white, or perfectly black, have sprung from the union of the European and the negro. Sometimes the colour, or other peculiarities of one parent, after having failed to show themselves in the immediate progeny, reappear in a subsequent generation, as where a white child is born of two black parents, the grandfather having been a white.

The same author judiciously observes, that if different species mixed their breed, and hybrid races were often propagated, the animal world would soon present a scene of confusion; its tribes would be everywhere blended together, and we should, perhaps, find more hybrid creatures than genuine and uncorrupted races.

The history of the vegetable kingdom has been thought to afford more decisive evidence in favour of the theory of the formation of new and permanent species from hybrid stocks. The first accurate experiments in illustration of this curious subject appear to have been made by Kölreuter, who obtained a hybrid from two species of Tobacco, *Nicotiana rustica* and *N. paniculata*, which differ greatly in the shape of their leaves, the colour of the corolla, and the height of the stem. The stigma of a female plant of *N. rustica* was impregnated with the pollen of a male plant of *N. paniculata*. The seed ripened and produced a hybrid which was intermediate between the two parents, and which, like all the hybrids which this botanist brought up, had imperfect stamens. He afterwards impregnated this hybrid with the pollen of *N. paniculata*, and obtained plants which much more resembled the last. This he continued through several generations, until, by due perseverance, he actually changed the *Nicotiana rustica* into the *Nicotiana paniculata*.

The plan of impregnation adopted, was the cutting off of the anthers of the plant intended for fructification before they had shed pollen, and then laying on foreign pollen upon the stigma. The same experiment has since been repeated, with success, by Wiegmann, who found that he could bring back the hybrids to the exact likeness of either parent, by crossing them a sufficient number of times.

The blending of the characters of the parent stocks, in many other of Weigmann's experiments, was complete; the colour and shape of the leaves and flowers, and even the scent, being intermediate, as in the offspring of the two species of verbascum. An intermarriage, also, between the common onion and the leek (*Allium cepa* and *A. porrum*) gave a mule plant, which, in the character of its leaves and flowers, approached most nearly to the garden onion, but had the elongated bulbous root and smell of the leek.

The same botanist remarks, that vegetable hybrids, when not strictly intermediate, more frequently approach the female than the male parent species, *but they never exhibit characters foreign to both*. A re-cross with one of the original stocks, generally causes the mule plant to revert towards that stock; but this is not always the case, the offspring sometimes continuing to exhibit the character of a full hybrid.

In general, the success attending the production and perpetuity of hybrids among plants, depends, as in the animal kingdom, on the degree of proximity between the species intermarried. If their organization be very remote, impregnation never takes place; if somewhat less distant, seeds are formed, but always imperfect and steril. The next degree of relationship yields hybrid seedlings, but these are barren; and it is only when the parent species are very nearly allied, that the hybrid race may be perpetuated for several generations. Even in this case the best authenticated examples seem confined to the crossing of hybrids with individuals of pure breed. In none of the experiments most accurately detailed does it appear that both the parents were mules.

Wiegmann diversified, as much as possible, his mode of bringing about these irregular unions among plants. He often sowed parallel rows, near to each other, of the species from which he desired to breed, and instead of mutilating, after Kölreuter's fashion, the plants of one of the parent stocks, he merely washed the pollen off their anthers. The branches of the plants, in each row, were then gently bent towards each other and intertwined, so that the wind, and numerous insects as they passed from the flowers of one to those of the other species, carried the pollen and produced fecundation.

The same observer saw a good exemplification of the manner in which hybrids may be formed in a state of nature. Some wallflowers and pinks had been growing in a garden, in a dry sunny situation, and their stigmas had been ripened so as to be moist, and to absorb pollen with avidity, although their anthers were not yet developed. These stigmas became impregnated by pollen, blown from some other adjacent plants of the same species, but had they been of different species, and not too remote in their organization, mule races must have resulted.

When, indeed, we consider how busily some insects have been shown to be engaged in conveying anther-dust from flower to flower, especially bees, flower-eating beetles, and the like, it seems a most enigmatical problem how it can happen, that promiscuous alliances between distinct species are not perpetually occurring.

How continually do we observe the bees diligently employed in col-lecting the red and yellow powder by which the stamens of flowers are covered, loading it on their hind legs, and carrying it to their hive for the purpose of feeding their young! In thus providing for their own progeny, these insects assist materially the process of fructification. Few of our readers need be reminded, that the stamens in certain plants grow on different blossoms from the pistils, and unless the summit of the pistil be touched with the fertilizing dust, the fruit does not swell, nor the seed arrive at maturity. It is by the help of bees chiefly, that the development of the fruit of many such species is secured, the powder which they have collected from the stamens being unconsciously left by them in visiting the pistils.

How often, during the heat of a summer's day, do we see the males of dioecious plants, such as the yew-tree, standing separate from the females, and sending off into the air, upon the slightest breath of wind, clouds of buoyant pollen! That the zephyr should so rarely intervene to fecundate the plants of one species with the anther-dust of others, seems almost to realize the converse of the miracle believed by the credulous herdsmen of the Lusitanian mares –

The Mares to Cliffs of rugged Rocks repair,
And with wide Nostrils sniff the Western Air:
When (wondrous to relate) the Parent Wind,
Without the Stallion, propagates the kind.[18]

But, in the first place, it appears that there is a natural aversion in plants, as well as in animals, to irregular sexual unions; and in most of the successful experiments in the animal and vegetable world, some violence has been used, in order to procure impregnation. The stigma imbibes, slowly and reluctantly, the granules of the pollen of another species, even when it is abundantly covered with it; and if it happen that, during this period, ever so slight a quantity of the anther-dust of its own species alight upon it, this is instantly absorbed, and the effect of the foreign pollen destroyed. Besides, it does not often happen that the male and female organs of fructification, in different species, arrive at a state of maturity at precisely the same time. Even where such synchronism does prevail, so that a cross impregnation is effected, the chances are very numerous against the establishment of a hybrid race.

If we consider the vegetable kingdom generally, it must be recollected, that even of the seeds which are well ripened, the greater part are either eaten by insects, birds, and other animals, or decay for want of room and opportunity to germinate. Unhealthy plants are the first which are cut off by causes prejudicial to the species, being usually stifled by more vigorous individuals of their own kind. If, therefore, the relative fecundity or hardiness of hybrids be in the least degree inferior, they cannot maintain their footing for many generations, even if they were ever produced beyond one generation in a wild state. In the universal struggle for existence, the right of the strongest eventually prevails; and the strength and durability of a race depends mainly on its prolificness, in which hybrids are acknowledged to be deficient.

Centaurea hybrida, a plant which never bears seed, and is supposed to be produced by the frequent intermixture of two well-known species of Centaurea, grows wild upon a hill near Turin. *Ranunculus lacerus*, also steril, has been produced accidentally at Grenoble, and near Paris, by the union of two Ranunculi; but this occurred in gardens.

Mr. Herbert, in one of his ingenious papers on mule plants, endeavours to account for their non-occurrence in a state of nature, from the circumstance that all the combinations that were likely to occur, have already been made many centuries ago, and have formed the various species of botanists; but in our gardens, he says, whenever species, having a certain degree of affinity to each other, are transported from different countries, and brought for the first time into contact, they give rise to hybrid species. But we have no data, as yet, to warrant the conclusion, that a single permanent hybrid race has ever been formed, even in gardens, by the intermarriage of two allied species brought from distant habitations. Until some fact of this kind is fairly established, and a new species, capable of perpetuating itself in a state of perfect independence of man, can be pointed out, we think it reasonable to call in question entirely this hypothetical source of new species. That varieties do sometimes spring up from cross-breeds, in a natural way, can hardly be doubted, but they probably die out even more rapidly than races propagated by grafts or layers.

Decandolle, whose opinion on a philosophical question of this kind deserves the greatest attention, has observed, in his Essay on Botanical Geography, that the *varieties* of plants range themselves under two general heads: those produced by external circumstances, and those formed by hybridity. After adducing various arguments to show that neither of these causes can explain the permanent diversity of plants indigenous in different regions, he says, in regard to the crossing of races, 'I can perfectly comprehend, without altogether sharing the opinion, that where many species of the same genera occur near together, hybrid species may be formed, and I am aware that the great number of species of certain genera which are found in particular regions, may be explained in this manner; but I am unable to conceive how any one can regard the same explanation as applicable to species which live naturally at great distances. If the three larches, for example, now known in the world, lived in the same localities, I might then believe that one of them was the produce of the crossing of the two others; but I never could admit that the Siberian species has been produced by the crossing of those of Europe and America. I see, then, that there exist, in organized beings, permanent differences which cannot be referred to any one of the actual causes of variation, and these differences are what constitute *species*.'[19]

The most decisive arguments, perhaps, amongst many others, against the probability of the derivation of permanent species from cross-breeds, are to be drawn from the fact alluded to by Decandolle, of species having a close affinity to each other occurring in distinct botanical provinces, or countries inhabited by groups of distinct species of indigenous plants. For in this case naturalists, who are not prepared to go the whole length of the transmutationists, are under the necessity of admitting, that in some cases species which approach very near to each other in their characters, were so created from their origin; an admission fatal to the idea of its being a general law of nature, that a few original types only should be formed, and that all intermediate races should spring from the intermixture of those stocks.

This notion, indeed, is wholly at variance with all that we know of hybrid generation; for the phenomena entitle us to affirm, that had the types been at first somewhat distant, *no cross-breeds would ever have been produced*, much less those prolific races which we now recognise as distinct species.

In regard, moreover, to the permanent propagation of hybrid races among animals, insuperable difficulties present themselves, when we endeavour to conceive the blending together of the different instincts and propensities of two species, so as to insure the preservation of the intermediate race. The common mule, when obtained by human art, may be protected by the power of man; but in a wild state, it would neither have precisely the same wants as the horse or the ass: and if, in consequence of some difference of this kind, it strayed from the herd, it would soon be hunted down by beasts of prey and destroyed.

If we take some genus of insects, such as the bee, we find that each of the numerous species has some difference in its habits, its mode of collecting honey, or constructing its dwelling, or providing for its young, and other particulars. In the case of the common hive-bee, the workers are described, by Kirby and Spence, as being endowed with no less than thirty distinct instincts. So also we find that amongst a most numerous class of spiders, there are nearly as many different modes of spinning their webs as there are species. When we recollect how complicated are the relations of these instincts with co-existing species, both of the animal and vegetable kingdoms, it is scarcely possible to imagine that a bastard race could spring from the union of two of these species, and retain just so much of the qualities of each parent-stock as to preserve its ground in spite of the dangers which surround it.

We should also ask, if a few generic types alone have been *created* among insects, and the intermediate species have proceeded from hybridity, where are those original types, combining, as they ought to do, the elements of all the instincts which have made their appearance in the numerous derivative races? So also in regard to animals of all classes, and of plants; if species in general are of hybrid origin, where are the stocks which combine in themselves the habits, properties, and organs, of which all the intervening species ought to afford us mere modifications?

We shall now conclude this subject by summing up, in a few words, the results to which the consideration of the phenomena of hybrids has led us. It appears that the aversion of individuals of distinct species to the sexual union is common to animals and plants, and that it is only when the species approach near to each other, in their organization and habits, that any offspring are produced from their connexion. Mules are of extremely rare occurrence in a state of nature, and no examples are yet known of their having procreated in a wild state. But it has been proved, that hybrids are not universally steril, provided the parent stocks have a near affinity to each other, although the continuation of the mixed race, for several generations, appears hitherto to have been obtained only by crossing the hybrids with individuals of pure species, an experiment which by no means bears out the hypothesis that a true hybrid race could ever be permanently established.

Hence we may infer, that aversion to sexual intercourse is, in general, a good test of the distinctness of original stocks, or of *species*, and the procreation of hybrids is a proof of the very near affinity of species. Perhaps, hereafter, the number of generations for which hybrids may be continued, before the race dies out (for it seems usually to degenerate rapidly), may afford the zoologist and botanist an experimental test of the difference in the degree of affinity of allied species.

We may also remark, that if it could have been shown that a single permanent species had ever been produced by hybridity (of which there is no satisfactory proof), it might certainly have lent some countenance to the notions of the ancients respecting the gradual deterioration of created things, but none whatever to Lamarck's theory of their progressive perfectibility; for observations have hitherto shown that there is a tendency, in mule animals and plants, to degenerate in organization.

We have already remarked, that the theory of progressive development arose from an attempt to ingraft the doctrines of the transmutationists upon one of the most popular generalizations in geology. But modern geological researches have almost destroyed every appearance of that gradation in the successive groups of animate beings, which was supposed to indicate the slow progress of the organic world from the more simple to the more compound structure. In the more modern formations, we find clear indications that the highest orders of the terrestrial mammalia were fully represented during several successive epochs; but, in the monuments which we have hitherto examined of more remote eras, in which there are as yet discovered few fluviatile, and perhaps no lacustrine formations, and, therefore, scarcely any means of obtaining an insight into the zoology of the then existing continents, we have only as yet found one example of a mammiferous quadruped. The recent origin of man, and the absence of all signs of any rational being holding an analogous relation to former states of the animate world, affords one, and the only reasonable argument, in support of the hypothesis of a progressive scheme, but none whatever in favour of the fancied evolution of one species out of another.

When the celebrated anatomist, Camper, first attempted to estimate the degrees of sagacity of different animals, and of the races of man, by the measurement of the facial angle, some speculators were bold enough to affirm, that certain simiæ differed as little from the more savage races of men, as do these from the human race in general; and that a scale might be traced from 'apes with foreheads villanous low,'[20] to the African variety of the human species, and from that to the European. The facial angle was measured by drawing a line from the prominent centre of the forehead to the most advanced part of the lower jaw-bone, and observing the angle which it made with the horizontal line; and it was affirmed, that there was a regular series from birds to the mammalia.

The gradation from the dog to the monkey was said to be perfect, and from that again to man. One of the ape tribe has a facial angle of 42°, and another, which approximated nearest to man in figure, an angle of 50°. To this succeeds (next but by a long interval[21]) the head of the African negro, which, as well as that of the Kalmuc, forms an angle of 70°, while that of the European contains 80°. The Roman painters preferred the angle of 95°, and the character of beauty and sublimity, so striking in some works of Grecian sculpture, as in the head of Apollo, and in the Medusa of Sisocles, is given by an angle which amounts to 100°.

A great number of valuable facts and curious analogies in comparative anatomy, were brought to light during the investigations which were made by Camper, John Hunter, and others, to illustrate this scale of organization; and their facts and generalizations must not be confounded with the fanciful systems which White and others deduced from them.

That there is some connexion between an elevated and capacious forehead in certain races of men, and a large development of the intellectual faculties, seems highly probable; and that a low facial angle is frequently accompanied with inferiority of mental powers, is certain; but the attempt to trace a graduated scale of intelligence through the different species of animals accompanying the modifications of the form of the skull, is a mere visionary speculation. It has been found necessary to exaggerate the sagacity of the ape tribe at the expense of the dog, and strange contradictions have arisen in the conclusions deduced from the structure of the elephant, some anatomists being disposed to deny the quadruped the intelligence which he really possesses, because they found that the volume of his brain was small in comparison to that of the other mammalia, while others were inclined to magnify extravagantly the superiority of its intellect, because the vertical height of its skull is so great when compared to its horizontal length.

It would be irrelevant to our subject if we were to enter into a farther discussion on these topics, because, even if a graduated scale of organization and intelligence could have been established, it would prove nothing in favour of a tendency, in each species, to attain a higher state of perfection. We may refer the reader to the writings of Blumenbach, Prichard, Lawrence, and others, for convincing proofs that the varieties of form, colour, and organization of different races of men, are perfectly consistent with the generally received opinion, that all the individuals of the species have originated from a single pair; and while they exhibit in man as many diversities of a physiological nature, as appear in any other species, they confirm also the opinion of the slight deviation from a common standard of which a species is capable.

The power of existing and multiplying in every latitude, and in every variety of situation and climate, which has enabled the great human family to extend itself over the habitable globe, is partly, says Lawrence, the result of physical constitution, and partly of the mental prerogative of man. If he did not possess the most enduring and flexible corporeal frame, his arts would not enable him to be the inhabitant of all climates, and to brave the extremes of heat and cold, and the other destructive influences of local situation. Yet, notwithstanding this flexibility of bodily frame, we find no signs of indefinite departure from a common standard, and the intermarriages of individuals of the most remote varieties are not less fruitful than between those of the same tribe.

There is yet another department of anatomical discovery, to which we must not omit some allusion, because it has appeared to some persons to afford a distant analogy, at least, to that progressive development by which some of the inferior species may have been gradually perfected into those of more complex organization. Tieddemann found, and his discoveries have been most fully confirmed and elucidated by M. Serres, that the brain of the fœtus, in the highest class of vertebrated animals, assumes, in succession, the various forms which belong to fishes, reptiles, and birds, before it acquires those additions and modifications which are peculiar to the mammiferous tribe. So that in the passage from the embryo to the perfect mammifer, there is a typical representation, as it were, of all those transformations which the primitive species are supposed to have undergone, during a long series of generations, between the present period and the remotest geological era.

If you examine the brain of the mammalia, says M. Serres, at an early stage of uterine life, you perceive the cerebral hemispheres consolidated, as in fish, in two vesicles isolated one from the other; at a later period, you see them affect the configuration of the cerebral hemispheres of reptiles; still later again, they present you with the forms of those of birds; finally, they acquire, at the era of birth, and sometimes later, the permanent forms which the adult mammalia present.

The cerebral hemispheres, then, only arrive at the state which we observe in the higher animals by a series of successive metamorphoses. If we reduce the whole of these evolutions to four periods, we shall see that in the first are born the cerebral lobes of fishes, and this takes place homogeneously in all classes. The second period will give us the organization of reptiles; the third the brain of birds; and the fourth the complex hemispheres of mammalia.

If we could develop the different parts of the brain of the inferior classes, we should make in succession a reptile out of a fish, a bird out of a reptile, and a mammiferous quadruped out of a bird. If, on the contrary, we could starve this organ in the mammalia, we might reduce it successively to the condition of the brain of the three inferior classes.

Nature often presents us with this last phenomenon in monsters, but never exhibits the first. Among the various deformities which organized beings may experience, they never pass the limits of their own classes to put on the forms of the class above them. Never does a fish elevate itself so as to assume the form of the brain of a reptile; nor does the latter ever attain that of birds; nor the bird that of the mammifer. It may happen that a monster may have two heads, but the conformation of the brain always remains circumscribed narrowly within the limits of its class.

It will be observed, that these curious phenomena disclose, in a highly interesting manner, the unity of plan that runs through the organization of the whole series of vertebrated animals; but they lend no support whatever to the notion of a gradual transmutation of one species into another, least of all of the passage, in the course of many generations, from an animal of a more simple, to one of a more complex structure. On the contrary, were it not for the sterility imposed on monsters, as well as on hybrids in general, the argument to be derived from Tieddemann's discovery, like that deducible from experiments respecting hybridity, would be in favour of the successive *degeneracy*, rather than the perfectibility, in the course of ages, of certain classes of organic beings.

For the reasons, therefore, detailed in this and the two preceding chapters, we draw the following inferences, in regard to the reality of *species* in nature.

First, That there is a capacity in all species to accommodate themselves, to a certain extent, to a change of external circumstances, this extent varying greatly according to the species.

2dly. When the change of situation which they can endure is great, it is usually attended by some modifications of the form, colour, size, structure, or other particulars; but the mutations thus superinduced are governed by constant laws, and the capability of so varying forms part of the permanent specific character.

3dly. Some acquired peculiarities of form, structure, and instinct, are transmissible to the offspring; but these consist of such qualities and attributes only as are intimately related to the natural wants and propensities of the species.

4thly. The entire variation from the original type, which any given kind of change can produce, may usually be effected in a brief period of time, after which no farther deviation can be obtained by continuing to alter the circumstances, though ever so gradually, – indefinite divergence, either in the way of improvement or deterioration, being prevented, and the least possible excess beyond the defined limits being fatal to the existence of the individual.

5thly. The intermixture of distinct species is guarded against by the aversion of the individuals composing them to sexual union, or by the sterility of the mule offspring. It does not appear that true hybrid races have ever been perpetuated for several generations, even by the assistance of man; for the cases usually cited relate to the crossing of mules with individuals of pure species, and not to the intermixture of hybrid with hybrid.

6thly. From the above considerations, it appears that species have a real existence in nature, and that each was endowed, at the time of its creation, with the attributes and organization by which it is now distinguished.

Geographical Distribution
of Species

Next to determining the question whether species have a real existence, the consideration of the laws which regulate their geographical distribution is a subject of primary importance to the geologist. It is only by studying these laws with attention, by observing the position which groups of species occupy at present, and inquiring how these may be varied in the course of time by migrations, by changes in physical geography, and other causes, that we can hope to learn whether the duration of species be limited, or in what manner the state of the animate world is affected by the endless vicissitudes of the inanimate.

That different regions of the globe are inhabited by entirely distinct animals and plants is a fact which has been familiar to all naturalists since Buffon first pointed out the want of *specific* identity between the land quadrupeds of America and those of the Old World. The same phenomenon has, in later times, been forced, in a striking manner, upon our attention, by the examination of New Holland, where the indigenous species of animals and plants were found to be, almost without exception, distinct from those known in other parts of the world.

But the extent of this parcelling out of the globe amongst different *nations*, as they have been termed, of plants and animals, – the universality of a phenomenon so extraordinary and unexpected, may be considered as one of the most interesting facts clearly established by the advance of modern science.

Scarcely fourteen hundred species of plants appear to have been known and described by the Greeks, Romans, and Arabians. At present, more than three thousand species are enumerated, as natives of our own island. In other parts of the world there have been collected, perhaps, upwards of seventy thousand species. It was not to be supposed, therefore, that the ancients should have acquired any correct notions respecting what may be called the geography of plants, although the influence of climate on the character of the vegetation could hardly have escaped their observation.

Antecedently to investigation, there was no reason for presuming that the vegetable productions, growing wild in the eastern hemisphere, should be unlike those of the western, in the same latitude; nor that the plants of the Cape of Good Hope should be unlike those of the South of Europe; situations where the climate is little dissimilar. The contrary supposition would have seemed more probable, and we might have anticipated an almost perfect identity in the animals and plants which inhabit corresponding parallels of latitude. The discovery, therefore, that each separate region of the globe, both of the land and water, is occupied by distinct groups of species, and that most of the exceptions to this general rule may be referred to disseminating causes now in operation, is eminently calculated to excite curiosity, and to stimulate us to seek some hypothesis respecting the first introduction of species which may be reconcileable with such phenomena.

A comparison of the *plants* of different regions of the globe affords results more to be depended upon in the present state of our knowledge, than those relating to the animal kingdom, because the science of botany is more advanced, and probably comprehends a great proportion of the total number of the vegetable productions of the whole earth.

Humboldt, in several eloquent passages of his Personal Narrative, was among the first to promulgate philosophical views on this subject. Every hemisphere, says this traveller, produces plants of different species; and it is not by the diversity of climates that we can attempt to explain why equinoctial Africa has no lauriniæ, and the New World no heaths; why the calceolariæ are found only in the southern hemisphere; why the birds of the continent of India glow with colours less splendid than the birds of the hot parts of America; finally, why the tiger is peculiar to Asia, and the ornithorhynchus to New Holland.

'We can conceive, he adds, that a small number of the families of plants, for instance the musaceæ and the palms, cannot belong to very cold regions, on account of their internal structure and the importance of certain organs; but we cannot explain why no one of the family of melastomas vegetates north of the parallel of thirty degrees; or why no rose-tree belongs to the southern hemisphere. Analogy of climates is often found in the two continents without identity of productions.'[22]

The luminous essay of Decandolle on 'Botanical Geography' presents us with the fruits of his own researches and those of Humboldt, Brown, and other eminent botanists, so arranged, that the principal phenomena of the distribution of plants are exhibited in connexion with the causes to which they are chiefly referrible. 'It might not, perhaps, be difficult,' observes this writer, 'to find two points, in the United States and in Europe, or in equinoctial America and Africa, which present all the same circumstances: as for example, the same temperature, the same height above the sea, a similar soil, an equal dose of humidity, yet nearly all, *perhaps all*, the plants in these two similar localities shall be distinct. A certain degree of analogy, indeed, of aspect, and even of structure, might very possibly be discoverable between the plants of the two localities in question, but the *species* would in general be different. Circumstances, therefore, different from those which now determine the *stations*, have had an influence on the *habitations* of plants.'[23]

As we shall frequently have occasion to speak of the *stations* and *habitations* of plants in the technical sense in which the terms are used in the above passage, we may remind the geological reader that station indicates the peculiar nature of the locality where each species is accustomed to grow, and has reference to climate, soil, humidity, light, elevation above the sea, and other analogous circumstances; whereas by habitation is meant a general indication of the country where a plant grows wild. Thus the *station* of a plant may be a salt-marsh, in a temperate climate, a hill-side, the bed of the sea, or a stagnant pool. Its *habitation* may be Europe, North America, or New Holland between the tropics. The study of stations has been styled the topography, that of habitations the geography of botany. The terms thus defined, express each a distinct class of ideas, which have been often confounded together, and which are equally applicable in zoology.

In further illustration of the principle above alluded to, that difference of longitude, independently of any influence of temperature, is accompanied by a great, and sometimes a complete diversity in the species of plants, Decandolle observes, that out of two thousand eight hundred and ninety-one species of phanerogamic plants described by Pursh, in the United States, there are only three hundred and eighty-five which are found in northern or temperate Europe. MM. Humboldt and Bonpland, in all their travels through equinoctial America, found only twenty-four species (these being all cyperacea and graminea) common to America and any part of the Old World. On comparing New Holland with Europe, Mr. Brown ascertained that out of four thousand one hundred species, discovered in Australia, there were only one hundred and sixty-six common to Europe, and of this small number there were some which may have been transported thither by man. Most of the others belong to those classes which are provided with the most ample means of dispersion to vast distances.

But it is still more remarkable, that in the more widely separated parts of the ancient continent, notwithstanding the existence of an uninterrupted land communication, the diversity in the specific character of the respective vegetations is almost as striking. Thus there is found one assemblage of species in China, another in the countries bordering the Black Sea and the Caspian, a third in those surrounding the Mediterranean, a fourth in the great platforms of Siberia and Tartary, and so forth.

The distinctness of the groups of indigenous plants, in the same parallel of latitude, is greatest where continents are disjoined by a wide expanse of ocean. In the northern hemisphere, near the Pole, where the extremities of Europe, Asia and America unite or approach near to one another, a considerable number of the same species of plants are found, common to the three continents. But it has been remarked, that these plants, which are thus so widely diffused in the Arctic regions, are also found in the chain of the Aleutian islands, which stretch almost across from America to Asia, and which may probably have served as the channel of communication for the partial blending of the Floras of the adjoining regions. It has, indeed, been found to be a general rule, that plants found at two points very remote from each other, occur also in places intermediate.

In islands very distant from continents, the total number of plants is comparatively small; but a large proportion of the species are such as occur nowhere else. In so far as the Flora of such islands is not peculiar to them, it contains, in general, species common to the nearest main lands.

The islands of the great southern ocean exemplify these rules; the easternmost containing more American, and the western more Indian plants. Madeira and Teneriffe contain many species, and even entire genera, peculiar to them; but they have also plants in common with Portugal, Spain, the Azores, and the north-west coast of Africa.

In the Canaries, out of five hundred and thirty-three species of phanerogamous plants, it is said that three hundred and ten are peculiar to these isles, and the rest identical with those of the African continent; but in the Flora of St. Helena, which is so far distant, even from the western shores of Africa, there have been found, out of sixty-one native species, only *two or three* which are to be found in any other part of the globe.

Decandolle has enumerated twenty great botanical provinces inhabited by indigenous or aboriginal plants; and although many of these contain a variety of species which are common to several others, and sometimes to places very remote, yet the lines of demarcation are, upon the whole, astonishingly well defined. Nor is it likely that the bearing of the evidence on which these general views are founded will ever be materially affected, since they are already confirmed by the examination of seventy or eighty thousand species of plants.

The entire change of opinion which the contemplation of these phenomena has brought about is worthy of remark. The first travellers were persuaded that they should find, in distant regions, the plants of their own country, and they took a pleasure in giving them the same names. It was some time before this illusion was dissipated; but so fully sensible did botanists at last become of the extreme smallness of the number of phænogamous plants common to different continents, that the ancient Floras fell into disrepute. All grew diffident of the pretended identifications, and we now find that every naturalist is inclined to examine each supposed exception with scrupulous severity. If they admit the fact, they begin to speculate on the mode whereby the seeds may have been transported from one country into the other, or inquire on which of two continents the plant was indigenous, assuming that a species, like an individual, cannot have two birth-places.

The marine vegetation is less known, but we learn from Lamouroux, that it is divisible into different systems, apparently as distinct as those on the land, notwithstanding that the uniformity of temperature is so much greater in the ocean. For on that ground we might have expected the phenomenon of partial distribution to have been far less striking, since climate is, in general, so influential a cause in checking the dispersion of species from one zone to another.

The number of hydrophytes, as they are termed, is very considerable, and their stations are found to be infinitely more varied than could have been anticipated; for while some plants are covered and uncovered daily by the tide, others live in abysses of the ocean, at the extraordinary depth of one thousand feet; and although in such situations there must reign darkness more profound than night, at least to our organs, many of these vegetables are highly coloured. From the analogy of terrestrial plants we should have inferred that the colouring of the algæ was derived from the influence of the solar rays; yet we are compelled to doubt when we reflect how feeble must be the rays which penetrate to these great depths.

The subaqueous vegetation of the Mediterranean is, upon the whole, distinct from that of the Atlantic on the west, and that part of the Arabian gulf which is immediately contiguous on the south. Other botanical provinces are found in the West-Indian seas, including the gulf of Mexico; in the ocean which washes the shores of South America, in the Indian ocean and its gulfs, in the seas of Australia, and in the Atlantic basin, from the 40° of north lat. to the pole. There are very few species common to the coast of Europe and the United States of North America, and none common to the Straits of Magellan and the shores of Van Diemen's Land.

It must not be overlooked, that the distinctness alluded to between the vegetation of these several countries relates strictly to *species* and not to forms. In regard to the numerical preponderance of certain forms, and many peculiarities of internal structure, there is a marked agreement in the vegetable productions of districts placed in corresponding latitudes, and under similar physical circumstances, however remote their position. Thus there are innumerable points of analogy between the vegetation of the Brazils, equinoctial Africa, and India; and there are also points of difference wherein the plants of these regions are distinguishable from all extra-tropical groups. But there are very few species common to the three continents. The same may be said, if we compare the plants of the Straits of Magellan with those of Van Diemen's Land, or the vegetation of the United States with that of the middle of Europe: the species are distinct, but the forms are in a great degree analogous.

Let us now consider what means of diffusion, independently of the agency of man, are possessed by plants, whereby, in the course of ages, they may be enabled to stray from one of the botanical provinces above mentioned to another, and to establish new colonies at a great distance from their birth-place.

The principal of the inanimate agents, provided by nature for scattering the seeds of plants over the globe, are the movements of the atmosphere and of the ocean, and the constant flow of water from the mountains to the sea. To begin with the winds: a great number of seeds are furnished with downy and feathery appendages, enabling them, when ripe, to float in the air, and to be wafted easily to great distances by the most gentle breeze. Other plants are fitted for dispersion by means of an attached wing, as in the case of the fir-tree, so that they are caught up by the wind as they fall from the cone, and are carried to a distance. Amongst the comparatively small number of plants known to Linnæus, no less than one hundred and thirty-eight genera are enumerated as having winged seeds.

As winds often prevail for days, weeks, or even months together, in the same direction, these means of transportation may sometimes be without limits; and even the heavier grains may be borne through considerable spaces, in a very short time, during ordinary tempests; for strong gales, which can sweep along grains of sand, often move at the rate of about forty miles an hour, and if the storm be very violent, at the rate of fifty-six miles. The hurricanes of tropical regions, which root up trees and throw down buildings, sweep along at the rate of ninety miles an hour, so that, for however short a time they prevail, they may carry even the heavier fruits and seeds over friths and seas of considerable width, and, doubtless, are often the means of introducing into islands the vegetation of adjoining continents. Whirlwinds are also instrumental in bearing along heavy vegetable substances to considerable distances. Slight ones may frequently be observed in our fields, in summer, carrying up haycocks into the air, and then letting fall small tufts of hay far and wide over the country; but they are sometimes so powerful as to dry up lakes and ponds, and to break off the boughs of trees, and carry them up in a whirling column of air.

Franklin tells us, in one of his letters, that he saw, in Maryland, a whirlwind which began by taking up the dust which lay in the road, in the form of a sugar-loaf with the pointed end downwards, and soon after grew to the height of forty or fifty feet, being twenty or thirty in diameter. It advanced in a direction contrary to the wind, and although the rotatory motion of the column was surprisingly rapid, its onward progress was sufficiently slow to allow a man to keep pace with it on foot. Franklin followed it on horseback, accompanied by his son, for three-quarters of a mile, and saw it enter a wood, where it twisted and turned round large trees with surprising force. These were carried up in a spiral line, and were seen flying in the air, together with boughs and innumerable leaves, which, from their height, appeared reduced to the apparent size of flies. As this cause operates at different intervals of time throughout a great portion of the earth's surface, it may be the means of bearing not only plants but insects, land-testacea and their eggs, with many other species of animals, to points which they could never otherwise have reached, and from which they may then begin to propagate themselves again as from a new centre.

The seeds of some aquatic fresh-water plants are of the form of shells, or small canoes, and on this account they swim on the surface, and are carried along by the wind and stream. Others are furnished with fibres, which serve the purpose of masts and sails, so that they are impelled along by the winds, even where there is no current. They cannot take root until the water stagnates, or till they reach some sheltered corner, where they may live without being exposed to too much agitation from winds and currents. The above-mentioned contrivances may enable aquatic plants to diffuse themselves gradually to considerable distances wherever there is a great chain of lakes, or a river which traverses a large continent.

It has been found that a great numerical proportion of the exceptions to the limitation of species to certain quarters of the globe, occur in the various tribes of cryptogamic plants. Linnæus observed, that as the germs of plants of this class, such as mosses, fungi, and lichens, consist of an impalpable powder, the particles of which are scarcely visible to the naked eye, there is no difficulty to account for their being dispersed throughout the atmosphere, and carried to every point of the globe, where there is a station fitted for them. Lichens in particular ascend to great elevations, sometimes growing two thousand feet above the line of perpetual snow, at the utmost limits of vegetation, and where the mean temperature is nearly at the freezing point. This elevated position must contribute greatly to facilitate the dispersion of those buoyant particles of which their fructification consists.

Some have inferred, from the springing up of mushrooms whenever particular soils and decomposed organic matter are mixed together, that the production of fungi is accidental, and not analogous to that of perfect plants. But Fries, whose authority on these questions is entitled to the highest respect, has shown the fallacy of this argument in favour of the old doctrine of equivocal generation. 'The sporules of fungi,' says this naturalist, 'are so infinite, that in a single individual of *Reticularia maxima*, I have counted above ten millions, and so subtile as to be scarcely visible, often resembling thin smoke; so light that they may be raised perhaps by evaporation into the atmosphere, and dispersed in so many ways by the attraction of the sun, by insects, wind, elasticity, adhesion, &c., that it is difficult to conceive a place from which they may be excluded.'[24]

In turning our attention, in the next place, to the instrumentality of the aqueous agents of dispersion, we cannot do better than cite the words of one of our ablest botanical writers. 'The mountain-stream or torrent,' observes Keith, 'washes down to the valley the seeds which may accidentally fall into it, or which it may happen to sweep from its banks when it suddenly overflows them. The broad and majestic river, winding along the extensive plain, and traversing the continents of the world, conveys to the distance of many hundreds of miles the seeds that may have vegetated at its source. Thus the southern shores of the Baltic are visited by seeds which grew in the interior of Germany; and the western shores of the Atlantic by seeds that have been generated in the interior of America.'[25] Fruits, moreover, indigenous to America and the West Indies, such as that of the *Mimosa scandens*, the cashew-nut, and others, have been known to be drifted across the Atlantic by the Gulf-stream, on the western coasts of Europe, in such a state that they might have vegetated had the climate and soil been favourable. Among these the *Guilandina Bonduc*, a leguminous plant, is particularly mentioned, as having been raised from a seed found on the west coast of Ireland. Sir Hans Sloane informs us that the *lenticula marina*, or sargasso, a bean which is frequently cast ashore on the Orkney isles, and coast of Ireland, grows on the rocks about Jamaica, where the surface of the sea is sometimes strewed with it, and from whence it is known to be carried by the winds and currents towards the coast of Florida.

The absence of liquid matter in the composition of seeds renders them comparatively insensible to heat and cold, so that they may be carried, without detriment, through climates where the plants themselves would instantly perish. Such is their power of resisting the effects of heat, that Spallanzani mentions some seeds that germinated after having been boiled in water. When, therefore, a strong gale, after blowing violently off the land for a time, dies away, and the seeds alight upon the surface of the waters, or wherever the ocean, by eating away the sea-cliffs, throws down into its waves plants which would never otherwise approach the shores, the tides and currents become active instruments in assisting the dissemination of almost all classes of the vegetable kingdom.

In a collection of six hundred plants from the neighbourhood of the river Zaire, in Africa, Mr. Brown found that thirteen species were also met with on the opposite shores of Guiana and Brazil. He remarked, that most of these plants were only found on the lower parts of the river Zaire, and were chiefly such as produced seeds capable of retaining their vitality a long time in the currents of the ocean.

Islands, moreover, and even the smallest rocks, play an important part in aiding such migrations, for when seeds alight upon them from the atmosphere, or are thrown up by the surf, they often vegetate and supply the winds and waves with a repetition of new and uninjured crops of fruits and seeds, which may afterwards pursue their course through the atmosphere, or along the surface of the sea, in the same direction. The number of plants found at any given time on an islet affords no test whatever of the extent to which it may have co-operated towards this end, since a variety of species may first thrive there and then perish, and be followed by other chance-comers like themselves.

Currents and winds, in the arctic regions, drift along icebergs covered with an alluvial soil on which herbs and pine saplings are seen growing, which often continue to vegetate on some distant shore where the ice-island is stranded.

With respect to marine vegetation, the seeds being in their native element, may remain immersed in water without injury for indefinite periods, so that there is no difficulty in conceiving the diffusion of species wherever uncongenial climates, contrary currents, and other causes, do not interfere. All are familiar with the sight of the floating sea-weed

> Flung from the rock on ocean's foam to sail,
> Where'er the surge may sweep, the tempest's breath prevail.[26]

Remarkable accumulations of drift weed occur on each side of the equator in the Atlantic, Pacific, and Indian Oceans. Columbus and other navigators who first encountered these banks of algæ in the Northern Atlantic, compared them to vast inundated meadows, and state that they retarded the progress of their vessels. The most extensive bank is a little west of the meridian of Fayal, one of the Azores, between latitude 25° and 36°; violent north winds sometimes prevail in this space, and drive the sea-weed to low latitudes, as far as the 24th or even the 20th degree.

The hollow pod-like receptacles in which the seeds of many algæ are lodged, and the filaments attached to the seed-vessels of others, seem intended to give buoyancy, and we may observe that these hydrophytes are in general *proliferous*, so that the smallest fragment of a branch can be developed into a perfect plant. The seeds, moreover, of the greater number of species are enveloped with a mucous matter like that which surrounds the eggs of some fish, and which not only protects them from injury, but serves to attach them to floating bodies or to rocks.

But we have as yet considered part only of the fertile resources of nature for conveying seeds to a distance from their place of growth. The various tribes of animals are busily engaged in furthering an object whence they derive such important advantages. Sometimes an express provision is found in the structure of seeds to enable them to adhere firmly by prickles, hooks, and hairs, to the coats of animals, or feathers of the winged tribe, to which they remain attached for weeks, or even months, and are borne along into every region whither birds or quadrupeds may migrate. Linnæus enumerates fifty genera of plants, and the number now known to botanists is much greater, which are armed with hooks by which, when ripe, they adhere to the coats of animals. Most of these vegetables, he remarks, require a soil enriched with dung. Few have failed to mark the locks of wool hanging on the thorn-bushes, wherever the sheep pass, and it is probable that the wolf or lion never give chace to herbivorous animals without being unconsciously subservient to this part of the vegetable economy.

A deer has strayed from the herd, when browsing on some rich pasture, when he is suddenly alarmed by the approach of his foe. He instantly plunges through many a thicket, and swims through many a river and lake. The seeds of the herbs and shrubs adhere to his smoking flanks, and are washed off again by the streams. The thorny spray is torn off and fixes itself in his hairy coat, until brushed off again in other thickets and copses. Even on the spot where the victim is devoured, many of the seeds which he had swallowed immediately before the pursuit may be left on the ground uninjured.

The passage, indeed, of undigested seeds through the stomachs of animals is one of the most efficient causes of the dissemination of plants, and is of all others, perhaps, the most likely to be overlooked. Few are ignorant that a portion of the oats eaten by a horse preserve their germinating faculty in the dung. The fact of their being still nutritious is not lost on the sagacious rook. To many, says Linnæus, it seems extraordinary, and something of a prodigy, that when a field is well tilled and sown with the best wheat, it frequently produces darnel or the wild oat, especially if it be manured with new dung: they do not consider that the fertility of the smaller seeds is not destroyed in the ventricles of animals.

Some of the order of the Passeres, says Ekmarck, devour the seeds of plants in great quantities, which they eject again in very distant places, without destroying its faculty of vegetation; thus a flight of larks will fill the cleanest field with a great quantity of various kinds of plants, as the melilot trefoil (*Medicago lupulina*), and others whose seeds are so heavy that the wind is not able to scatter them to any distance. In like manner, the blackbird and missel-thrush, when they devour berries in too great quantities, are known to consign them to the earth undigested in their excrement.

Pulpy fruits serve quadrupeds and birds as food, while their seeds, often hard and indigestible, pass uninjured through the intestines, and are deposited far from their original place of growth in a condition peculiarly fit for vegetation. So well are our farmers, in some parts of England, aware of this fact, that when they desire to raise a quick-set hedge in the shortest possible time, they feed turkeys with the haws of the common white-thorn (*Cratægus oxyacantha*), and then sow the stones which are ejected in their excrement, whereby they gain an entire year in the growth of the plant. Birds when they pluck cherries, sloes, and haws, fly away with them to some convenient place, and when they have devoured the fruit drop the stone into the ground. Captain Cook, in his account of the volcanic island of Tanna, one of the New Hebrides, which he visited in his second voyage, makes the following interesting observation. 'Mr. Forster, in his botanical excursion this day, shot a pigeon, in the craw of which was a wild nutmeg. He took some pains to find the tree on this island, but his endeavours were without success.'[27] It is easy, therefore, to perceive, that birds in their migrations to great distances, and even across seas, may transport seeds to new isles and continents.

The sudden deaths to which great numbers of frugivorous birds are annually exposed, must not be omitted as auxiliary to the transportation of seeds to new habitations. When the sea retires from the shore, and leaves fruits and seeds on the beach, or in the mud of estuaries, it might, by the returning tide, wash them away again, or destroy them by long immersion; but when they are gathered by land birds which frequent the sea-side, or by waders and water-fowl, they are often borne inland, and if the bird to whose crop they have been consigned is killed, they may be left to grow up far from the sea. Let such an accident happen but once in a century, or a thousand years, it will be sufficient to spread many of the plants from one continent to another; for, in estimating the activity of these causes, we must not consider whether they act slowly in relation to the period of our observation, but in reference to the duration of species in general.

Let us trace the operation of this cause in connexion with others. A tempestuous wind bears the seeds of a plant many miles through the air, and then delivers them to the ocean; the oceanic current drifts them to a distant continent; by the fall of the tide they become the food of numerous birds, and one of these is seized by a hawk or eagle, which, soaring across hill and dale to a place of retreat, leaves, after devouring its prey, the unpalatable seeds to spring up and flourish in a new soil.

The machinery before adverted to is so capable of disseminating seeds over almost unbounded spaces, that were we more intimately acquainted with the economy of nature, we might probably explain all the instances which occur of the aberration of plants to great distances from their native countries. The real difficulty which must present itself to every one who contemplates the present geographical distribution of species, is the small number of exceptions to the rule of the non-intermixture of different groups of plants. Why have they not, supposing them to have been ever so distinct originally, become more blended and confounded together in the lapse of ages?

But in addition to all the agents already enumerated as instrumental in diffusing plants over the globe, we have still to consider man – one of the most important of all. He transports with him, into every region, the vegetables which he cultivates for his wants, and is the involuntary means of spreading a still greater number which are useless to him, or even noxious. 'When the introduction of cultivated plants is of recent date, there is no difficulty in tracing their origin; but when it is of high antiquity, we are often ignorant of the true country of the plants on which we feed. No one contests the American origin of the maize or the potato, nor the origin, in the old world, of the coffee-tree and of wheat. But there are certain objects of culture, of very ancient date, between the tropics, such, for example, as the banana, of which the origin cannot be verified. Armies, in modern times, have been known to carry, in all directions, grain and cultivated vegetables from one extremity of Europe to the other, and thus have shown us how, in more ancient times, the conquests of Alexander, the distant expeditions of the Romans, and afterwards the Crusades, may have transported many plants from one part of the world to the other.'[28]

But besides the plants used in agriculture, the number which have been naturalized by accident, or which man has spread unintentionally, is considerable. One of our old authors, Josselyn, gives a catalogue of such plants as had, in his time, sprung up in the colony since the English planted and kept cattle in New England. They were two and twenty in number. The common nettle was the first which the settlers noticed, and the plantain was called by the Indians, 'English man's foot,' as if it sprung from their footsteps.

'We have introduced everywhere,' observes Decandolle, 'some weeds which grow among our various kinds of wheat, and which have been received, perhaps, originally from Asia with them. Thus, together with the Barbary wheat, the inhabitants of the south of Europe have sown, for many ages, the plants of Algiers and Tunis. With the wools and cottons of the East, or of Barbary, there are often brought into France, the grains of exotic plants, some of which naturalize themselves. Of this I will cite a striking example. There is at the gate of Montpelier, a meadow set apart for drying foreign wool *after it has been washed*. There hardly passes a year without some foreign plants being found naturalized in this drying ground. I have gathered there *Centaurea parviflora*, *Psoralea palæstina*, and *Hypericum crispum*.'[29] This fact is not only illustrative of the aid which man lends inadvertently to the propagation of plants, but it also demonstrates the multiplicity of seeds which are borne about in the woolly and hairy coats of wild animals.

The same botanist mentions instances of plants naturalized in sea-ports by the ballast of ships, and several examples of others which have spread through Europe from botanical gardens, so as to have become more common than many indigenous species.

It is scarcely a century, says Linnæus, since the Canadian erigiron, or flea-bane, was brought from America to the botanical garden at Paris, and already the seeds have been carried by the winds, so that it is diffused over France, the British islands, Italy, Sicily, Holland, and Germany. Several others are mentioned by the Swedish naturalist, as having been dispersed by similar means. The common thorn-apple, *Datura stramonium*, observes Willdenow, now grows as a noxious weed throughout all Europe, with the exception of Sweden, Lapland, and Russia. It came from the East Indies and Abyssinia to us, and was so universally spread by certain quacks who used its seed as an emetic.

In hot and ill-cultivated countries, such naturalizations take place more easily. Thus the *Chenopodium ambrosioides*, sown by Mr. Burchell on a point of St. Helena, multiplied so in four years as to become one of the commonest weeds in the island.

The most remarkable proof, says Decandolle, of the extent to which man is unconsciously the instrument of dispersing and naturalizing species, is found in the fact, that in New Holland, America, and the Cape of Good Hope, the aboriginal European species exceed in number all the others which have come from any distant regions, so that, in this instance, the influence of man has surpassed that of all the other causes which tend to disseminate plants to remote districts.

Although we are but slightly acquainted, as yet, with the extent of our instrumentality in naturalizing species, yet the facts ascertained afford no small reason to suspect that the number which we introduce unintentionally, exceeds all those transported by design. Nor is it unnatural to suppose that the functions, which the inferior beings extirpated by man once discharged in the economy of nature, should devolve upon the human race. If we drive many birds of passage from different countries, we are probably required to fulfil their office of carrying seeds, eggs of fish, insects, molluscs, and other creatures, to distant regions; if we destroy quadrupeds, we must replace them, not merely as consumers of the animal and vegetable substances which they devoured, but as disseminators of plants, and of the inferior classes of the animal kingdom. We do not mean to insinuate that the same changes which man brings about, would have taken place by means of the agency of other species, but merely that he supersedes a certain number of agents, and so far as he disperses plants unintentionally, or against his will, his intervention is strictly analogous to that of the species so extirpated.

We may observe, moreover, that if, at former periods, the animals inhabiting any given district have been partially altered by the extinction of some species, and the introduction of others, whether by new creations or by immigration, a change must have taken place in regard to the particular plants conveyed about with them to foreign countries. As for example, when one set of migratory birds is substituted for another, the countries from and to which seeds are transported are immediately changed. Vicissitudes, therefore, analogous to those which man has occasioned, may have previously attended the springing up of new relations between species in the vegetable and animal worlds.

It may also be remarked, that if man is the most active agent in enlarging, so also is he in circumscribing the geographical boundaries of particular plants. He promotes the migration of some, he retards that of other species, so that while in many respects he appears to be exerting his power to blend and confound the various provinces of indigenous species, he is, in other ways, instrumental in obstructing the fusion into one group of the inhabitants of contiguous provinces.

Thus, for example, when two botanical regions exist in the same great continent, such as *the European region*, comprehending the central parts of Europe and those surrounding the Mediterranean, and *the Oriental region*, as it has been termed, embracing the countries adjoining the Black Sea and the Caspian, the interposition between these of thousands of square miles of cultivated lands, opposes a new and powerful barrier against the mutual interchange of indigenous plants. Botanists are well aware that garden plants naturalize and diffuse themselves with great facility in comparatively unreclaimed countries, but spread themselves slowly and with difficulty in districts highly cultivated. There are many obvious causes for this difference; by drainage and culture the natural variety of stations is diminished, and those stray individuals by which the passage of a species from one fit station to another is effected, are no sooner detected by the agriculturist, than they are uprooted as weeds. The larger shrubs and trees, in particular, can scarcely ever escape observation, when they have attained a certain size, and will rarely fail to be cut down if unprofitable.

The same observations are applicable to the interchange of the insects, birds, and quadrupeds of two regions situated like those above alluded to. No beasts of prey are permitted to make their way across the intervening arable tracts. Many birds, and hundreds of insects, which would have found some palatable food amongst the various herbs and trees of the primeval wilderness, are unable to subsist on the olive, the vine, the wheat, and a few trees and grasses favoured by man. In addition, therefore, to his direct intervention, man, in this case, operates indirectly to impede the dissemination of plants, by intercepting the migrations of animals, many of which would otherwise have been active in transporting seeds from one province to another.

Whether in the vegetable kingdom the influence of man will tend, after a considerable lapse of ages, to render the geographical range of *species in general* more extended, as Decandolle seems to anticipate, or whether the compensating agency above alluded to will not counterbalance the exceptions caused by our naturalizations, admits at least of some doubt. In the attempt to form an estimate on this subject, we must be careful not to underrate, or almost overlook, as some appear to have done, the influence of man in checking the diffusion of plants, and restricting their distribution to narrower limits.

[*The analysis of geographical distribution is continued in Chapter 6, which discusses mammals, birds and reptiles. Chapter 7 traces the diffusion of fish, shells, corals, sponges and other aquatic animals, and ends with a brief account of the spread of man across the face of the earth, 'in a manner singularly analogous to that in which many plants and animals are diffused'. Similarly, Lyell concludes that the role of human agency in extending the range and population of other species is 'strictly analogous to that of the inferior animals'.*]

CHAPTER 8

Changes in the Animate World, which Tend to the Extinction of Species

It would be superfluous to examine the various attempts which were made to explain the phenomena of the distribution of species alluded to in the preceding chapters, in the infancy of the sciences of botany, zoology, and physical geography. The theories or rather conjectures then indulged, now stand refuted by a simple statement of facts; and if Linnæus were living, he would be the first to renounce the notions which he promulgated. For he imagined the habitable world to have been for a certain time limited to one small tract, the only portion of the earth's surface that was as yet laid bare by the subsidence of the primæval ocean. In this fertile spot he supposed the originals of all the species of plants which exist on this globe to have been congregated, together with the first ancestors of all animals and of the human race. 'In this commodious habitat lived all animals, and all plants were produced in the greatest luxuriance.'[30] In order to accommodate the various habitudes of so many creatures, and to provide a diversity of climate suited to their several natures, the tract in which the creation took place was supposed to have been situated in some warm region of the earth, but to have contained a lofty mountain range, on the heights and in the declivities of which were to be found all temperatures and every clime, from the torrid to the frozen zone.

That there never was a universal ocean since the planet was inhabited, or rather since the oldest groups of strata yet known to contain organic remains were formed, is proved by the presence of terrestrial plants in all the older formations; and if this conclusion was not established, yet no geologist could deny that since the first small portion of the earth was laid dry, there have been many entire changes in the species of plants and animals inhabiting the land.

But without dwelling on the above and other refuted theories, let us inquire whether we can substitute some hypothesis as simple as that of Linnæus, to which the phenomena now ascertained in regard to the distribution both of aquatic and terrestrial species may be referred. The following may, perhaps, be reconcileable with known facts: – Each species may have had its origin in a single pair, or individual, where an individual was sufficient, and species may have been created in succession at such times and in such places as to enable them to multiply and endure for an appointed period, and occupy an appointed space on the globe.

In order to explain this theory, let us suppose every living thing to be destroyed in the western hemisphere, both on the land and in the ocean, and permission to be given to man to people this great desert, by transporting into it animals and plants from the eastern hemisphere, a strict prohibition being enforced against introducing two original stocks of the same species.

Now the result we conceive of such a mode of colonizing would correspond exactly, so far as regards the grouping of animals and plants, with that now observed throughout the globe. It would be necessary for naturalists, before they imported species into particular localities, to study attentively the climate and other physical conditions of each spot. It would be no less requisite to introduce the different species in succession, so that each plant and animal might have time and opportunity to multiply before the species destined to prey upon it was admitted. Many herbs and shrubs, for example, must spread far and wide before the sheep, the deer, and the goat could be allowed to enter, lest they should devour and annihilate the original stocks of many plants, and then perish themselves for want of food. The above-mentioned herbivorous animals in their turn must be permitted to make considerable progress before the entrance of the first pair of wolves or lions. Insects must be allowed to swarm before the swallow could be permitted to skim through the air and feast on thousands at one repast.

It is evident that, however equally in this case our original stocks were distributed over the whole surface of land and water, there would nevertheless arise distinct botanical and zoological provinces, for there are a great many natural barriers which oppose common obstacles to the advance of a variety of species. Thus, for example, almost all the animals and plants naturalized by us towards the extremity of South America, would be unable to spread beyond a certain limit, towards the east, west, and south, because they would be stopped by the ocean, and a few of them only would succeed in reaching the cooler latitudes of the northern hemisphere, because they would be incapable of bearing the heat of the tropics, through which they must pass. In the course of ages, undoubtedly, exceptions would arise, and some species might become common to the temperate and polar regions, or both sides of the equator; for we have before shown that the powers of diffusion conferred on some classes are very great. But we should confidently predict that these exceptions would never become so numerous as to invalidate the general rule.

Some of the plants and animals transplanted by us to the coast of Chili or Peru would never be able to cross the Andes, so as to reach the Eastern plains; nor, for a similar reason, would those first established in the Pampas, or the valleys of the Amazon and the Orinoco, ever arrive at the shores of the Pacific.

In the ocean an analogous state of things would prevail; for there, also, climate would exert a great influence in limiting the range of species, and the land would stop the migrations of aquatic tribes as effectually as the sea arrests the dispersion of the terrestrial. As certain birds, insects, and the seeds of plants, can never cross the direction of prevailing winds, so currents form natural barriers to the dissemination of many oceanic races. A line of shoals may be as impassable to pelagian species, as are the Alps and the Andes to plants and animals peculiar to plains, while deep abysses may prove insuperable obstacles to the migrations of the inhabitants of shallow waters.

It is worthy of observation, that one effect of the introduction of single pairs of each species must be the confined range of certain groups in spots which, like small islands, or solitary inland lakes, have few means of interchanging their inhabitants with adjoining regions. Now this congregating, in a small space, of many peculiar species, would give an appearance of *centres* or *foci* of creation, as they have been termed, as if there were favourite points where the creative energy has been in greater action than in others, and where the numbers of peculiar organic beings have consequently become more considerable.

We do not mean to call in question the soundness of the inferences of some botanists, as to the former existence of certain limited spots whence species of plants have been propagated, radiating, as it were, in all directions from a common centre. On the contrary, we conceive these phenomena to be the necessary consequences of the plan of nature before suggested, operating during the successive mutations of the surface, some of which the geologist can prove to have taken place subsequently to the period when many species now existing were created. In order to exemplify how this arrangement of plants may have been produced, let us imagine that, about three centuries before the discovery of St. Helena (itself of submarine volcanic origin), a multitude of new isles had been thrown up in the surrounding sea, and that these had each become clothed with plants emigrating from St. Helena, in the same manner as the wild plants of Campania have diffused themselves over Monte Nuovo. Whenever the first botanist investigated the new archipelago, he would, in all probability, find a different assemblage of plants in each of the isles of recent formation; but in St. Helena itself, he would meet with individuals of every species belonging to all parts of the archipelago, and some, in addition, peculiar to itself, viz., those which had not been able to obtain a passage into any one of the surrounding new-formed lands. In this case, it might be truly said that the original isle was the primitive focus, or centre, of a certain type of vegetation, whereas, in the surrounding isles, there would be a smaller number of species, yet all belonging to the same group.

But this peculiar distribution of plants would not warrant the conclusion that, in the space occupied by St. Helena, there had been a greater exertion of creative power than in the spaces of equal area occupied by the new adjacent lands, because, within the period in which St. Helena had acquired its peculiar vegetation, each of the spots supposed to be subsequently converted into land, may have been the birth-places of a great number of *marine* animals and plants, which may have had time to scatter themselves far and wide over the southern Atlantic.

Perhaps it may be objected to some part of the foregoing train of reasoning, that during the lapse of past ages, especially during many partial revolutions of the globe of comparatively modern date, different zoological and botanical provinces ought to have become more confounded and blended together – that the distribution of species approaches too nearly to what might have been expected, if animals and plants had been introduced into the globe when its physical geography had already assumed the features which it now wears; whereas we know that, in certain districts, considerable geographical changes have taken place since species identical with those now in being were created.

These, and many kindred topics, cannot be fully discussed until we have considered, not merely the general laws which may regulate the first introduction of species, but those which may limit their *duration* on the earth. Brocchi, whose untimely death in Egypt is deplored by all who have the progress of geology at heart, has remarked, when hazarding some interesting conjectures respecting 'the loss of species,' that a modern naturalist had no small assurance, who declared 'that individuals alone were capable of destruction, and that species were so perpetuated that nature could not annihilate them, so long as the planet lasted, or at least that nothing less than the shock of a comet, or some similar disaster, could put an end to their existence.'[31] The Italian geologist, on the contrary, had satisfied himself, that many species of testacea, which formerly inhabited the Mediterranean, had become extinct, although a great number of others, which had been the contemporaries of those lost races, still survived. He came to the opinion, that about half the species which peopled the waters when the Subapennine strata were deposited, had gone out of existence; and in this inference he does not appear to have been far wrong.

But instead of seeking a solution of this problem, like some other geologists of his time, in a violent and general catastrophe, Brocchi endeavoured to imagine some regular and constant law by which species might be made to disappear from the earth gradually and in succession. The death, he suggested, of a species might depend, like that of individuals, on certain peculiarities of constitution conferred upon them at their birth, and as the longevity of the one depends on a certain force of vitality, which, after a period, grows weaker and weaker, so the duration of the other may be governed by the quantity of prolific power bestowed upon the species, which, after a season, may decline in energy, so that the fecundity and multiplication of individuals may be gradually lessened from century to century, 'until that fatal term arrives, when the embryo, incapable of extending and developing itself, abandons, almost at the instant of its formation, the slender principle of life by which it was scarcely animated, – and so all dies with it.'[32]

Now we might coincide in opinion with the Italian naturalist, as to the gradual extinction of species one after another, by the operation of regular and constant causes, without admitting an inherent principle of deterioration in their physiological attributes. We might concede 'that many species are on the decline, and that the day is not far distant when they will cease to exist;'[33] yet deem it consistent with what we know of the nature of organic beings, to believe that the last individuals of each species retain their prolific powers in their full intensity.

Brocchi has himself speculated on the share which a change of climate may have had in rendering the Mediterranean unfit for the habitation of certain testacea, which still continued to thrive in the Indian ocean, and of others which were now only represented by analogous forms within the tropics. He must also have been aware that other extrinsic causes, such as the progress of human population, or the increase of some one of the inferior animals, might gradually lead to the extirpation of a particular species, although its fecundity might remain to the last unimpaired. If, therefore, amid the vicissitudes of the animate and inanimate world, there are known causes capable of bringing about the decline and extirpation of species, it became him thoroughly to investigate the full extent to which these might operate, before he speculated on any cause of so purely hypothetical a kind, as 'the diminution of the prolific virtue.'[34]

If it could have been shown that some wild plant had insensibly dwindled away and died out, as sometimes happens to cultivated varieties propagated by cuttings, even though climate, soil, and every other circumstance should continue identically the same – if any animal had perished while the physical condition of the earth, and the number and force of its foes, with every other extrinsic cause, remained unaltered, then might we have some ground for suspecting that the infirmities of age creep on as naturally on species as upon individuals. But in the absence of such observations, let us turn to another class of facts, and examine attentively the circumstances which determine the *stations* of particular animals and plants, and perhaps we shall discover, in the vicissitudes to which these stations are exposed, a cause fully adequate to explain the phenomena under consideration.

Stations comprehend all the circumstances, whether relating to the animate or inanimate world, which determine whether a given plant or animal can exist in a given locality, so that if it be shown that stations can become essentially modified by the influence of known causes, it will follow that species, as well as individuals, are mortal.

Every naturalist is familiar with the fact, that although in a particular country, such as Great Britain, there may be more than three thousand species of plants, ten thousand insects, and a great variety in each of the other classes, yet there will not be more than a hundred, perhaps not half that number, inhabiting any given locality. There may be no want of space in the supposed tract; it may be a large mountain, or an extensive moor, or a great river-plain, containing room enough for individuals of every species in our island; yet the spot will be occupied by a few to the exclusion of many, and these few are enabled, throughout long periods, to maintain their ground successfully against every intruder, notwithstanding the facilities which species enjoy, by virtue of their powers of diffusion, of invading adjacent territories.

The principal causes which enable a certain assemblage of plants thus to maintain their ground against all others depend, as is well known, on the relations between the physiological nature of each species, and the climate, exposure, soil, and other physical conditions of the locality. Some plants live only on rocks, others in meadows, a third class in marshes. Of the latter, some delight in a fresh-water morass, – others in salt marshes, where their roots may copiously absorb saline particles. Some prefer an alpine region in a warm latitude, where, during the heat of summer, they are constantly irrigated by the cool waters of melting snows. To others loose sand, so fatal to the generality of species, affords the most proper station. The *Carex arenaria* and the *Elymus arenarius* acquire their full vigour on a sandy dune, obtaining an ascendency over the very plants which in a stiff clay would immediately stifle them.

Where the soil of a district is of so peculiar a nature that it is extremely favourable to certain species, and agrees ill with every other, the former get exclusive possession of the ground, and, as in the case of heaths, live in societies. In like manner, the Bog moss (*Sphagnum palustre*) is fully developed in peaty swamps, and becomes, like the heath, in the language of botanists, a social plant. Such monopolies would be very frequent, if the powers of a great number of species were not equally balanced, and if animals did not interfere most actively to preserve an equilibrium in the vegetable kingdom.

'All the plants of a given country,' says Decandolle in his usual spirited style, 'are at war one with another. The first which establish themselves by chance in a particular spot, tend, by the mere occupancy of space, to exclude other species – the greater choke the smaller, the longest livers replace those which last for a shorter period, the more prolific gradually make themselves masters of the ground, which species multiplying more slowly would otherwise fill.'[35]

In this continual strife, it is not always the resources of the plant itself which enable it to maintain or extend its ground. Its success depends, in a great measure, on the number of its foes or allies among the animals and plants inhabiting the same region. Thus, for example, a herb which loves the shade may multiply, if some tree with spreading boughs and dense foliage flourish in the neighbourhood. Another, which, if unassisted, would be overpowered by the rank growth of some hardy competitor, is secure, because its leaves are unpalatable to cattle, which, on the other hand, annually crop down its antagonist, and rarely suffer it to ripen its seed.

Oftentimes we see some herb which has flowered in the midst of a thorny shrub, when all the other individuals of the same species, in the sunny fields around, are eaten down, and cannot bring their seed to maturity. In this case, the shrub has lent his armour of spines and prickles to protect the defenceless herb against the mouths of the cattle, and thus a few individuals which occupied, perhaps, the most unfavourable station in regard to exposure, soil, and other circumstances, may nevertheless, by the aid of an ally, become the principal source whereby the winds are supplied with seeds which perpetuate the species throughout the surrounding tract.

In the above example we see one plant shielding another from the attacks of animals; but instances are, perhaps, still more numerous, where some animal defends a plant against the enmity of some other subject of the vegetable kingdom.

Scarcely any beast, observes a Swedish naturalist, will touch the nettle, but fifty different kinds of insects are fed by it. Some of these seize upon the root, others upon the stem; some eat the leaves, others devour the seeds and flowers: but for this multitude of enemies, the nettle would annihilate a great number of plants. Linnæus tells us, in his Tour in Scania, that goats were turned into an island which abounded with the *Agrostis arundinacea*, where they perished by famine; but horses, which followed them, grew fat on the same plant. The goat, also, he says, thrives on the meadow-sweet and water hemlock, plants which are injurious to cattle.

Every plant, observes Wilcke, has its proper insect allotted to it to curb its luxuriancy, and to prevent it from multiplying to the exclusion of others. 'Thus grass in meadows sometimes flourishes so as to exclude all other plants: here the Phalæna graminis (*Bombyx gram.*), with her numerous progeny, find a well-spread table; they multiply in immense numbers, and the farmer for some years laments the failure of his hay crop; but the grass being consumed, the moths die with hunger, or remove to another place. Now the quantity of grass being greatly diminished, the other plants, which were before choked by it, spring up, and the ground becomes variegated with a multitude of different species of flowers. Had not nature given a commission to this minister for that purpose, the grass would destroy a number of species of vegetables, of which the equilibrium is now kept up.'[36]

In the above passage allusion is made to the ravages committed in 1740, and the two following years, in many provinces of Sweden, by a most destructive insect. The same moth is said never to touch the fox-tail grass, so that it may be classed as a most active ally and benefactor of that species, and as peculiarly instrumental in preserving it in its present abundance. A discovery of Rolander, cited in the treatise of Wilcke above-mentioned, affords a good illustration of the checks and counter-checks which nature has appointed to preserve the balance of power amongst species. 'The *Phalæna strobilella* has the fir cone assigned to it to deposit its eggs upon; the young caterpillars coming out of the shell consume the cone and superfluous seed; but lest the destruction should be too general, the *Ichneumon strobilellæ* lays its eggs in the caterpillar, inserting its long tail in the openings of the cone till it touches the included insect, for its body is too large to enter. Thus it fixes its minute egg upon the caterpillar, which being hatched destroys it.'[37]

Entomologists enumerate many parallel cases where insects, appropriated to certain plants, are kept down by other insects, and these again by parasites expressly appointed to prey on them. Few, perhaps, are in the habit of duly appreciating the extent to which insects are active in preserving the balance of species among plants, and thus regulating indirectly the relative numbers of many of the higher orders of terrestrial animals.

The peculiarity of their agency consists in their power of suddenly multiplying their numbers, to a degree which could only be accomplished in a considerable lapse of time in any of the larger animals, and then as instantaneously relapsing, without the intervention of any violent disturbing cause, into their former insignificance.

If for the sake of employing, on different but rare occasions, a power of many hundred horses, we were under the necessity of feeding all these animals at great cost in the intervals when their services were not required, we should greatly admire the invention of a machine, such as the steam-engine, which was capable, at any moment, of exerting the same degree of strength without any consumption of food during periods of inaction. The same kind of admiration is strongly excited when we contemplate the powers of insect life, in the creation of which nature has been so prodigal. A scanty number of minute individuals, only to be detected by careful research, are ready in a few days, weeks, or months, to give birth to myriads which may repress any degree of monopoly in another species, or remove nuisances, such as dead carcasses, which might taint the air. But no sooner has the destroying commission been executed, than the gigantic power becomes dormant – each of the mighty host soon reaches the term of its transient existence, and the season arrives when the whole species passes naturally into the egg, and thence into the larva and pupa state. In this defenceless condition it may be destroyed either by the elements, or by the augmentation of some of its numerous foes which may prey upon it in these stages of its transformation; or it often happens that, in the following year, the season proves unfavourable to the hatching of the eggs or the development of the pupæ.

Thus the swarming myriads depart which may have covered the vegetation like the aphides, or darkened the air like locusts. In almost every season there are some species which in this manner put forth their strength, and then, like Milton's spirits which thronged the spacious hall, 'reduce to smallest forms their shapes immense' –

> ————————— So thick the aëry crowd
> Swarm'd and were straiten'd; till, the signal given,
> Behold a wonder! they but now who seem'd
> In bigness to surpass earth's giant sons,
> Now less than smallest dwarfs.[38]

A few examples will illustrate the mode in which this force operates. It is well known that among the countless species of the insect creation, some feed on animal, others on vegetable matter, and, upon considering a catalogue of eight thousand British insects and arachnidæ, Mr. Kirby found that these two divisions were nearly a counterpoise to each other, the carnivorous being somewhat preponderant. There are also distinct species, some appointed to consume living, others dead or putrid animal and vegetable substances. One female, of *Musca carnaria*, will give birth to twenty thousand young; and the larvæ of many flesh-flies devour so much food in twenty-four hours, and grow so quickly, as to increase their weight two hundredfold! In five days after being hatched they arrive at their full growth and size, so that there was ground, says Kirby, for the assertion of Linnæus, that three flies of *M. vomitoria* could devour a dead horse as quickly as a lion; and another Swedish naturalist remarks, that so great are the powers of propagation of a single species, even of the smallest insects, that each can commit, when required, more ravages than the elephant.

Next to locusts, the aphides, perhaps, exert the greatest power over the vegetable world, and, like them, are sometimes so numerous as to darken the air. The multiplication of these little creatures is without parallel, and almost every plant has its peculiar species. Reaumur has proved, that in five generations one aphis may be the progenitor of 5,904,900,000 descendants; and it is supposed that in one year there may be twenty generations. Mr. Curtis observes, that as among caterpillars we find some that are constantly and unalterably attached to one or more particular species of plants, and others that feed indiscriminately on most sorts of herbage, so it is precisely with the aphides; some are particular, others more general feeders; and as they resemble other insects in this respect, so they do also in being more abundant in some years than others. In 1793 they were the chief, and in 1798 the sole cause of the failure of the hops. In 1794, a season almost unparalleled for drought, the hop was perfectly free from them, while peas and beans, especially the former, suffered very much from their depredations.

The ravages of the caterpillars of some of our smaller moths afford a good illustration of the temporary increase of a species. The oak-trees of a considerable wood have been stripped of their leaves as bare as in winter, by the caterpillars of a small green moth (*Tortrix irridana,*) which has been observed the year following not to abound. The Gamma moth (*Plusia gamma*), although one of our common species, is not dreaded by us for its devastations, but legions of their caterpillars have, at times, created alarm in France, as in 1735. Reaumur observes, that the female moth lays about four hundred eggs; so that if twenty caterpillars were distributed in a garden, and all lived through the winter and became moths in the succeeding May, the eggs laid by these, if all fertile, would produce eight hundred thousand. A modern writer, therefore, justly observes, that did not Providence put causes in operation to keep them in due bounds, the caterpillars of this moth alone, leaving out of consideration the two thousand other British species, would soon destroy more than half of our vegetation.

In the latter part of the last century an ant, most destructive to the sugar-cane (*Formica saccharivora*), appeared in such infinite hosts, in the island of Grenada, as to put a stop to the cultivation of that vegetable. Their numbers were incredible. The plantations and roads were filled with them; many domestic quadrupeds, together with rats, mice, and reptiles, and even birds, perished in consequence of this plague. It was not till 1780 that they were at length annihilated by torrents of rain, which accompanied a dreadful hurricane.

We may conclude by mentioning some instances of the devastations of locusts in various countries. Among other parts of Africa, Cyrenaica has been at different periods infested by myriads of these creatures, which have consumed nearly every green thing. The effect of the havoc committed by them may be estimated by the famine they occasioned. St. Augustin mentions a plague of this kind in Africa which destroyed no less than eight hundred thousand men in the kingdom of Masanissa alone, and many more upon the territories bordering upon the sea. It is also related, that in the year 591 an infinite army of locusts migrated from Africa into Italy, and, after grievously ravaging the country, were cast into the sea, when there arose a pestilence from their stench which carried off nearly a million of men and beasts.

In the Venetian territory also, in 1478, more than thirty thousand persons are said to have perished in a famine, occasioned by this scourge; and other instances are recorded of their devastations in France, Spain, Italy, Germany, &c. In different parts of Russia also, Hungary, and Poland, – in Arabia and India, and other countries, their visitations have been periodically experienced. Although they have a preference for certain plants, yet, when these are consumed, they will attack almost all the remainder. In the accounts of the invasions of locusts, the statements which appear most marvellous relate to the prodigious mass of matter which encumbers the sea wherever they are blown into it, and the pestilence arising from its putrefaction. Their dead bodies are said to have been, in some places, heaped one upon another, to the depth of four feet, in Russia, Poland, and Lithuania; and when in southern Africa they were driven into the sea by a north-west wind, they formed, says Barrow, along the shore, for fifty miles, a bank three or four feet high. But when we consider that forests are stripped of their foliage, and the earth of its green garment, for thousands of square miles, it may well be supposed that the volume of animal matter produced may equal that of great herds of quadrupeds and flights of large birds suddenly precipitated into the sea.

The occurrence of such events at certain intervals, in hot countries, like the severe winters and damp summers returning after a series of years in the temperate zone, affect the proportional numbers of almost all classes of animals and plants, and are probably fatal to the existence of many which would otherwise thrive there, while, on the contrary, they must be favourable to certain species which, if deprived of such aid, might not maintain their ground.

Although it may usually be remarked that the extraordinary increase of some one species is immediately followed and checked by the multiplication of another, yet this is not always the case, partly because many species feed in common on the same kinds of food, and partly because many kinds of food are often consumed indifferently by one and the same species. In the former case, where a variety of different animals have precisely the same taste, as, for example, when many insectivorous birds and reptiles devour alike some particular fly or beetle, the unusual numbers of the latter may only cause a slight and almost imperceptible augmentation of each of those species of bird and reptile. In the other instance, where one animal preys on others of almost every class, as, for example, where our English buzzards devour not only small quadrupeds, as rabbits and field-mice, but also birds, frogs, lizards, and insects, the profusion of any one of these last may cause all such general feeders to subsist more exclusively upon the species thus in excess, and the balance may thus be restored.

The number of species which are nearly omnivorous is considerable; and although every animal has, perhaps, a predilection for some one description of food rather than another, yet some are not even confined to one of the great kingdoms of the organic world. Thus when the racoon of the West Indies can neither procure fowls, fish, snails, nor insects, it will attack the sugar-canes, and devour various kinds of grain. The civets, when animal food is scarce, maintain themselves on fruits and roots.

Numerous birds, which feed indiscriminately on insects and plants, are perhaps more instrumental than any other of the terrestrial tribes in preserving a constant equilibrium between the relative numbers of different classes of animals and vegetables. If the insects become very numerous and devour the plants, these birds will immediately derive a larger portion of their subsistence from insects, just as the Arabians, Syrians, and Hottentots feed on locusts, when the locusts devour their crops.

The intimate relation of the inhabitants of the water to those of the land, and the influence exerted by each on the relative number of species, must not be overlooked amongst the complicated causes which determine the existence of animals and plants in certain regions. A large proportion of the amphibious quadrupeds and reptiles prey partly on aquatic plants and animals, and in part on terrestrial; and a deficiency of one kind of prey causes them to have immediate recourse to the other. The voracity of certain insects, as the dragon-fly, for example, is confined to the water during one stage of their transformations, and in their perfect state to the air. Innumerable water-birds both of rivers and seas derive in like manner their food indifferently from either element; so that the abundance or scarcity of prey in one induces them either to forsake or more constantly to haunt the other. Thus an intimate connexion between the state of the animate creation in a lake or river, and in the adjoining dry land, is maintained; or between a continent, with its lakes and rivers, and the ocean. It is well known that many birds migrate, during stormy seasons, from the sea-shore into the interior, in search of food; while others, on the contrary, urged by like wants, forsake their inland haunts, and live on substances rejected by the tide.

The migrations of fish into rivers during the spawning season supplies another link of the same kind. Suppose the salmon to be reduced in numbers by some marine foes, as by seals and grampuses, the consequence must often be, that in the course of a few years the otters at the distance of several hundred miles inland will be lessened in number from the scarcity of fish. On the other hand, if there be a dearth of food for the young fry of the salmon in rivers and estuaries, so that few return to the sea, the sand-eels and other marine species, which are usually kept down by the salmon, will swarm in greater profusion.

It is unnecessary to accumulate a greater number of illustrations in order to prove that the stations of different plants and animals depend on a great complication of circumstances, – on an immense variety of relations in the state of the animate and inanimate worlds. Every plant requires a certain climate, soil, and other conditions, and often the aid of many animals, in order to maintain its ground. Many animals feed on certain plants, being often restricted to a small number, and sometimes to one only; other members of the animal kingdom feed on plant-eating species, and thus become dependent on the conditions of the *stations* not only of their prey, but of the plants consumed by them.

Having duly reflected on the nature and extent of these mutual relations in the different parts of the organic and inorganic worlds, we may next proceed to examine the results which may be anticipated from the fluctuations now continually in progress in the state of the earth's surface, and in the geographical distribution of its living productions.

CHAPTER 9

Changes in the Animate World,
which Tend to the Extinction
of Species, continued

We have seen that the stations of animals and plants depend not merely on the influence of external agents in the inanimate world, and the relations of that influence to the structure and habits of each species, but also on the state of the contemporary living beings which inhabit the same part of the globe. In other words, the possibility of the existence of a certain species in a given locality, or of its thriving more or less therein, is determined not merely by temperature, humidity, soil, elevation, and other circumstances of the like kind, but also by the existence or non-existence, the abundance or scarcity, of a particular assemblage of other plants and animals in the same region.

If we show that both these classes of circumstances, whether relating to the animate or inanimate creation, are perpetually changing, it will follow that species are subject to incessant vicissitudes; and if the result of these mutations, in the course of ages, be so great as materially to affect the general condition of *stations*, it will follow that the successive destruction of species must now be part of the regular and constant order of Nature.

It will be desirable, first, to consider the effects which every extension of the numbers or geographical range of one species must produce on the condition of others inhabiting the same regions. When the necessary consequences of such extensions have been fully explained, the reader will be prepared to appreciate the important influence which slight modifications in the physical geography of the globe may exert on the condition of organic beings.

In the first place it is clear, that when any region is stocked with as great a variety of animals and plants as the productive powers of that region will enable it to support, the addition of any new species, or the *permanent* numerical increase of one previously established, must always be attended either by the local extermination or the numerical decrease of some other species.

There may undoubtedly be considerable fluctuations from year to year, and the equilibrium may be again restored without any permanent alteration; for in particular seasons a greater supply of heat, humidity, or other causes may augment the total quantity of vegetable produce, in which case all the animals subsisting on vegetable food, and others which prey on them, may multiply without any one species giving way; but whenever the aggregate quantity of vegetable produce remains unaltered, the progressive increase of one animal or plant implies the decline of another.

All agriculturists and gardeners are familiar with the fact, that when weeds intrude themselves into the space appropriated to cultivated species, the latter are starved in their growth or stifled. If we abandon for a short time a field or garden, a host of indigenous plants,

The darnel, hemlock, and rank fumitory,[39]

pour in and obtain the mastery, extirpating the exotics, or putting an end to the monopoly of some native plants.

If we inclose a park, and stock it with as many deer as the herbage will support, we cannot add sheep without lessening the number of the deer; nor can other herbivorous species be subsequently introduced, unless the individuals of each species in the park become fewer in proportion.

So if there be an island where leopards are the only beasts of prey, and the lion, tiger, and hyæna afterwards enter, the leopards, if they stand their ground, will be reduced in number. If the locusts then arrive and swarm greatly, it may deprive a large number of phytophagous animals of their food, and thereby cause a famine, not only among them, but among the beasts of prey; – certain species, perhaps, which had the weakest footing in the island will thus be annihilated.

We have seen how many distinct geographical provinces there are of aquatic and terrestrial species, and how great are the powers of migration conferred on different classes, whereby the inhabitants of one region may be enabled from time to time to invade another, and do actually so migrate and diffuse themselves over new countries. Now, although our knowledge of the history of the animate creation dates from so recent a period, that we can scarcely trace the advance or decline of any animal or plant, except in those cases where the influence of man has intervened, yet we can easily conceive what must happen when some new colony of wild animals or plants enters a region for the first time, and succeeds in establishing itself.

Let us consider how great are the devastations committed at certain periods by the Greenland bears, when they are drifted to the shores of Iceland in considerable numbers on the ice. These periodical invasions are formidable even to man; so that when the bears arrive, the inhabitants collect together, and go in pursuit of them with fire-arms – each native who slays one being rewarded by the king of Denmark. The Danes of old, when they landed in their marauding expeditions upon our coast, hardly excited more alarm; nor did our islanders muster more promptly for the defence of their lives and property against a common enemy, than the modern Icelanders against these formidable brutes. It frequently happens, says Henderson, that the natives are pursued by the bear when he has been long at sea, and when his natural ferocity has been strengthened by the keenness of hunger; if unarmed, it is frequently by stratagem only that they make their escape.

Let us cast our thoughts back to the period when the first polar bears reached Iceland, before it was colonized by the Norwegians in 874; – we may imagine the breaking up of an immense barrier of ice, like that which, in 1816 and the following year, disappeared from the east coast of Greenland, which it had surrounded for four centuries. By the aid of such means of transportation, a great number of these quadrupeds might effect a landing at the same time, and the havoc which they would make among the species previously settled in the island would be terrific. The deer, foxes, seals, and even birds, on which these animals sometimes prey, would be soon thinned down.

But this would be a part only, and probably an insignificant portion, of the aggregate amount of change brought about by the new invader. The plants on which the deer fed being less consumed in consequence of the lessened numbers of that herbivorous species, would soon supply more food to several insects, and probably to some terrestrial testacea, so that the latter would gain ground. The increase of these would furnish other insects and birds with food, so that the numbers of these last would be augmented. The diminution of the seals would afford a respite to some fish which they had persecuted; and these fish, in their turn, would then multiply and press upon their peculiar prey. Many water-fowls, the eggs and young of which are devoured by foxes, would increase when the foxes were thinned down by the bears; and the fish on which the water-fowls subsisted would then, in their turn, be less numerous. Thus the numerical proportions of a great number of the inhabitants, both of the land and sea, might be permanently altered by the settling of one new species in the region; and the changes caused indirectly might ramify through all classes of the living creation, and be almost endless.

An actual illustration of what we have here only proposed hypothetically, is in some degree afforded by the selection of small islands by the eider duck for its residence during the season of incubation; its nests being seldom, if ever, found on the shores of the main land, or even of a large island. The Icelanders are so well aware of this, that they have expended a great deal of labour in forming artificial islands, by separating from the main-land certain promontories, joined to it by narrow isthmuses. This insular position is necessary to guard against the destruction of the eggs and young birds, by foxes, dogs, and other animals. One year, says Hooker, it happened that, in the small island of Vidoc, adjoining the coast of Iceland, a fox got over *upon the ice*, and caused great alarm, as an immense number of ducks were then sitting on their eggs or young ones. It was long before he was taken, which was at last, however, effected by bringing another fox to the island, and fastening it by a string near the haunt of the former, by which he was allured within shot of the hunter.

It is usually the first appearance of an animal or plant, in a region to which it was previously a stranger, that gives rise to the chief alteration; since, after a time, an equilibrium is again established. But it must require ages before such a new adjustment of the relative forces of so many conflicting agents can be definitively settled. The causes in simultaneous action are so numerous, that they admit of an almost infinite number of combinations; and it is necessary that all these should have occurred once before the total amount of change, capable of flowing from any new disturbing force, can be estimated.

Thus, for example, suppose that once in two centuries a frost of unusual intensity, or a volcanic eruption of immense violence, accompanied by floods from the melting of glaciers, should occur in Iceland; or an epidemic disease, fatal to the larger number of individuals of some one species, and not affecting others, – these, and a variety of other contingencies, all of which may occur at once, or at periods separated by different intervals of time, ought to happen before it would be possible for us to declare what ultimate alteration the presence of any new comer, such as the bear before mentioned, might occasion in the animal population of the isle.

Every new condition in the state of the organic or inorganic creation, a new animal or plant, an additional snow-clad mountain, any permanent change, however slight in comparison to the whole, gives rise to a new order of things, and may make a material change in regard to some one or more species. Yet a swarm of locusts, or a frost of extreme intensity, may pass away without any great apparent derangement; no species may be lost, and all may soon recover their former relative numbers, because the same scourges may have visited the region, again and again, at some former periods. Every plant that was incapable of resisting such a degree of cold, every animal which was exposed to be entirely cut off by famine, in consequence of the consumption of vegetation by the locusts, may have perished already, so that the subsequent recurrence of similar catastrophes is attended only by a temporary change.

We are best acquainted with the mutations brought about by the progress of human population, and the growth of plants and animals favoured by man. To these, therefore, we should, in the first instance, turn our attention. If we conclude, from the concurrent testimony of history and of the evidence yielded by geological data, that man is, comparatively speaking, of very modern origin, we must at once perceive how great a revolution in the state of the animate world the increase of the human race, considered merely as consumers of a certain quantity of organic matter, must necessarily cause.

It may, perhaps, be said, that man has, in some degree, compensated for the appropriation to himself of so much food, by artificially improving the natural productiveness of soils, by irrigation, manure, and a judicious intermixture of mineral ingredients conveyed from different localities. But it admits of reasonable doubt, whether, upon the whole, we fertilize or impoverish the lands which we occupy. This assertion may seem startling to many, because they are so much in the habit of regarding the sterility or productiveness of land in relation to the wants of man, and not as regards the organic world generally. It is difficult, at first, to conceive, if a morass is converted into arable land, and made to yield a crop of grain, even of moderate abundance, that we have not improved the capabilities of the habitable surface – that we have not empowered it to support a larger quantity of organic life. In such cases, a tract, before of no utility to man, may be reclaimed and become of high agricultural importance, but it may yield, at the same time, a scantier vegetation. If a lake be drained and turned into a meadow, the space will provide sustenance to man and many terrestrial animals serviceable to him, but not perhaps so much food as it previously yielded to the aquatic races.

If the pestiferous Pontine Marshes were drained and covered with corn, like the plains of the Po, they might, perhaps, feed a smaller number of animals than they do now; for these morasses are filled with herds of buffaloes and swine, and they swarm with birds, reptiles, and insects.

The felling of dense and lofty forests which covered, even within the records of history, a considerable space on the globe, now tenanted by civilized man, must usually have lessened the amount of vegetable food throughout the space where these woods grew. We must also take into our account the area covered by towns, and a still larger surface occupied by roads.

If we force the soil to bear extraordinary crops one year, we are, perhaps, compelled to let it lie fallow the next. But nothing so much counterbalances the fertilizing effects of human art as the extensive cultivation of foreign herbs and shrubs, which, although they are often more nutritious to man, seldom thrive with the same rank luxuriance as the native plants of a district. Man is, in truth, continually striving to diminish the natural diversity of the *stations* of animals and plants in every country, and to reduce them all to a small number fitted for species of economical use. He may succeed perfectly in attaining his object, even though the vegetation be comparatively meagre, and the total amount of animal life be greatly lessened.

Spix and Martius have given a lively description of the incredible number of insects which lay waste the crops in Brazil, besides swarms of monkeys, flocks of parrots and other birds, as well as the paca, agouti, and wild swine. They describe the torment which the planter and the naturalist suffer from the musquitoes, and the devastation of the ants and blattæ; they speak of the dangers to which they were exposed from the jaguar, the poisonous serpents, lizards, scorpions, centipedes, and spiders. But with the increasing population and cultivation of the country, observe these naturalists, these evils will gradually diminish; when the inhabitants have cut down the woods, drained the marshes, made roads in all directions, and founded villages and towns, man will by degrees triumph over the rank vegetation and the noxious animals, and all the elements will second and amply recompense his activity.

The number of human beings now peopling the earth is supposed to amount to eight hundred millions, so that we may easily understand how great a number of beasts of prey, birds, and animals of every class, this prodigious population must have displaced, independently of the still more important consequences which have followed from the derangement brought about by man in the relative numerical strength of particular species.

Let us make some inquiries into the extent of the influence which the progress of society has exerted, during the last seven or eight centuries, in altering the distribution of our indigenous British animals. Dr. Fleming has prosecuted this inquiry with his usual zeal and ability, and in a memoir on the subject has enumerated the best-authenticated examples of the decrease or extirpation of certain species during a period when our population has made the most rapid advances. We shall offer a brief outline of his results.

The stag, as well as the fallow deer and the roe, were formerly so abundant that, according to Lesley, from five hundred to a thousand were sometimes slain at a hunting-match; but the native races would already have been extinguished, had they not been carefully preserved in certain forests. The otter, the marten, and the polecat, were also in sufficient numbers to be pursued for the sake of their fur; but they have now been reduced within very narrow bounds. The wild cat and fox have also been sacrificed throughout the greater part of the country, for the security of the poultry-yard or the fold. Badgers have been expelled from nearly every district which at former periods they inhabited.

Besides these, which have been driven out from some haunts, and everywhere reduced in number, there are some which have been wholly extirpated; such as the ancient breed of indigenous horses, the wild boar, and the wild oxen, of which last, however, a few remains are still preserved in the parks of some of our nobility. The beaver, which was eagerly sought after for its fur, had become scarce at the close of the ninth century, and, by the twelfth century, was only to be met with, according to Giraldus de Barri, in one river in Wales, and another in Scotland. The wolf, once so much dreaded by our ancestors, is said to have maintained its ground in Ireland so late as the beginning of the eighteenth century (1710), though it had been extirpated in Scotland thirty years before, and in England at a much earlier period. The bear, which in Wales was regarded as a beast of the chace equal to the hare or the boar, only perished as a native of Scotland in the year 1057.

Many native birds of prey have also been the subjects of unremitting persecution. The eagles, larger hawks, and ravens, have disappeared from the more cultivated districts. The haunts of the mallard, the snipe, the redshank, and the bittern, have been drained equally with the summer dwellings of the lapwing and the curlew. But these species still linger in some portion of the British isles; whereas the large capercailzies, or wood grouse, formerly natives of the pine-forests of Ireland and Scotland, have been destroyed within the last fifty years. The egret and the crane, which appear to have been formerly very common in Scotland, are now only occasional visitants.

The bustard (*Otis tarda*), observes Graves in his British Ornithology, 'was formerly seen in the downs and heaths of various parts of our island, in flocks of forty or fifty birds; whereas it is now a circumstance of rare occurrence to meet with a single individual.'[40] Bewick also remarks, 'that they were formerly more common in this island than at present; they are now found only in the open counties of the south and east, in the plains of Wiltshire, Dorsetshire, and some parts of Yorkshire.'[41] In the few years that have elapsed since Bewick wrote, this bird has entirely disappeared from Wiltshire and Dorsetshire.

These changes, we may observe, are derived from very imperfect memorials, and relate only to the larger and more conspicuous animals inhabiting a small spot on the globe; but they cannot fail to exalt our conception of the enormous revolutions which, in the course of several thousand years, the whole human species must have effected.

The kangaroo and the emu are retreating rapidly before the progress of colonization in Australia; and it scarcely admits of doubt, that the general cultivation of that country must lead to the extirpation of both. The most striking example of the loss, even within the last two centuries, of a remarkable species, is that of the dodo – a bird first seen by the Dutch when they landed on the Isle of France, at that time uninhabited, immediately after the discovery of the passage to the East Indies by the Cape of Good Hope. It was of a large size and singular form; its wings short, like those of an ostrich, and wholly incapable of sustaining its heavy body even for a short flight. In its general appearance it differed from the ostrich, cassowary, or any known bird.

Many naturalists gave figures of the dodo after the commencement of the seventeenth century, and there is a painting of it in the British Museum, which is said to have been taken from a living individual. Beneath the painting is a leg, in a fine state of preservation, which ornithologists are agreed cannot belong to any other known bird. In the museum at Oxford, also, there is a foot and a head, in an imperfect state, but M. Cuvier doubts the identity of this species with that of which the painting is preserved in London.*

In spite of the most active search, during the last century, no information respecting the dodo was obtained, and some authors have gone so far as to pretend that it never existed; but amongst a great mass of satisfactory evidence in favour of the recent existence of this species, we may mention that an assemblage of fossil bones were recently discovered, under a bed of lava, in the Isle of France, and sent to the Paris museum by M. Desjardins. They almost all belonged to a large living species of land-tortoise, called *Testudo Indica*, but amongst them were the head, sternum, and humerus of the dodo. M. Cuvier showed me these valuable remains in Paris, and assured me that they left no doubt in his mind that the huge bird was one of the gallinaceous tribe.

Next to the direct agency of man, his indirect influence in multiplying the numbers of large herbivorous quadrupeds of domesticated races, may be regarded as one of the most obvious causes of the extermination of species. On this, and on several other grounds, the introduction of the horse, ox, and other mammalia, into America, and their rapid propagation over that continent within the last three centuries, is a fact of great importance in natural history. The extraordinary herds of wild cattle and horses which overran the plains of South America, sprung from a very few pairs first carried over by the Spaniards; and they prove that the wide geographical range of large species in great continents does not necessarily imply that they have existed there from remote periods. Humboldt observes, in his Travels, on the authority of Azzara, that it is believed there exist, in the Pampas of Buenos Ayres, twelve million cows and three million horses, without comprising in this enumeration the cattle that have no acknowledged proprietor. In the Llanos of Caraccas, the rich hateros, or proprietors of pastoral farms, are entirely ignorant of the

---

* Some have complained that inscriptions on tomb-stones convey no general information except that individuals were born and died, accidents which must happen alike to all men. But the death of a *species* is so remarkable an event in natural history, that it deserves commemoration, and it is with no small interest that we learn, from the archives of the University of Oxford, the exact day and year when the remains of the last specimen of the dodo, which had rotted in the Ashmolean museum, were cast away. The relics, we are told, were 'a Musæo subducta, annuente Vice-cancellario aliisque curatoribus, ad ea lustranda convocatis, die Januarii, 8vo. A.D., 1755' [taken out of the Museum, with the assent of the Vice-Chancellor and the other curators summoned to sacrifice them, 8 January 1755]. Zool. Journ., No. 12, p. 559 [-60].

number of cattle they possess. The young are branded with a mark peculiar to each herd, and some of the most wealthy owners mark as many as fourteen thousand a year. In the northern plains, from the Orinoco to the lake of Maracaybo, M. Depons reckoned that one million two hundred thousand oxen, one hundred and eighty thousand horses, and ninety thousand mules, wandered at large. In some parts of the valley of the Mississippi, especially in the country of the Osage Indians, wild horses are immensely numerous.

The establishment of black cattle in America dates from Columbus's second voyage to St. Domingo. They there multiplied rapidly; and that island presently became a kind of nursery from which these animals were successively transported to various parts of the continental coast, and from thence into the interior. Notwithstanding these numerous exportations, in twenty-seven years after the discovery of the island, herds of four thousand head, as we learn from Oviedo, were not uncommon, and there were even some that amounted to eight thousand. In 1587, the number of hides exported from St. Domingo alone, according to Acosta's report, was thirty-five thousand four hundred and forty-four; and in the same year there were exported sixty-four thousand three hundred and fifty from the ports of New Spain. This was in the sixty-fifth year after the taking of Mexico, previous to which event the Spaniards, who came into that country, had not been able to engage in anything else than war.

All our readers are aware that these animals are now established throughout the American continent, from Canada to Paraguay.

The ass has thriven very generally in the New World; and we learn from Ulloa, that in Quito they ran wild, and multiplied in amazing numbers, so as to become a nuisance. They grazed together in herds, and, when attacked, defended themselves with their mouths. If a horse happened to stray into the places where they fed, they all fell upon him, and did not cease biting and kicking till they left him dead.

The first hogs were carried to America by Columbus, and established in the island of St. Domingo the year following its discovery in November, 1493. In succeeding years they were introduced into other places where the Spaniards settled; and, in the space of half a century, they were found established in the New World, from the latitude of 25° north, to the 40th degree of south latitude. Sheep, also, and goats have multiplied enormously in the New World, as have also the cat and the rat, which last, as we before stated, has been imported unintentionally in ships. The dogs introduced by man, which have at different periods become wild in America, hunted in packs like the wolf and the jackal, destroying not only hogs, but the calves and foals of the wild cattle and horses.

Ulloa in his voyage, and Buffon on the authority of old writers, relate a fact which illustrates very clearly the principle before explained by us, of the check which the increase of one animal necessarily offers to that of another. The Spaniards had introduced goats into the island of Juan Fernandez, where they became so prolific as to furnish the pirates who infested those seas with provisions. In order to cut off this resource from the buccaneers, a number of dogs were turned loose into the island; and so numerous did they become in their turn, that they destroyed the goats in every accessible part, after which the number of the wild dogs again decreased.

As an example of the rapidity with which a large tract may become peopled by the offspring of a single pair of quadrupeds, we may mention, that in the year 1773 thirteen rein-deer were exported from Norway, only three of which reached Iceland. These were turned loose into the mountains of Guldbringè Syssel, where they multiplied so greatly, in the course of forty years, that it was not uncommon to meet with herds consisting of from forty to one hundred in various districts.

In Lapland, observes a modern writer, the rein-deer is a loser by his connexion with man, but Iceland will be this creature's paradise. There is, in the interior, a tract which Sir G. Mackenzie computes at not less than forty thousand square miles, without a single human habitation, and almost entirely unknown to the natives themselves. There are no wolves; the Icelanders will keep out the bears; and the rein-deer, being almost unmolested by man, will have no enemy whatever, unless it has brought with it its own tormenting gad-fly.

Besides the quadrupeds before enumerated by us, our domestic fowls have also succeeded in the West Indies and America, where they have the common fowl, the goose, the duck, the peacock, the pigeon, and the guinea-fowl. As these were often taken suddenly from the temperate to very hot regions, they were not reared at first without much difficulty; but after a few generations they became familiarized to the climate, which, in many cases, approached much nearer than that of Europe to the temperature of their original native countries.

The fact of so many millions of wild and tame individuals of our domestic species, almost all of them the largest quadrupeds and birds, having been propagated throughout the new continent within the short period that has elapsed since the discovery of America, while no appreciable improvement can have been made in the productive powers of that vast continent, affords abundant evidence of the extraordinary changes which accompany the diffusion and progressive advancement of the human race over the globe. That it should have remained for us to witness such mighty revolutions is a proof, even if there was no other evidence, that the entrance of man into the planet is, comparatively speaking, of extremely modern date, and that the effects of his agency are only beginning to be felt.

A modern writer has estimated, that there are in America upwards of four million square miles of useful soil, each capable of supporting two hundred persons; and nearly six million, each mile capable of supporting four hundred and ninety persons. If this conjecture be true, it will follow, as that author observes, that if the natural resources of America were fully developed, it would afford sustenance to five times as great a number of inhabitants as the entire mass of human beings existing at present upon the globe. The new continent, he thinks, though less than half the size of the old, contains an equal quantity of useful soil, and much more than an equal amount of productive power. Be this as it may, we may safely conclude that the amount of human population now existing, constitutes but a small proportion of that which the globe is capable of supporting, or which it is destined to sustain at no distant period, by the rapid progress of society, especially in America, Australia, and certain parts of the old continent.

But if we reflect that already many millions of square miles of the most fertile land, occupied originally by a boundless variety of animal and vegetable forms, have been already brought under the dominion of man, and compelled, in a great measure, to yield nourishment to him, and to a limited number of plants and animals which he has caused to increase, we must at once be convinced, that the annihilation of a multitude of species has already been effected, and will continue to go on hereafter, in certain regions, in a still more rapid ratio, as the colonies of highly-civilized nations spread themselves over unoccupied lands.

Yet, if we wield the sword of extermination as we advance, we have no reason to repine at the havoc committed, nor to fancy, with the Scotch poet, that 'we violate the social union of nature;'[42] or complain, with the melancholy Jaques, that we

Are mere usurpers, tyrants, and, what's worse,
To fright the animals, and to kill them up
In their assign'd and native dwelling-place.[43]

We have only to reflect, that in thus obtaining possession of the earth by conquest, and defending our acquisitions by force, we exercise no exclusive prerogative. Every species which has spread itself from a small point over a wide area, must, in like manner, have marked its progress by the diminution, or the entire extirpation, of some other, and must maintain its ground by a successful struggle against the encroachments of other plants and animals. That minute parasitic plant, called 'the rust' in wheat, has, like the Hessian fly, the locust, and the aphis, caused famines ere now amongst the 'lords of the creation.' The most insignificant and diminutive species, whether in the animal or vegetable kingdom, have each slaughtered their thousands, as they disseminated themselves over the globe, as well as the lion, when first it spread itself over the tropical regions of Africa.

We cannot conclude this division of our subject without observing, that although we have as yet considered one class only of the causes (the organic) whereby species may become exterminated, yet the continued action of these alone, throughout myriads of future ages, must work an entire change in the state of the organic creation, not merely on the continents and islands, where the power of man is chiefly exerted, but in the great ocean, where his control is almost unknown. The mind is prepared by the contemplation of such future revolutions to look for the signs of others, of an analogous nature, in the monuments of the past. Instead of being astonished at the proofs there manifested of endless mutations in the animate world, they will appear to one who has thought profoundly on the fluctuations now in progress, to afford evidence in favour of the uniformity of the system, unless, indeed, we are precluded from speaking of *uniformity* when we characterize a principle of endless variation.

CHAPTER 10

Changes in the Inorganic World, Tending to the Extinction of Species

Having shown in the last chapter how considerably the numerical increase or the extension of the geographical range of any one species must derange the numbers and distribution of others, let us now direct our attention to the influence which the inorganic causes described in our first volume are continually exerting on the habitations of species.

So great is the instability of the earth's surface, that if Nature were not continually engaged in the task of sowing seeds and colonizing animals, the depopulation of a certain portion of the habitable sea and land would in a few years be considerable. Whenever a river transports sediment into a lake or sea, the aquatic animals and plants which delight in deep water are expelled: the tract, however, is not allowed to remain useless, but is soon peopled by species which require more light and heat, and thrive where the water is shallow. Every addition made to the land by the encroachment of the delta of a river banishes many subaqueous species from their native abodes; but the new-formed plain is not permitted to lie unoccupied, being instantly covered with terrestrial vegetation. The ocean devours continuous lines of sea-coast, and precipitates forests or rich pasture-land into the waves; but this space is not lost to the animate creation, for shells and sea-weed soon adhere to the new-made cliffs, and numerous fish people the channel which the current has scooped out for itself. No sooner has a volcanic isle been thrown up than some lichens begin to grow upon it, and it is sometimes clothed with verdure, while smoke and ashes are still occasionally thrown from the crater. The cocoa, pandanus, and mangrove take root upon the coral reef before it has fairly risen above the waves. The burning stream of lava that descends from Etna rolls through the stately forest, and converts to ashes every tree and herb which stand in its way; but the black strip of land thus desolated, is covered again, in the course of time, with oaks, pines, and chestnuts, as luxuriant as those which the fiery torrent swept away.

Every flood and landslip, every wave which a hurricane or earthquake throws upon the shore, every shower of volcanic dust and ashes which buries a country far and wide to the depth of many feet, every advance of the sand-flood, every conversion of salt-water into fresh when rivers alter their main channel of discharge, every permanent variation in the rise or fall of tides in an estuary – these and countless other causes displace in the course of a few centuries certain plants and animals from stations which they previously occupied. If, therefore, the Author of Nature had not been prodigal of those numerous contrivances before alluded to, for spreading all classes of organic beings over the earth – if he had not ordained that the fluctuations of the animate and inanimate creation should be in perfect harmony with each other, it is evident that considerable spaces, now the most habitable on the globe, would soon be as devoid of life as are the Alpine snows, or the dark abysses of the ocean, or the moving sands of the Sahara.

The powers then of migration and diffusion conferred on animals and plants, are indispensable to enable them to maintain their ground, and would be necessary even though it were never intended that a species should gradually extend its geographical range. But a facility of shifting their quarters being once given, it cannot fail to happen that the inhabitants of one province should occasionally penetrate into some other, since the strongest of those barriers which we before described as separating distinct regions, are all liable to be thrown down one after the other, during the vicissitudes of the earth's surface.

The numbers and distribution of particular species are affected in two ways, by changes in the physical geography of the earth. First, these changes promote or retard the migrations of species; secondly, they alter the physical conditions of the localities which species inhabit. If the ocean should gradually wear its way through an isthmus, like that of Suez, it would open a passage for the intermixture of the aquatic tribes of two seas previously disjoined, and would, at the same time, close a free communication which the terrestrial plants and animals of two continents had before enjoyed. These would be, perhaps, the most important consequences in regard to the distribution of species, which would result from the breach made by the sea in such a spot; but there would be others of a distinct nature, such as the conversion of a certain tract of land which formed the isthmus into sea. This space previously occupied by terrestrial plants and animals would be immediately delivered over to the aquatic, a local revolution which might have happened in innumerable other parts of the globe, without being attended by any alteration in the blending together of species of two distinct provinces.

This observation leads us to point out one of the most interesting conclusions to which we are led by the contemplation of the vicissitudes of the inanimate world in relation to those of the animate. It is clear that if the agency of inorganic causes be uniform as we have supposed, they must operate very irregularly on the state of organic beings, so that the rate according to which these will change in particular regions will not be equal in equal periods of time.

We are not about to advocate the doctrine of general catastrophes recurring at certain intervals, as in the ancient oriental cosmogonies, nor do we doubt that if very considerable periods of equal duration could be taken into our consideration and compared one with another, the rate of change in the living, as well as in the inorganic world, would be nearly uniform; but if we regard each of the causes separately, which we know to be at present the most instrumental in remodelling the state of the surface, we shall find that we must expect each to be in action for thousands of years, without producing any extensive alterations in the habitable surface, and then to give rise, during a very brief period, to important revolutions.

We shall illustrate this principle by a few of the most remarkable examples which present themselves. In the course of the last century, as we have before pointed out, a considerable number of instances are recorded of the solid surface, whether covered by water or not, having been permanently sunk or upraised by the power of earthquakes. Most of these convulsions are only accompanied by temporary fluctuations in the state of limited districts, and a continued repetition of these events for thousands of years might not produce any decisive change in the state of many of those great zoological or botanical provinces of which we have sketched the boundaries.

When, for example, large parts of the ocean and even of inland seas are a thousand fathoms or upwards in depth it is a matter of no moment to the animate creation that vast tracts should be heaved up many fathoms at certain intervals, or should subside to the same amount. Neither can any material revolution be produced in South America either in the terrestrial or the marine plants or animals by a series of shocks on the coast of Chili, each of which, like that of Penco, in 1750, should uplift the coast about twenty-five feet. Nor if the ground sinks fifty feet at a time, as in the harbour of Port Royal, in Jamaica, in 1692, will such alterations of level work any general fluctuations in the state of organic beings inhabiting the West India islands, or the Caribbean Sea.

It is only when these subterranean powers, by shifting gradually the points where their principal force is developed, happen to strike upon some particular region where a slight change of level immediately affects the distribution of land and water, or the state of the climate, or the barriers between distinct groups of species over extensive areas, that the rate of fluctuation becomes accelerated, and may, in the course of a few years or centuries, work mightier changes than had been experienced in myriads of antecedent years.

Thus, for example, a repetition of subsidences causing the narrow isthmus of Panamá to sink down a few hundred feet, might in a few centuries bring about a great revolution in the state of the animate creation in the western hemisphere. Thousands of aquatic species would pass for the first time from the Caribbean Sea into the Pacific; and thousands of others, before peculiar to the Pacific ocean, would make their way into the Caribbean Sea, the Gulf of Mexico, and the Atlantic. A considerable modification would probably be occasioned by the same event in the direction or volume of the Gulf-stream, and thereby the temperature of the sea and the contiguous lands would be altered as far as the influence of that current extends. A change of climate might thus be produced in the ocean from Florida to Spitzbergen, and in many countries of North America, Europe, and Greenland. Not merely the heat, but the quantity of rain which falls would be altered in certain districts, so that many species would be excluded from tracts where they before flourished; others would be reduced in number; and some would thrive more and multiply. The seeds also and the fruits of plants would no longer be drifted in precisely the same directions, nor the eggs of aquatic animals; neither would species be any longer impeded in their migrations towards particular stations before shut out from them by their inability to cross the mighty current.

Let us take another example from a part of the globe which is at present liable to suffer by earthquakes, viz., the low sandy tract which intervenes between the sea of Azof and the Caspian. If there should occur a sinking down to a trifling amount, and such ravines should be formed as might be produced by a few earthquakes, not more considerable than have fallen within our limited observation during the last one hundred and forty years, the waters of the sea of Azof would pour rapidly into the Caspian, which, according to the lowest estimate, is fifty feet lower than the level of the Black Sea, and which, according to some writers of considerable authority, is one hundred and fifty feet, – according to others, three hundred feet below the level of the Sea of Azof. The latter sea would immediately borrow from the Euxine, the Euxine from the Mediterranean, and the Mediterranean from the Atlantic, so that an inexhaustible current would pour down into the low tracts of Asia bordering the Caspian, by which all the sandy salt steppes adjacent to that sea would be inundated.

The diluvial waters would reach the salt lake of Aral, nor stop until their eastern shores were bounded by the high land which in the steppe of the Kirghis connects the Altay with the Himalaya mountains. A few years, perhaps a few months might suffice for the accomplishment of this great revolution in the geography of the interior of Asia; and it is impossible for those who believe in the permanence of the energy with which existing causes now act, not to anticipate such events again and again in the course of future ages.

Let us next imagine a few cases of the elevation of land of small extent at certain critical points, as, for example, in the shallowest parts of the Straits of Gibraltar, where the soundings from the African to the European side give only two hundred and twenty fathoms. In proportion as this submarine barrier of rock was upheaved, to effect which would merely require the shocks of partial and confined earthquakes, the volume of water which pours in from the Atlantic into the Mediterranean would be lessened. But the loss of the inland sea by evaporation would remain the same, so that being no longer able to draw on the ocean for a supply sufficient to restore its equilibrium, it must sink, and leave dry a certain portion of land around its borders. The current which now flows constantly out of the Black Sea into the Mediterranean would then rush in more rapidly, and the level of the Mediterranean would be thereby prevented from falling so low; but the level of the Black Sea would, for the same reason, sink, so that when, by a continued series of elevatory movements, the Straits of Gibraltar had become completely closed up, we might expect large and level sandy steppes to surround both the Euxine and Mediterranean, like those occurring at present on the skirts of the Caspian, and the sea of Aral. The geographical range of hundreds of aquatic species would be thereby circumscribed, and that of hundreds of terrestrial plants and animals extended.

A line of submarine volcanos crossing the channel of some strait, and gradually choking it up with ashes and lava, might produce a new barrier as effectually as a series of earthquakes; especially if thermal springs, plentifully charged with carbonate of lime, silica, and other mineral ingredients, should promote the rapid multiplication of corals and shells, and cement them together with solid matter precipitated during the intervals between eruptions. Suppose in this manner a stoppage to be caused of the Bahama Channel between the bank of that name and the coast of Florida. This insignificant revolution, confined to a mere spot in the bottom of the ocean, would, by diverting the main current of the Gulf-stream, give rise more effectually than the opening of the Straits of Panamá before supposed, to extensive changes in the climate and distribution of animals and plants inhabiting the northern hemisphere.

A repetition of elevatory movements of earthquakes might continue over an area as extensive as Europe, for thousands of ages, at the bottom of the ocean in certain regions, and produce no visible effects; whereas, if they should operate in some shallow parts of the Pacific, amid the coral archipelagos, they would soon give birth to a new continent. Hundreds of volcanic islands may be thrown up and become covered with vegetation, without causing more than local fluctuations in the animate world; but if a chain like the Aleutian archipelago or the Kurile isles, run for a distance of many hundred miles, so as to form an almost uninterrupted commun-ication between two continents, or two distant islands, the migrations of plants, birds, insects, and even of some quadrupeds, may cause in a short time an extraordinary series of revolutions, tending to augment the range of some animals and plants, and to limit that of others. A new archipelago might be formed in the Mediterranean, the Bay of Biscay, and a thousand other localities, and might produce less important events than one rock which should rise up between Australia and Java, so placed that winds and currents might cause an interchange of the plants, insects, and birds, of the latter countries.

If we turn from the igneous to the aqueous agents, we find the same tendency to an irregular rate of change, naturally connected with the strictest uniformity in the energy of those causes. When the sea, for example, gradually encroaches upon both sides of a narrow isthmus, as that of Sleswick, separating the North Sea from the Baltic, where, as we stated, the cliffs on both the opposite coasts are wasting away, no material alteration results for thousands of years, save only that there is a progressive conversion of a small strip of land into water. A few feet only, or a few yards, are annually removed; but when at last the partition shall be broken down, and the tides of the ocean shall enter by a direct passage into the inland sea, instead of going by a circuitous route through the Cattegat, a body of salt-water will sweep up as far as the Gulfs of Bothnia and Finland, the waters of which are now brackish, or almost fresh; and this revolution will be attended by the local annihilation of many species.

Similar consequences must have resulted, on a small scale, when the sea opened its way through the isthmus of Staveren in the thirteenth century, forming an union between an inland lake and the ocean, and opening, in the course of one century, a shallow strait more than half as wide as the narrowest part of that which divides England from France.

It will almost seem superfluous, after we have thus traced the important modifications in the condition of living beings which flow from changes of trifling extent, to argue that entire revolutions might be brought about, if the climate and physical geography of the whole globe were greatly altered. Species we have stated are, in general, local, some being confined to extremely small spots, and depending for their existence on a combination of causes which, if they are to be met with elsewhere, occur only in some very remote region. Hence it must happen that when the nature of these localities is changed the species will perish; for it will rarely happen that the cause which alters the character of the district will afford new facilities to the species to establish itself elsewhere.

If we attribute the origin of a great part of the desert of Africa to the gradual progress of moving sands, driven eastward by the westerly winds, we may safely infer that a variety of species must have been annihilated by this cause alone. The sand-flood has been inundating, from time immemorial, the rich lands on the west of the Nile, and we have only to multiply this effect a sufficient number of times, in order to understand how, in the lapse of ages, a whole group of terrestrial animals and plants may become extinct.

This desert, without including Bornou and Darfour, extends, according to the calculation of Humboldt, over one hundred and ninety-four thousand square leagues, an area far more than double that of the Mediterranean, which occupies only seventy-nine thousand eight hundred square leagues. In a small portion of so vast a space, we may infer, from analogy, that there were many peculiar species of plants and animals which must have been banished by the sand, and their habitations invaded by the camel and by birds and insects formed for the arid sands.

There is evidently nothing in the nature of the catastrophe to favour the escape of the former inhabitants to some adjoining province; nothing to weaken, in the bordering lands, that powerful barrier against emigration – pre-occupancy. Nor, even if the exclusion of a certain group of species from a given tract were compensated by an extension of their range over a new country, would that circumstance tend to the conservation of species in general; for the extirpation would merely then be transferred to the region so invaded. If it be imagined, for example, that the aboriginal quadrupeds, birds, and other animals of Africa emigrated in consequence of the advance of drift-sand, and colonized Arabia, the indigenous Arabian species must have given way before them, and have been reduced in number or destroyed.

Let us next suppose that, in some central and more elevated parts of the great African desert, the upheaving power of earthquakes should be exerted throughout an immense series of ages, accompanied, at certain intervals, by volcanic eruptions such as gave rise at once, in 1755, to a mountain one thousand seven hundred feet high, on the Mexican plateau. When the continued repetition of these events had caused a mountain-chain, it is obvious that a complete transformation in the state of the climate would be brought about throughout a vast area.

We will imagine the summits of the new chain to rise so as to be covered, like Mount Atlas, for several thousand feet, with snow during a great part of the year. The melting of these snows, during the greatest heat, would cause the rivers to swell in the season when the greatest drought now prevails; the waters, moreover, derived from this source, would always be of lower temperature than the surrounding atmosphere, and would thus contribute to cool the climate. During the numerous earthquakes and volcanic eruptions which would attend the gradual formation of the chain, there would be many floods, caused by the bursting of temporary lakes and by the melting of snows by lava. These inundations would deposit alluvial matter far and wide over the original sands at all levels, as the country assumed various shapes, and was modified again and again by the moving power from below, and the aqueous erosion of the surface above. At length the Sahara would be fertilized, irrigated by rivers and streamlets intersecting it in every direction, and covered by jungle and morasses, so that the animals and plants which now people northern Africa would disappear, and the region would gradually become fitted for the reception of a population of species perfectly dissimilar in their forms, habits, and organization.

There are always some peculiar and characteristic features in the physical geography of each large division of the globe; and on these peculiarities the state of animal and vegetable life is dependent. If, therefore, we admit incessant fluctuations in the physical geography, we must, at the same time, concede the successive extinction of terrestrial and aquatic species to be part of the economy of our system. When some great class of stations is in excess of certain latitudes, as, for example, in wide savannahs, arid sands, lofty mountains, or inland seas, we find a corresponding development of species adapted for such circumstances. In North America, where there is a chain of vast inland lakes of fresh-water, we find an extraordinary abundance and variety of aquatic birds, fresh-water fish, testacea, and small amphibious reptiles, fitted for such a climate. The greater part of these would perish if the lakes were destroyed, – an event that might be brought about by some of the least of those important revolutions contemplated in geology. It might happen that no fresh-water lakes of corresponding magnitude might then exist on the globe; but if they occurred elsewhere, they might be situated in New Holland, Southern Africa, Eastern Asia, or some region so distant as to be quite inaccessible to the North American species; or they might be situated within the tropics, in a climate uninhabitable by species fitted for a temperate zone; or, finally, we may presume that they would be pre-occupied by *indigenous* tribes.

To pursue this train of reasoning farther is unnecessary; the reader has only to reflect on what we have said of the habitations and the stations of organic beings in general, and to consider them in relation to those effects which we have contemplated in our first volume as resulting from the igneous and aqueous causes now in action, and he will immediately perceive that, amidst the vicissitudes of the earth's surface, species cannot be immortal, but must perish one after the other, like the individuals which compose them. There is no possibility of escaping from this conclusion, without resorting to some hypothesis as violent as that of Lamarck, who imagined, as we have before seen, that species are each of them endowed with indefinite powers of modifying their organization, in conformity to the endless changes of circumstances to which they are exposed.

Some of the effects which must attend every general alteration of *climate* are sufficiently peculiar to claim a separate consideration before concluding the present chapter.

We have before stated that, during seasons of extraordinary severity, many northern birds, and, in some countries, many quadrupeds, migrate southwards. If these cold seasons were to become frequent, in consequence of a gradual and general refrigeration of the atmosphere, such migrations would be more and more regular, until, at length, many animals, now confined to the arctic regions, would become the tenants of the temperate zone; while the inhabitants of the latter would approach nearer to the equator. At the same time, many species previously established on high mountains, would begin to descend, in every latitude, towards the middle regions, and those which were confined to the flanks of mountains would make their way into the plains. Analogous changes would also take place in the vegetable kingdom.

If, on the contrary, the heat of the atmosphere be on the increase, the plants and animals of low grounds would ascend to higher levels, the equatorial species would migrate into the temperate zone, and those of the temperate into the arctic circle.

But although some species might thus be preserved, every great change of climate must be fatal to many which can find no place of retreat, when their original habitations become unfit for them. For if the general temperature be on the rise, then is there no cooler region whither the polar species can take refuge; if it be on the decline, then the animals and plants previously established between the tropics have no resource. Suppose the general heat of the atmosphere to increase, so that even the arctic region became too warm for the musk-ox and rein-deer, it is clear that they must perish; so, if the torrid zone should lose so much of its heat by the progressive refrigeration of the earth's surface, as to be an unfit habitation for apes, boas, bamboos, and palms, these tribes of animals and plants, or at least most of the species now belonging to them, would become extinct, for there would be no warmer latitudes for their reception.

It will follow, therefore, that as often as the climates of the globe are passing from the extreme of heat to that of cold – from the summer to the winter of the great year before alluded to by us – the migratory movement will be directed constantly from the poles towards the equator; and for this reason the species inhabiting parallel latitudes, in the northern and southern hemispheres, must become widely different. For we assume,

on grounds before stated, that the original stock of each species is introduced into one spot of the earth only, and, consequently, no species can
be at once indigenous in the arctic and antarctic circles.

But when, on the contrary, a series of changes in the physical geography
of the globe, or any other supposed cause, occasions an elevation of the
general temperature, – when there is a passage from the winter to one of
the vernal or summer seasons of the great cycle of climates, then the order
of the migratory movement is inverted. The different species of animals
and plants direct their course from the equator towards the poles; and
the northern and southern hemispheres may become peopled, to a great
degree, by identical species. Such is not the actual state of the inhabited
earth, as we have already shown in our sketch of the geographical distribution of its living productions; and this fact adds one more additional
proof to a great body of evidence, derived from independent sources, that
the general temperature has been cooling down during the epochs which
immediately preceded our own.

· We do not mean to speculate on the entire transposition of a group of
animals and plants from tropical to polar latitudes, or the reverse, as a
probable, or even possible, event; for although we believe the mean annual
temperature of one zone to be transferrible to another, we know that
the same climate cannot be so transferred. Whatever be the general
temperature of the earth's surface, comparative equability of heat will
characterize the tropical regions, while great periodical variations will
belong to the temperate, and still more to the polar latitudes. These, and
many other peculiarities connected with heat and light, depend on fixed
astronomical causes, such as the motion of the earth and its position in
relation to the sun, and not on those fluctuations of its surface, which may
influence the general temperature.

Among many obstacles to such extensive transferences of habitations,
we must not forget the immense lapse of time required, according to any
hypothesis yet suggested, especially that which has appeared to us most
feasible, to bring about a considerable change in climate. During a period
so vast, the other causes of extirpation, before enumerated by us, would
exert so powerful an influence as to prevent all, save a very few hardy
species, from passing from equatorial to polar regions, or from the tropics
to the pole.

But the power of accommodation to new circumstances is great in
certain species, and might enable many to pass from one zone to another,

if the mean annual heat of the atmosphere and the ocean were greatly altered. To the marine tribes, especially, such a passage would be possible, for they are less impeded in their migrations, by barriers of land, than are the terrestrial by the ocean. Add to this, that the temperature of the ocean is much more uniform than that of the atmosphere investing the land, so that we may easily suppose that most of the testacea, fish, and other classes, might pass from the equatorial into the temperate regions, if the mean temperature of those regions were transposed, although a second expatriation of these species of tropical origin into the arctic and antarctic circles would probably be impossible.

On the principles above explained, if we found that at some former period, as when, for example, our carboniferous strata were deposited, the same tree-ferns and other plants inhabited the regions now occupied by Europe and Van Diemen's Land, we should suspect that the species in question had, at some antecedent period, inhabited lands within the tropics, and that an increase of the mean annual heat had caused them to emigrate into both the temperate zones. There are no geological data, however, as yet obtained, to warrant the opinion that such identity of species existed in the two hemispheres in the era in question.

Let us now consider more particularly the effect of vicissitudes of climate in causing one species to give way before the increasing numbers of some other.

When temperature forms the barrier which arrests the progress of an animal or plant in a particular direction, the individuals are fewer and less vigorous as they approach the extreme confines of the geographical range of the species. But these stragglers are ready to multiply rapidly on the slightest increase or diminution of heat that may be favourable to them, just as particular insects increase during a hot summer, and certain plants and animals gain ground after a series of congenial seasons.

In almost every district, especially if it be mountainous, there are a variety of species the limits of whose habitations are conterminous, some being unable to proceed farther without encountering too much heat, others too much cold. Individuals, which are thus on the borders of the regions proper to their respective species, are like the out-posts of hostile armies, ready to profit by every slight change of circumstances in their favour, and to advance upon the ground occupied by their neighbours and opponents.

The proximity of distinct climates, produced by the inequalities of the earth's surface, brings species possessing very different constitutions into such immediate contact, that their naturalizations are very speedy whenever opportunities of advancing present themselves. Many insects and plants, for example, are common to low plains within the arctic circle, and to lofty mountains in Scotland and other parts of Europe. If the climate, therefore, of the polar regions were transferred to our own latitudes, the species in question would immediately descend from these elevated stations to overrun the low grounds. Invasions of this kind, attended by the expulsion of the pre-occupants, are almost instantaneous, because the change of temperature not only places the one species in a more favourable position, but renders the others sickly and almost incapable of defence.

Lamarck appears to have speculated on the modifications to which every variation of external circumstances might give rise in the form and organization of species, as if he had indefinite periods of time at his command, not sufficiently reflecting that revolutions in the state of the habitable earth, whether by changes of climate or any other condition, are attended by still greater fluctuations in the relative condition of contemporary species. They can avail themselves of these alterations in their favour instantly, and augment their numbers to the injury of some other species; whereas the supposed transmutations are only assumed to be brought about by slow and insensible degrees, and in a lapse of ages, the duration of which is beyond the reach of human conception. Even if we thought it possible that the palm or the elephant, which now flourish in equatorial regions, could ever learn to bear the variable seasons of our temperate zone, or the rigours of an arctic winter, we should, with no less confidence, affirm, that they must perish before they had time to become habituated to such new circumstances. That they would be supplanted by other species at each variation of climate, may be inferred from what we have before said of the known local exterminations of species which have resulted from the multiplication of others. Some minute insect, perhaps, might be the cause of destruction to the huge and powerful elephant.

Suppose the climate of the highest part of the woody zone of Etna to be transferred to the sea-shore at the base of the mountain, no botanist would anticipate that the olive, lemon-tree, and prickly pear (*Cactus opuntia*), would be able to contend with the oak and chestnut, which would begin forthwith to descend to a lower level, or that these last would be able to stand their ground against the pine, which would also, in the space of a few years, begin to occupy a lower position. We might form some kind of estimate of the time which might be required for the migrations of these plants; whereas we have no data for concluding that any number of thousands of years would be sufficient for one step in the pretended metamorphosis of one species into another, possessing distinct attributes and qualities.

This argument is applicable not merely to *climate*, but to any other cause of mutation. However slowly a lake may be converted into a marsh, or a marsh into a meadow, it is evident that before the lacustrine plants can acquire the power of living in marshes, or the marsh-plants of living in a less humid soil, other species, already existing in the region, and fitted for these several stations, will intrude and keep possession of the ground. So if a tract of salt-water becomes fresh by passing through every intermediate degree of brackishness, still the marine molluscs will never be permitted to be gradually metamorphosed into fluviatile species; because long before any such transformation can take place by slow and insensible degrees, other tribes, which delight in brackish or fresh-water, will avail themselves of the change in the fluid, and will, each in their turn, monopolize the space.

It is idle to dispute about the abstract possibility of the conversion of one species into another, when there are known causes so much more active in their nature, which must always intervene and prevent the actual accomplishment of such conversions. A faint image of the certain doom of a species less fitted to struggle with some new condition in a region which it previously inhabited, and where it has to contend with a more vigorous species, is presented by the extirpation of savage tribes of men by the advancing colony of some civilized nation. In this case the contest is merely between two different *races*, each gifted with equal capacities of improvement – between two varieties, moreover, of a species which exceeds all others in its aptitude to accommodate its habits to the most extraordinary variations of circumstances. Yet few future events are more certain than the speedy extermination of the Indians of North America and the savages of New Holland in the course of a few centuries, when these tribes will be remembered only in poetry and tradition.

CHAPTER 11

Whether the Extinction and Creation of Species can Now be in Progress

We have pointed out in the preceding chapters the strict dependence of each species of animal and plant on certain physical conditions in the state of the earth's surface, and on the number and attributes of other organic beings inhabiting the same region. We have also endeavoured to show that all these conditions are in a state of continual fluctuation, the igneous and aqueous agents remodelling, from time to time, the physical geography of the globe, and the migrations of species causing new relations to spring up successively between different organic beings. We have deduced as a corollary, that the species existing at any particular period must, in the course of ages, become extinct one after the other. 'They must die out,' to borrow an emphatical expression from Buffon, 'because Time fights against them.'[44]

If the views which we have taken are just, there will be no difficulty in explaining why the habitations of so many species are now restrained within exceedingly narrow limits. Every local revolution, such as those contemplated in the preceding chapter, tends to circumscribe the range of some species, while it enlarges that of others; and as we have been led to infer that new species originate in one spot only, each must require time to diffuse itself over a wide area. The recent origin, therefore, of some species, and the high antiquity of others, may be equally consistent with the general fact of their limited distribution, some being local, because they have not existed long enough to admit of their wide dissemination; others, because circumstances in the animate or inanimate world have occurred to restrict the range which they may once have obtained.

As considerable modifications in the relative levels of land and sea have taken place, in certain regions, since the existing species were in being, we can feel no surprise that the zoologist and botanist have hitherto found it difficult to refer the geographical distribution of species to any clear and determinate principles, since they have usually speculated on the phenomena, upon the assumption that the physical geography of the globe had undergone no material alteration since the introduction of the species now living. So long as this assumption was made, the facts relating to the geography of plants and animals appeared capricious in the extreme, and by many the subject was pronounced to be so full of mystery and anomalies, that the establishment of a satisfactory theory was hopeless.

Some botanists conceived, in accordance with the hypothesis of Willdenow, that mountains were the centres of creation from which the plants now inhabiting large continents have radiated, to which Decandolle and others, with much reason, objected, that mountains, on the contrary, are often the barriers between two provinces of distinct vegetation. The geologist who is acquainted with the extensive modifications which the surface of the earth has undergone in very recent geological epochs, may be able, perhaps, to reconcile both these theories in their application to different regions.

A lofty range of mountains, which is so ancient as to date from a period when the species of animals and plants differed from those now living, will naturally form a barrier between contiguous provinces; but a chain which has been raised, in great part, within the epoch of existing species, and around which new lands have arisen from the sea within that period, will be a centre of peculiar vegetation.

'In France,' observes Decandolle, 'the Alps and Cevennes prevent a great number of the plants of the south from spreading themselves to the northward; but it has been remarked that some species have made their way through the gorges of these chains, and are found on their northern sides, principally in those places where they are lower and more interrupted.'[45] Now the chains here alluded to have probably been of considerable height, ever since the era when the existing vegetation began to appear, and were it not for the deep fissures which divide them, they might have caused much more abrupt terminations to the extension of distinct assemblages of species.

Parts of the Italian peninsula, on the other hand, have gained a considerable portion of their present height since a majority of the marine species now inhabiting the Mediterranean, and probably, also, since the terrestrial plants of the same region, were in being. Large tracts of land have been added, both on the Adriatic and Mediterranean side, to what originally constituted a much narrower range of mountains, if not a chain of islands running nearly north and south, like Corsica and Sardinia. It may, therefore, be presumed, that the Apennines have been a centre whence species have diffused themselves over the contiguous *lower* and *newer* regions. In this and all analogous situations, the doctrine of Willdenow, that species have radiated from the mountains as from centres, may be well founded.

It appears from Mr. Brown's remarks on the vegetation of New Holland, that there are two groups of plants occurring between the thirty-third and thirty-fifth degrees of southern latitude, and principally at the two opposite extremities of this tract, that is, near the eastern and western coasts. These points have been termed the two principal *foci* of Australian vegetation, each of them possessing certain genera which are almost peculiar to it. If, when this continent has been more thoroughly investigated, we do not discover some physical barriers, such as a great marsh, or a desert, or a lofty mountain-chain, now intervening between these districts, there may, perhaps, be geological evidence hereafter discovered, that a sea was interposed up to a modern period separating two large islands. Sufficient time may not have elapsed since the union of such isles, to allow of a complete intermixture by mutual immigrations.

If the reader should infer, from the facts laid before him in the preceding chapters, that the successive extinction of animals and plants may be part of the constant and regular course of nature, he will naturally inquire whether there are any means provided for the repair of these losses? Is it part of the economy of our system that the habitable globe should, to a certain extent, become depopulated both in the ocean and on the land; or that the variety of species should diminish until some new era arrives when a new and extraordinary effort of creative energy is displayed? Or is it possible that new species can be called into being from time to time, and yet that so astonishing a phenomenon can escape the observation of naturalists?

Humboldt has characterized these subjects as among the mysteries which natural science cannot reach; and he observes, that the investigation of the origin of beings does not belong to zoological or botanical geography. To geology, however, these topics do strictly appertain; and this science is only interested in inquiries into the state of the animate creation as it now exists, with a view of pointing out its relations to antecedent periods when its condition was different.

Before offering any hypothesis towards the solution of so difficult a problem, let us consider what kind of evidence we ought to expect, in the present state of science, of the first appearance of new animals or plants, if we could imagine the successive creation of species to constitute, like their gradual extinction, a regular part of the economy of nature.

In the first place it is obviously more easy to prove that a species, once numerously represented in a given district, has ceased to be, than that some other which did not pre-exist has made its appearance – assuming always, for reasons before stated, that single stocks only of each animal and plant are originally created, and that individuals of new species do not suddenly start up in many different places at once.

So imperfect has the science of Natural History remained down to our own times, that within the memory of persons now living, the numbers of known animals and plants have been doubled, or even quadrupled, in many classes. New and often conspicuous species are annually discovered in parts of the old continent, long inhabited by the most civilized nations. Conscious, therefore, of the limited extent of our information, we always infer, when such discoveries are made, that the beings in question had previously eluded our research; or had at least existed elsewhere, and only migrated at a recent period into the territories where we now find them. It is difficult even in contemplation to anticipate the time when we shall be entitled to make any other hypothesis in regard to all the marine tribes, and to by far the greater number of the terrestrial; – such as birds, which possess such unlimited powers of migration; insects which, besides their numbers, are also so capable of being diffused to vast distances; and cryptogamous plants, to which, as to many other classes, both of the animal and vegetable kingdom, similar observations are applicable.

What kind of proofs, therefore, could we reasonably expect to find of the origin at a particular period of a new species?

Perhaps it may be said in reply, that within the last two or three centuries some forest tree or new quadruped might have been observed to appear suddenly in those parts of England or France which had been most thoroughly investigated; – that naturalists might have been able to shew that no such being inhabited any other region of the globe, and that there was no tradition of anything similar having before been observed in the district where it had made its appearance.

Now although this objection may seem plausible, yet its force will be found to depend entirely on the rate of fluctuation which we suppose to prevail in the animate world, and on the proportion which such conspicuous subjects of the animal and vegetable kingdoms bear to those which are less known, and escape our observation. There are probably more than a million, perhaps two millions of species of plants and animals, exclusive of the microscopic and infusory animalcules, now inhabiting the terraqueous globe. The terrestrial plants, it is supposed, may amount, if fully known, to about one hundred thousand, and the insects to four times that number. To these we have still to add for the remainder of the terrestrial classes, many of the invertebrated and all the vertebrated animals. As to the aquatic tribes, it remains at present in a great degree mere matter of conjecture what proportion they bear to the denizens of the land; but the habitable surface beneath the waters can hardly be estimated at less than double that of the continents and islands, even admitting that a very considerable area is destitute of life, in consequence of great depth, cold, darkness, and other circumstances. In the late polar expedition it was found that in some regions, as in Baffin's Bay, there were marine animals inhabiting the bottom at great depths, where the temperature of the water was below the freezing point. That there is life at much greater profundities in warmer regions may be confidently inferred. We have before stated that marine plants not only exist but acquire vivid colours at depths where, to our senses, there would be darkness deep as night.

The ocean teems with life – the class of *polyps* alone are conjectured by Lamarck to be as strong in individuals as insects. Every tropical reef is described as bristling with corals, budding with sponges, and swarming with crustacea, echini, and testacea; while almost every tide-washed rock is carpeted with fuci and studded with corallines, actiniæ, and mollusca. There are innumerable forms in the seas of the warmer zones, which have scarcely begun to attract the attention of the naturalist; and there are parasitic animals without number, three or four of which are sometimes appropriated to one genus, as to the *Balæna*, for example. Even though we concede, therefore, that the geographical range of marine species is more extensive in general than that of the terrestrial, (the temperature of the sea being more uniform, and the land impeding less the migrations of the oceanic than the ocean those of the terrestrial,) yet we think it most probable that the aquatic species far exceed in number the inhabitants of the land.

Without insisting on this point, we may safely assume, as we before stated, that, exclusive of microscopic beings, there are between one and two millions of species now inhabiting the terraqueous globe; so that if only one of these were to become extinct annually, and one new one were to be every year called into being, more than a million of years would be required to bring about a complete revolution in organic life.

We are not hazarding at present any hypothesis as to the probable rate of change, but none will deny that when we propose as a mere speculation the *annual* birth and the *annual* death of one species on the globe, we imagine no slight degree of instability in the animate creation. If we divide the surface of the earth into twenty regions of equal area, one of these might comprehend a space of land and water about equal in dimensions to Europe, and might contain a twentieth part of the million of species which we will suppose to exist. In this region one species only would, according to the rate of mortality before assumed, perish in twenty years, or only five out of fifty thousand in the course of a century. But as a considerable proportion of the whole would belong to the aquatic classes, with which we have a very imperfect acquaintance, we may exclude them from our consideration, and thus one only might be lost in about forty years among the terrestrial tribes. Now the mammiferous quadrupeds in Great Britain are only to other terrestrial species of organic beings, both plants and animals, in the proportion of about one to two hundred and eighty; and taking this as a rude approximation to a general standard, it would require more than eight thousand years before it would come to the turn of this conspicuous class to lose one of their number even in a region of the dimensions of Europe.

It is easy, therefore, to conceive, that in a small portion of such an area, in countries, for example, of the size of England and France, periods of much greater duration must elapse before it would be possible to authenticate the first appearance of one of the larger plants and animals, assuming the annual birth and death of one species to be the rate of vicissitude in the animate creation throughout the world.

The observations of naturalists may, in the course of future centuries, accumulate positive data, from which an insight into the laws which govern this part of our terrestrial system may be derived; but, in the present deficiency of historical records, we have traced up the subject to that point where geological monuments alone are capable of leading us on to the discovery of ulterior truths. To these, therefore, we must now appeal, carefully examining the strata of recent formation wherein the remains of *living* species, both animal and vegetable, are known to occur. We must study these strata in strict reference to their chronological order as deduced from their superposition, and other relations. From these sources we may learn which of the species, now our contemporaries, have survived the greatest revolutions of the earth's surface; which of them have co-existed with the greatest number of animals and plants now extinct, and which have made their appearance only when the animate world had nearly attained its present condition.

From such data we may be enabled to infer whether species have been called into existence in succession or all at one period; whether singly, or whether by groups simultaneously; whether the antiquity of man be as high as that of any of the inferior beings which now share the planet with him, or whether the human species is one of the most recent of the whole.

To some of these questions we can even now return a satisfactory answer; and with regard to the rest, we have some data to guide conjecture, and to enable us to speculate with advantage: but it would be premature to anticipate such discussions until we have laid before the reader an ample body of materials amassed by the industry of modern geologists.

CHAPTER 12

Modifications in Physical Geography
Caused by Plants, the Inferior Animals,
and Man

The second branch of our inquiry, respecting the changes of the organic world, relates to the effects produced by the powers of vitality on the state of the earth's surface, and on the material constituents of its crust.

By the effects produced on the surface, we mean those modifications in physical geography of which the existence of organic beings is the direct cause, – as when the growth of certain plants covers the slope of a mountain with peat, or converts a swamp into dry land; or when vegetation prevents the soil, in certain localities, from being washed away by running water.

By the agency of the powers of vitality on the material constituents of the earth's crust, we mean those permanent modifications in the composition and structure of new strata, which result from the imbedding therein of animal and vegetable remains. In this case, organic beings may not give rise immediately to any new features in the physical geography of certain tracts, which would not equally have resulted from the mere operation of inorganic causes; as, for example, if a lake be filling up with sediment, held in suspension by the waters of some river, and with mineral matter precipitated from the waters of springs, the character of the deposits may be modified by aquatic animals and plants, which may convert the earthy particles into shell, peat, and other substances: but the lake may, nevertheless, be filled up in the same time, and the new strata may be deposited in nearly the same order as would have prevailed if its waters had never been peopled by living beings.

In treating of the first division of our subject we may remark, that when we talk of alterations in physical geography, we are apt to think too exclusively of that part of the earth's surface which has emerged from beneath the waters, and with which alone, as terrestrial beings, we are familiar. Here the direct power of animals and plants to cause any important variations is, of necessity, very limited, except in checking the progress of that decay of which the land is the chief theatre. But if we extend our views, and instead of contemplating the dry land we consider that larger portion which is assigned to the aquatic tribes, we discover the immediate influence of the living creation, in imparting varieties of conformation to the solid exterior which the sole agency of inanimate causes would not produce, to be very great.

Thus, when timber is floated into the sea, it is often drifted to vast distances and subsides in spots where there might have been no deposit, at that time and place, if the earth had not been tenanted by living beings. If, therefore, in the course of ages, a hill of wood, or lignite, be thus formed in the subaqueous regions, a change in the submarine geography may be said to have resulted from the action of organic powers. So in regard to the growth of coral reefs: it is probable that almost all the matter of which they are composed is supplied by mineral springs, which we know often rise up at the bottom of the sea, and which, on land, abound throughout volcanic regions thousands of miles in extent. The matter thus constantly given out could not go on accumulating for ever in the waters, but would be precipitated in the abysses of the sea, even if there were no polyps and testacea; but these animals arrest and secrete the carbonate of lime on the summits of submarine mountains, and form reefs many hundred feet in thickness, and hundreds of leagues in length, where, but for them, none might ever have existed.

If no such voluminous masses are formed on the land, it is not from the want of solid matter in the structure of terrestrial animals and plants, but merely because, as we have so often stated, the continents are those parts of the globe where accessions of matter can scarcely ever take place, – where, on the contrary, the most solid parts already formed are, each in their turn, exposed to gradual degradation. The quantity of timber and vegetable matter which grows in a tropical forest in the course of a century is enormous, and multitudes of animal skeletons are scattered there in the same period, besides innumerable land-shells and other organic substances. The aggregate of these materials might constitute, perhaps, a mass greater in volume than that which is produced in any coral-reef during the same lapse of years; but, although this process should continue on the land for ever, no mountains of wood or bone would be seen stretching far and wide over the country, or pushing out bold promontories into the sea.

The whole solid mass is either devoured by animals, or decomposes, as does a portion of the rock and soil on which the animals and plants are supported. For the decomposition of the strata themselves, especially of their alkaline ingredients and of the organic remains which they so frequently include, is one source from whence running water and the atmosphere may derive the materials which are absorbed by the roots and leaves of plants. Another source is the passage into a gaseous form of even the hardest parts of animals and plants which die and are exposed to putrefy in the air, where they are soon resolved into the elements of which they are composed; and while a portion of these parts is volatilized, the rest is taken up by rain-water and sinks into the earth or flows towards the sea, so that they enter again and again into the composition of different organic beings.

The principal elements found in plants are hydrogen, carbon, and oxygen, so that water and the atmosphere contain all of them, either in their own composition or in solution. The constant supply of these elements is maintained not only by the putrefaction of animal and vegetable substances, and the decay of rocks before mentioned, but also by the copious evolution of carbonic acid and other gases from volcanos and mineral springs, and by the effects of ordinary evaporation, whereby aqueous vapours are made to rise from the ocean and to circulate round the globe.

It is well known that when two gases of different specific gravity are brought into contact, even though the heavier be the lowermost, they become uniformly diffused by mutual absorption through the whole space which they occupy. By virtue of this law, the heavy carbonic acid finds its way upwards through the lighter air, and conveys nourishment to the lichen which covers the mountain top.

The fact, therefore, that the vegetable mould which covers the earth's surface does not decrease in thickness, will not altogether bear out the argument which was founded upon it by Playfair. This vegetable soil, he observes, consists partly of loose earthy materials easily removed, in the form of sand and gravel, partly of finer particles suspended in the waters, which tinge those of some rivers continually, and those of all occasionally, when they are flooded. The soil, although continually diminished from this cause, 'remains the same in quantity, or at least nearly the same, and must have done so ever since the earth was the receptacle of animal or vegetable life. The soil, therefore, is augmented from other causes, just as much, at an average, as it is diminished by that now mentioned; and this augmentation evidently can proceed from nothing but the constant and slow disintegration of the rocks.'[46]

That the repair of the *earthy* portion of the soil can only proceed, as Playfair suggests, from the decomposition of rocks, may be admitted; but the *vegetable* matter may be supplied, and is actually furnished in a great degree, by absorption from the atmosphere, as we before mentioned, so that in level situations, such as in platforms that intervene between valleys where the action of running water is very trifling, the fine vegetable particles carried off by the rain may be perpetually restored, not by the waste of the rock below, but from the air above.

If we supposed the quantity of food consumed by terrestrial animals, together with the matter absorbed by them in breathing, and the elements imbibed by the roots and leaves of plants, to be derived entirely from that supply of hydrogen, carbon, oxygen, azote, and other elements, given out into the atmosphere and the waters by the putrescence of organic substances, then we might imagine that the vegetable mould would, after a series of years, neither gain nor lose a single particle by the action of organic beings. This conclusion is not far from the truth; but the operation which renovates the vegetable and animal mould is by no means so simple as that here supposed. Thousands of carcasses of terrestrial animals are floated down every century into the sea, and, together with forests of drift-timber, are imbedded in subaqueous deposits, where their elements are imprisoned in solid strata, and may there remain throughout whole geological epochs before they again become subservient to the purposes of life.

On the other hand, fresh supplies are derived by the atmosphere, and by running water, as we before stated, from the disintegration of rocks and their organic contents, and from the interior of the earth, from whence all the elements before-mentioned, which enter principally into the composition of animals and vegetables, are continually evolved. Even nitrogen has been recently found to be contained very generally in the waters of mineral springs.

If we suppose that the copious discharge from the nether regions, by springs and volcanic vents, of carbonic acid and other gases, together with the decomposition of rocks, may be just sufficient to counterbalance that loss of matter which, having already served for the nourishment of animals and plants, is annually carried down in organized forms, and buried in subaqueous strata, we believe that we concede the utmost that is consistent with probability. When more is required by a theorist – when we are told that a counterpoise is derived from the same source to that enormous disintegration of solid rock and its transportation to lower levels, which is the annual result of the action of rivers and marine currents, we must entirely withhold our assent. Such an opinion has been recently advanced by an eminent geologist, or we should have deemed it unnecessary to dwell on propositions which appear to us so clear and obvious.

The descriptions which we gave of the degradation yearly going on through the eastern shores of England, and of the enormous weight of solid matter hourly rolled down by the Ganges or the Mississippi, have been represented as extreme cases, calculated to give a partial view of the changes now in progress, especially as we omitted, it is said, to point out the silent but universal action of a great antagonist power, whereby the destructive operations before alluded to are neutralized, and even, in a great degree, counterbalanced.

'Are there,' says Professor Sedgwick, 'no *antagonist* powers in nature to oppose these mighty ravages – no conservative principle to meet this vast destructive agency? The forces of degradation very often of themselves produce their own limitation. The mountain-torrent may tear up the solid rock and bear its fragments to the plain below; but there its power is at an end, and the rolled fragments are left behind to a new action of material elements. And what is true of a single rock, is true of a mountain-chain; and vast regions on the surface of the earth, now only the monuments of spoliation and waste, may hereafter rest secure under the defence of a thick vegetable covering, and become a new scene of life and animation.'

'It well deserves remark that the destructive powers of nature act only upon lines, while some of the grand principles of conservation act upon the whole surface of the land. By the processes of vegetable life an incalculable mass of solid matter is absorbed, year after year, from the elastic and non-elastic fluids circulating round the earth, and is then thrown down upon its surface. In this *single* operation there is a *vast counterpoise to all* the agents of destruction. And the deltas of the Ganges and the Mississippi are not solely formed at the expense of the solid materials of our globe, but in part, and I believe also in a considerable part, by one of the great conservative operations by which the elements are made to return into themselves.'[47]

This is splendid eloquence, full of the energy and spirit that breathes through the whole address: –

... a mighty raging Flood,
That from some Mountain flows,
Rapid, and warm, and deep, and loud,
Whose Force no Limits knows.[48]

but we must pause for a moment, lest we be hurried away by its tide. Let us endeavour calmly to consider whither it would carry us.

If by the elements returning *into themselves* be meant their return to higher levels, it is certainly possible that a fraction of the organic matter which is intermixed with the mud and sand deposited in alternate strata in the delta of the Ganges, may have been derived by the leaves and roots of plants from such aqueous vapour, carbonic acid, and other gases, as had ascended into the atmosphere from *lower* regions, and which were not, therefore, derived from the waste of rocks and their organic contents, or from the putrescence of vegetables previously nourished from these sources. This fraction, and this alone, may then be deducted from the mass of solid matter annually transported into the Bay of Bengal, and what remains, whether organic or inorganic, will be the measure of the degradation which thousands of torrents in the Himalaya mountains, and many rivers of other parts of India, bring down in a single year. Even in this case it will be found that the sum of the force of vegetation can merely be considered as having been in a slight degree *conservative*, retarding the waste of land, and not acting as an antagonist power.

But the untenable nature of the doctrine which we are now controverting may be set in a clearer light by examining the present state of the earth's surface, on which it is declared that 'an incalculable mass of solid matter is thrown down year after year,' in such a manner as to form a counterpoise to the agents of decay.[49] Is it not a fact that the vegetable mould is seldom more than a few feet in thickness, and that it often does not exceed a few inches? Do we find that its volume is more considerable on those parts of our continents which we can prove, by geological data, to have been elevated at more ancient periods, and where there has been the greatest time for the accumulation of vegetable matter, produced throughout successive zoological epochs? On the contrary, are not these higher and older regions more frequently denuded, so as to expose the bare rock to the action of the sun and air?

Do we find in the torrid zone, where the growth of plants is most rank and luxurious, that accessions of matter due to their agency are most conspicuous on the surface of the land? On the contrary, is it not there where the vegetation is most active that, for reasons to be explained in the next chapter, even those superficial peat mosses are unknown which cover a large area in some parts of our temperate zone? If the operation of animal and vegetable life could restore to the general surface of the continents a portion of the elements of those disintegrated rocks, of which such enormous masses are swept down annually into the sea, along particular river-courses and lines of coast, the effects would have become ere now most striking; and would have constituted one of the most leading features in the structure and composition of our continents. All the great steppes and table-lands of the world, where the action of running water is feeble, would have become the grand repositories of organic matter, accumulated without that intermixture of sediment which so generally characterizes the subaqueous strata.

Even the formation of peat in certain districts where the climate is cold and moist, the only case, perhaps, which affords the shadow of a support to the theory under consideration, has not in every instance a conservative tendency. A peat-moss often acts like a vast sponge, absorbing water in large quantities, and swelling to the height of many yards above the surrounding country. The turfy covering of the bog serves, like the skin of a bladder, to retain for a while the fluid within, and a violent inundation sometimes ensues when that skin bursts, as has often happened in Ireland, and many parts of the continent. Examples will be mentioned by us in a subsequent chapter, where the Stygian torrent has hollowed out ravines and borne along rocks and sand, in countries where such ravages could not have happened but for the existence of peat. Here, therefore, the force of vegetation accelerates the rate of decay of land, and the solid matter swept down to lower levels during such floods, counterbalances, to a certain degree, the accessions of vegetable mould which may accrue to the land by the growth of peat.

We may explain more clearly the kind of force which we imagine vegetation to exert, by comparing it to the action of frost, which augments the height of some few Alpine summits by causing a mass of perpetual snow to lodge thereon, or fills up some valleys with glaciers; but although by this process of congelation the rain-water that has risen by evaporation from the sea, is retained for awhile in a solid form upon the land, and although some elevated spots may be protected from waste by a constant covering of ice, yet by the sudden melting of snow and ice, the degradation of rocks is often accelerated. Although every year fresh snow and ice are formed, as also more vegetable and animal matter, yet there is no increase; the one melts, the other putrifies, or is drifted down to the sea by rivers. If this were not the case, frost might be considered as an antagonist power, as well as the action of animal and vegetable life, and these by their combined energy might restore to continents a portion of that solid matter which is swept down into the sea from mountains and wasting cliffs. By the aid of such machinery might a theorist repair the losses of the solid land, sand and rocky fragments being carried down annually to the subaqueous regions from hills of granite, limestone, and shale, while vegetation and frost might raise new mountains, which, like the cliffs in Eschscholtz's Bay, might consist of icebergs, intermixed with vegetable mould.

We have stated in a former volume that, in the known operation of the *igneous* causes, a real antagonist power is found which may counterbalance the levelling action of running water; and there seems no good reason for presuming that the upheaving and depressing force of earthquakes, together with the heaping up of ejected matter by volcanos, may not be fully adequate to restore the superficial inequalities which rivers and oceanic currents annually tend to lessen. If a counterpoise be derived from this source, the quantity and elevation of land above the sea may for ever remain the same, in spite of the action of the aqueous causes, which, if thus counteracted, may never be able to reduce the surface of the earth more nearly to a state of equilibrium than that which it has now attained; and, on the other hand, the force of the aqueous agents themselves might thus continue for ever unimpaired. This permanence of the intensity of the powers now in operation would account for any amount of disturbance or degradation of the earth's crust, so far as the *mere quantity* of movement or decay is concerned; provided only that indefinite periods of time are contemplated.

As to the *intensity* of the disturbing causes at particular epochs, their effects have as yet been studied for too short a time to enable us fully to compare the signs of ancient convulsions with the permanent monuments left in the earth's crust by the events of the last few thousand years. But notwithstanding the small number of changes which have been witnessed and carefully recorded, observation has at least shown that our knowledge of the extent of the subterranean agency, as now developed from time to time, is in its infancy; and there can be no doubt that great partial mutations in the structure of the earth's crust are brought about in volcanic regions, without any interruption to the general tranquillity of the habitable surface.

Some geologists point to particular cases of enormous dislocation of ancient date, and confessedly not of frequent occurrence, where shifts in the strata of two thousand feet and upwards appear to have been produced suddenly and at one effort. But they have been at no pains to prove that similar consequences could not result from earthquakes such as have happened within the last three thousand years. They have usually proceeded on *à priori* reasoning to assume that such convulsions were paroxysmal, and attended by catastrophes such as have never occurred in modern times. It would be irrelevant to the subject immediately under consideration to enter into a long digression on these topics, but we may remind the reader, that the subsidence of the quay at Lisbon to the depth of six hundred feet only gave rise to a slight whirlpool; and we may thence infer the possibility of a sinking down or elevation four or five times as great, especially in deeper seas, without any superficial disturbance unparalleled in the events of the last century.

If a certain sect of geologists were as anxious to reconcile the actual and former course of nature as they are eager to contrast them, they would perceive that the effects witnessed by us of subterranean action on supramarine land, may not be a type of those which the submerged rocks undergo, and they would proceed with more caution when reasoning from a comparison between the accumulated results of disturbing causes in the immensity of past time, and those which are recorded in the meagre annals of a brief portion of the human era.

The same rash generalizations which are now made respecting eras of paroxysmal violence and chaotic derangement, led formerly to the doctrines of universal formations, the improbability of which might have been foreseen by a slight reference to the causes now in operation.

To the same modes of philosophising we may ascribe the unwillingness of some naturalists to admit, that all the fossil species are not the same as those now living on the globe; whereas, if the facts and reasoning set forth in a former part of this volume, respecting the present instability of the organic creation be just, it might always *à priori* have been seen that the species inhabiting the planet at two periods very remote could hardly be identical.

In our view of the Huttonian theory, we pointed out as one of its principal defects, the assumed want of synchronism in the action of the great antagonist powers – the introduction, first, of periods when continents gradually wasted away, and then of others when new lands were elevated by violent convulsions. In order to have a clear conception of the working of such a system, let the reader suppose the earthquakes and volcanic eruptions of the Andes to be suspended for a million of years, and sedimentary deposits to accumulate throughout the whole of that period, as they now accumulate at the mouths of the Orinoco and Amazon, and along the intervening coast. Then let a period arrive when the subterranean power, which had obtained no vent during those ten thousand centuries, should escape suddenly in one tremendous explosion.

It is natural that geologists who reject such portions of the Huttonian theory as we embrace, should cling fondly to those parts which we deem unsound and unphilosophical. They have accordingly selected the distinctness of the periods when the antagonist forces are developed, as a principle peculiarly worthy of implicit faith. For this reason they have declined making any strenuous effort to account for those violations of continuity in the series of geological phenomena which are exhibited in large but limited regions; and which we have hinted may admit of explanation by the shifting of the volcanic foci, without the necessity of calling in to our aid any hypothetical eras of convulsion.

In the Oriental cosmogonies, as we have seen, both the physical and moral worlds were represented to be subject to gradual deterioration, until a crisis arrived when they were annihilated, or reverted to a state of chaos; – there had been alternating periods of tranquillity and disorder – an endless vicissitude of destructions and renovations of the globe.

In the spirit of this antique philosophy, some modern geologists conceive that nature, after long periods of repose, is agitated by fits of 'feverish spasmodic energy, during which her very frame-work is torn asunder;' – these paroxysms of internal energy are accompanied by the sudden elevation of mountain chains, 'followed by mighty waves desolating whole regions of the earth';[50] and, according to some authors, whole races of organic beings are thus suddenly annihilated.

It was to be expected that when, in opposition to these favourite dogmas, we enumerated the subterranean catastrophes of the last one hundred and forty years, pointing out how defective were our annals, and called on geologists to multiply the amount of disturbances arising from this source by myriads of ages during the existence of successive races of organic beings, that we should provoke some vehement expostulation. We could not hope that the self-appointed guardians of Nature's slumber would allow us with impunity thus suddenly to intrude upon her rest, or that they would fail to resent so rude an attempt to rouse her from the torpor into which she had been lulled by their hypothesis. We were prepared to see our proofs and authorities severely sifted, our inferences rigorously scrutinized; but we never supposed it possible that our adversaries would set up 'as a vast counterpoise to all the agents of destruction,' a cause so nugatory as 'the single operation of vegetable life.'[51]

As it will appear from what we before said, that vegetation cannot act as an antagonist power amid the mighty agents of change which are always modifying the surface of the globe, let us next inquire how far its influence is conservative, – how far it may retard the levelling power of running water, which it cannot oppose, much less counterbalance.

It is well known that a covering of herbage and shrubs may protect a loose soil from being carried away by rain, or even by the ordinary action of a river, and may prevent hills of loose sand from being blown away by the wind. For the roots bind together the separate particles into a firm mass, and the leaves intercept the rain-water, so that it dries up gradually instead of flowing off in a mass and with great velocity. The old Italian hydrographers make frequent mention of the increased degradation which has followed the clearing away of natural woods in several parts of Italy. A remarkable example was afforded in the Upper Val d'Arno, in Tuscany, on the removal of the woods clothing the steep declivities of the hills by which that valley is bounded. When the ancient forest laws were abolished by the Grand Duke Joseph, during the last century, a considerable tract of surface in the Cassentina (the Clausentinium of the Romans) was denuded, and, immediately, the quantity of sand and soil washed down into the Arno increased enormously. Frisi, alluding to such occurrences, observes, that as soon as the bushes and plants were removed, the waters flowed off more rapidly, and, in the manner of floods, swept away the vegetable soil.

This effect of vegetation is of high interest to the geologist, when he is considering the formation of those valleys which have been principally due to the action of rivers. The spaces intervening between valleys, whether they be flat or ridgy, when covered with vegetation, may scarcely undergo the slightest waste, as the surface may be protected by the green sward of grass; and this may be renewed, in the manner before described, from elements derived from rain-water and the atmosphere. Hence, while the river is continually bearing down matter in the alluvial plain, and undermining the cliffs on each side of every valley, the height of the intervening rising grounds may remain stationary.

In this manner a cone of loose scoriæ, sand and ashes, such as Monte Nuovo, may, when it has once become densely clothed with herbage and shrubs, suffer scarcely any farther dilapidation; and the perfect state of the cones of hundreds of extinct volcanos in France, Campania, Sicily, and elsewhere, may prove nothing whatever, either as to their relative or absolute antiquity. We may be enabled to infer from the integrity of such conical hills of incoherent materials, that no flood can have passed over the countries where they are situated since their formation; but the atmospheric action alone in spots where there happen to be no torrents, and where the surface was clothed with vegetation, could scarcely in any lapse of ages have destroyed them.

During a late tour in Spain I was surprized to see a district of gently undulating ground in Catalonia, consisting of red and grey sandstone, and in some parts of red marl, almost entirely denuded of herbage, while the roots of the pines, holm oaks, and some other trees were half exposed, as if the soil had been washed away by a flood. Such is the state of the forests, for example, between Orista and Vich, and near San Lorenzo. Being at length overtaken by a violent thunderstorm, in the month of August, I saw the whole surface, even the highest levels of some flat-topped hills, streaming with mud, while on every declivity the devastation of torrents was terrific. The peculiarities in the physiognomy of the district were at once explained, and I was taught that in speculating on the greater effects which the direct action of rain may once have produced on the surface of certain parts of England, we need not revert to periods when the heat of the climate was *tropical*.

In the torrid zone the degradation of land is generally more rapid, but the waste is by no means proportioned to the superior quantity of rain or the suddenness of its fall, the transporting power of water being counteracted by a greater luxuriance of vegetation. A geologist who is no stranger to tropical countries observes, that the softer rocks would speedily be washed away in such regions, if the numerous roots of plants were not matted together in such a manner as to produce considerable resistance to the destructive power of the rains. The parasitical and creeping plants also entwine in every possible direction so as to render the forests nearly impervious, and the trees possess forms and leaves best calculated to shoot off the heavy rains, which when they have thus been broken in their fall are quickly absorbed by the ground beneath, or when thrown into the drainage depressions give rise to furious torrents.

The felling of forests has been attended, in many countries, by a diminution of rain, as in Barbadoes and Jamaica. For in tropical countries, where the quantity of aqueous vapour in the atmosphere is very great, but where, on the other hand, the direct rays of the sun have immense power, any impediment to the free circulation of air, or any screen which shades the earth from the solar rays, becomes a powerful cause of humidity, and wherever dampness and cold have begun to be generated by such causes, the condensation of vapour continues. The leaves moreover of all plants are alembics, and some of those in the torrid zone have a remarkable power of distilling water, thus contributing to prevent the earth from becoming parched up.

There can be no doubt that the state of the climate, especially the humidity of the atmosphere, influences vegetation, and that, in its turn, vegetation reacts upon the climate; but some writers seem to have attributed too much importance to the influence of forests, particularly those of America, as if they were the primary cause of the moisture of the climate.

The theory of a modern author on this subject, 'that forests exist in those parts of America only where the predominant winds carry with them a considerable quantity of moisture from the ocean,' seems far more rational. In all countries, he says, 'having a summer heat exceeding 70°, the presence or absence of natural woods, and their greater or less luxuriance, may be taken as a measure of the amount of humidity, and of the fertility of the soil. Short and heavy rains, in a warm country, will produce grass, which, having its roots near the surface, springs up in a few days, and withers when the moisture is exhausted; but transitory rains, however heavy, will not nourish trees, because, after the surface is saturated with water, the rest runs off, and the moisture lodged in the soil neither sinks deep enough, nor is in sufficient quantity, to furnish the giants of the forest with the necessary sustenance. It may be assumed, that twenty inches of rain falling moderately, or at intervals, will leave a greater permanent supply in the soil than forty inches falling, as it sometimes does in the torrid zone, in as many hours.'

'In all regions,' he continues, 'where ranges of mountains intercept the course of the constant or predominant winds, the country on the windward side of the mountains will be moist, and that on the leeward dry, and hence parched deserts will generally be found on the west side of countries within the tropics, and on the east side of those beyond them, the prevailing winds in these cases being generally in opposite directions. On this principle, the position of forests in North and South America may be explained. Thus, for example, in the region within the thirtieth parallel, the moisture swept up by the trade-wind from the Atlantic is precipitated in part upon the mountains of Brazil, which are but low and so distributed as to extend far into the interior. The portion which remains is borne westward, and, losing a little as it proceeds, it is at length arrested by the Andes, where it falls down in showers on their summits. The aërial current, now deprived of all the humidity with which it can part, arrives in a state of complete exsiccation at Peru, where, consequently, no rain falls. In the same manner the Ghauts in India, a chain only three or four thousand feet high, intercept the whole moisture of the atmosphere, having copious rains on their windward side, while on the other the weather remains clear and dry. The rains in this case change regularly from the west side to the east, and vice versâ, *with the monsoons*. But in the region of America, beyond the thirtieth parallel, the Andes serve as a screen to intercept the moisture brought by the prevailing winds from the Pacific Ocean; rains are copious on their summits, and in Chili on their *western* declivities; but none falls on the plains to the *eastward*, except occasionally when the wind blows from the Atlantic.'[52]

We have been more particular in explaining these views, because they appear to us to place in a true light the dependence of vegetation on climate, notwithstanding the reciprocal action which each exerts on the other, the humidity being increased, and more uniformly diffused throughout the year, by the gradual spreading of wood.

Before concluding this chapter, we must offer a few observations on the influence of man in modifying the physical geography of the globe, for we must class his agency among the powers of organic nature.

The modifications of the surface, resulting from human agency, are only on a considerable scale when we have obtained so much knowledge of the working of the laws of nature as to be able to use them as instruments to effect our purposes. We must command nature by obeying her laws, according to the saying of the philosopher, and for this reason we can never materially interfere with any of the great changes which either the aqueous or igneous causes are bringing about on the earth. In vain would the inhabitants of Italy strive to prevent the tributaries of the Po and Adige from bearing down, annually, an immense volume of sand and mud from the Alps and Apennines; in vain would they toil to re-convey to the mountains the mass torn from them year by year, and deposited in the form of sediment in the Adriatic. But they have, nevertheless, been able to vary the distribution of this sediment over a considerable area, by embanking the rivers, and preventing the sand and mud from being spread, by annual inundations, over the plains.

We have explained how the form of the delta of the Po has been altered by this system of embankment, and how much more rapid, in consequence of these banks, have been the accessions of land at the mouths of the Po and Adige within the last twenty centuries. There is a limit, however, to these modifications, since the danger of floods augments with the increasing height of the river-beds, while the expense of maintaining the barrier is continually enhanced, as well as the difficulty of draining the low surrounding country.

In the Ganges, says Major R. H. Colebrooke, no sooner is a slight covering of soil observed on a new sand-bank, than the island is cultivated; water-melons, cucumbers, and mustard, become the produce of the first year, and rice is often seen growing near the water's edge, where the mud is in large quantity. Such islands may be swept away before they have acquired a sufficient degree of stability to resist permanently the force of the stream; but if, by repeated additions of soil, they acquire height and firmness, the natives take possession, and bring over their families, cattle and effects. They choose the highest spots for the sites of villages, where they erect their dwellings with as much confidence as they would do on the main land; for although the foundation is sandy, the uppermost soil being interwoven with the roots of grass and other plants, and hardened by the sun, is capable of withstanding all attacks of the river. These islands often grow to a considerable size, and endure for the lives of the new possessors, being only at last destroyed by the same gradual process of undermining and encroachment to which the banks of the Ganges are subject.

If Bengal were inhabited by a nation more advanced in opulence and agricultural skill, they might, perhaps, succeed in defending these possessions against the ravages of the stream for much longer periods; but no human power could ever prevent the Ganges, or the Mississippi, from making and unmaking islands. By fortifying one spot against the set of the current, its force is only diverted against some other point; and, after a vast expense of time and labour, the property of individuals may be saved, but no addition would thus be made to the sum of productive land. It may be doubted, whether any system could be devised so conducive to *national* wealth, as the simple plan pursued by the peasants of Hindostan, who, wasting no strength in attempts to thwart one of the great operations of nature, permit the alluvial surface to be perpetually renovated, and find their losses in one place compensated in some other, so that they continue to reap an undiminished harvest from a virgin soil.

To the geologist, the Gangetic islands, and their migratory colonies, may present an epitome of the globe as tenanted by man. For during every century we cede some territory which the earthquake has sunk, or the volcano has covered by its fiery products, or which the ocean has devoured by its waves. On the other hand, we gain possession of new lands, which rivers, tides, or volcanic ejections have formed, or which subterranean causes have upheaved from the deep. Whether the human species will outlast the whole, or a great part of the continents and islands now seen above the waters, is a subject far beyond the reach of our conjectures; but thus much may be inferred from geological data, – that if such should be its lot, it will be no more than has already fallen to pre-existing species, some of which have, ere now, outlived the form and distribution of land and sea which prevailed at the era of their birth.

We have before shown, when treating of the excavation of new estuaries in Holland by inroads of the ocean, as also of the changes on our own coasts, that although the conversion of sea into land by artificial labours may be great, yet it must always be in subordination to the great movements of the tides and currents. If, in addition to the assistance obtained by parliamentary grants for defending Dunwich from the waves, all the resources of Europe had been directed to the same end, the existence of that port might possibly have been prolonged for many centuries. But, in the meantime, the current would have continued to sweep away portions from the adjoining cliffs on each side, rounding off the whole line of coast into its present form, until at length the town must have projected as a narrow promontory, becoming exposed to the irresistible fury of the waves.

It is scarcely necessary to observe, that the control which man can exert over the igneous agents is less even than that which he may obtain over the aqueous. He cannot modify the upheaving or depressing force of earthquakes, or the periods or degree of violence of volcanic eruptions; and on these causes the inequalities of the earth's surface, and, consequently, the shape of the sea and land, appear mainly to depend. The utmost that man can hope to effect in this respect, is occasionally to divert the course of a lava-stream, and to prevent the burning matter, for a season at least, from overwhelming a city, or other fruit of human industry.

No application, perhaps, of human skill and labour tends so greatly to vary the state of the habitable surface, as that employed in the drainage of lakes and marshes, since not only the *stations* of many animals and plants, but the general climate of a district, may thus be modified. It is also a kind of alteration to which it is difficult, if not impossible, to find anything analogous in the agency of inferior beings. For we ought always, before we decide that any part of the influence of man is novel and anomalous, carefully to consider all the powers of other animate agents which may be limited or superseded by him. Many who have reasoned on these subjects seem to have forgotten that the human race often succeeds to the discharge of functions previously fulfilled by other species; a topic on which we have already offered some hints, when explaining how the distribution and numbers of each species are dependent on the state of contemporary beings.

Suppose the growth of some of the larger terrestrial plants, or, in other words, the extent of forests, to be diminished by man, and the climate to be thereby modified, it does not follow that this kind of innovation is unprecedented. It is a change in the state of the vegetation, and such may often have been the result of the entrance of new species into the earth. The multiplication, for example, of certain insects in parts of Germany, during the last century, destroyed more trees than man, perhaps, could have felled during an equal period.

It is a curious fact, to which we shall again advert, that the sites of many European forests, cut down since the time of the Romans, have become peat-mosses; and thus a permanent change has been effected in these regions. But other woods, blown down by winds, in the same countries, have also become peat-bogs; so that, although man may have accelerated somewhat the change, yet it may be doubted whether other animate and inanimate causes might not, without his interference, have produced similar results. The atmosphere of our latitudes may have been slowly and insensibly cooling down since the ancient forests began to grow, and the time may have arrived when slight accidents were sufficient to cause the decrease of trees, and the usurpation of their site by other plants.

We do not pretend to decide how far the power of man, to modify the surface, may differ in kind or degree from that of other living beings, but we suspect that the problem is more complex than has been imagined by many who have speculated on such topics. If new land be raised from the sea, the greatest alteration in its physical condition, which could ever arise from the influence of organic beings, would probably be produced by the first immigration of terrestrial plants, whereby the tract would become covered with vegetation. The change next in importance would seem to be when animals enter, and modify the proportionate numbers of certain species of plants. If there be any anomaly in the intervention of man, in farther varying the relative numbers in the vegetable kingdom, it may not so much consist in the kind or absolute quantity of alteration, as in the circumstance that *a single species*, in this case, would exert, by its superior power and universal distribution, an influence equal to that of hundreds of other terrestrial animals.

If we inquire whether man, by his direct removing power, or by the changes which he may give rise to indirectly, tends, upon the whole, to lessen or increase the inequalities of the earth's surface, we shall incline, perhaps, to the opinion that he is a levelling agent. He conveys upwards a certain quantity of materials from the bowels of the earth in mining operations; but, on the other hand, much rock is taken annually from the land, in the shape of ballast, and afterwards thrown into the sea, whereby, in spite of prohibitory laws, many harbours, in various parts of the world, have been blocked up. We rarely transport heavy materials to higher levels, and our pyramids and cities are chiefly constructed of stone brought down from more elevated situations. By ploughing up thousands of square miles, and exposing a surface for part of the year to the action of the elements, we assist the abrading force of rain, and destroy the conservative effects of vegetation.

But the aggregate force exerted by man is truly insignificant, when we consider the operations of the great physical causes, whether aqueous or igneous, in the inanimate world. If all the nations of the earth should attempt to quarry away the lava which flowed during one eruption from the Icelandic volcanoes in 1783 and the two following years, and should attempt to consign it to the deepest abysses of the ocean, wherein it might approach most nearly to the profundities from which it rose in the volcanic vent, they might toil for thousands of years before their task was accomplished. Yet the matter borne down by the Ganges and Burrampooter, in a single year, probably exceeds, in weight and volume, the mass of Icelandic lava produced by that great eruption.

How the Remains of Man and his Works are becoming Fossil beneath the Waters

We now come to the second subdivision of the inquiry explained in the preceding chapter, – the consideration of the permanent modifications produced in the material constituents of the earth's crust, by the action of animal and vegetable life.

New mineral compounds, such as might never have existed in this globe but for the action of the powers of vitality, are annually formed, and made to enter into deposits accumulated both above and beneath the waters. Although we can neither explain nor imitate the processes of animal and vegetable life whereby those substances are produced, yet we can investigate the laws by virtue of which organic matter becomes imbedded in new strata, – sometimes imparting to them a peculiar mineral composition, – sometimes leaving durable impressions and casts of the forms of animate beings in rocks, so as to modify their structure and appearance.

It has been well remarked by M. Constant Prevost, that the effects of geological causes are divisible into two great classes; those produced on the surface during the immersion of land beneath the waters, and those which take place after its emersion. Agreeably to this classification we shall consider, first, in what manner animal and vegetable remains become included and preserved in solid deposits on emerged land, or that part of the surface which is not *permanently* covered by water, whether of the sea or lakes; secondly, the manner in which organic remains become imbedded in sub-aqueous deposits.

* * *

[*Before fossils can be used to reconstruct the record of life on earth, the conditions through which the remains of once-living organisms are preserved must be understood. In the next five chapters (13–17), Lyell gives a sustained analysis of the processes of fossilization of terrestrial animals and plants. Chapters 13 and 14 show how they become fossilized on dry land; Chapters 15 to 17 examine processes taking place under water. Chapter 16, included here, shows how human remains are preserved:*]

We shall now proceed to inquire in what manner the mortal remains of man and the works of his hands may be permanently preserved in subaqueous strata. Of the many hundred million human beings which perish in the course of every century on the land, every vestige is usually destroyed in the course of a few thousand years, but of the smaller number that perish in the waters, a considerable proportion must frequently be entombed, under such circumstances, that parts of them may endure throughout entire geological epochs.

We have already seen how the bodies of men, together with those of the inferior animals, are occasionally washed down during river-inundations into seas and lakes, of which we shall now enumerate some additional examples.

Belzoni witnessed a flood on the Nile in September, 1818, where, although the river only rose three feet and a half above its ordinary level, several villages, with some hundreds of men, women, and children, were swept away. We mentioned in a former volume that a rise of six feet of water in the Ganges in 1763, was attended with a much greater loss of lives.

In the year 1771, at the time of the bursting of the Solway moss before alluded to, when the inundations in the north of England appear to have equalled the recent floods in Morayshire, a great number of houses and their inhabitants were swept away by the rivers Tyne, Can, Wear, Tees, and Greta; and no less than twenty-one bridges were destroyed in the courses of these rivers. At the village of Bywell the flood tore the dead bodies and coffins out of the churchyard, and bore them away, together with many of the living inhabitants. During the same tempest an immense number of cattle, horses, and sheep, were also transported to the sea, while the whole coast was covered with the wreck of ships. Four centuries before (in 1338), the same district had been visited by a similar continuance of heavy rains followed by disastrous floods, and it is not improbable that these catastrophes may recur periodically. As the population increases, and buildings and bridges are multiplied, we must expect that the loss of lives and property will rather augment.

If to the hundreds of human bodies committed to the deep in the way of ordinary burial, we add those of individuals lost by shipwreck, we shall find that, in the course of a single year, a great number of human remains are consigned to the subaqueous regions. We shall hereafter advert to a calculation by which it appears that more than five hundred *British* vessels alone, averaging each a burden of about one hundred and twenty tons, are wrecked, and sink to the bottom, *annually*. Of these the crews for the most part escape, although it sometimes happens that all perish. In one great naval action several thousand individuals sometimes share a watery grave.

Many of these corpses are instantly devoured by predaceous fish, sometimes before they reach the bottom; still more frequently when they rise again to the surface and float in a state of putrefaction. Many decompose on the floor of the ocean where no sediment is thrown down upon them, but if they fall upon a reef where corals and shells are becoming agglutinated into a solid rock, or subside where the delta of a river is advancing, they may be preserved for an incalculable series of ages in these deposits.

Often at the distance of a few hundred feet from a coral reef there are no soundings at the depth of many hundred fathoms. Here if a ship strike and be wrecked, it may soon be covered by calcareous sand and fragments of coral detached by the breakers from the summit of a submarine mountain, and which may roll down to its base. Wrecks are known to have been common for centuries near certain reefs, so that canoes, merchant vessels, and ships of war may have sunk and have been enveloped in these situations in calcareous sand and breccia. Suppose a volcanic eruption to cover such remains with ashes and sand, and that over the tufaceous strata resulting from these ejections, a current of lava is afterwards poured, the ships and human skeletons might then remain uninjured beneath the superincumbent rock, like the houses and works of art in the subterranean cities of Campania. That cases may have already occurred where human remains have been thus preserved in a fossil state beneath masses more than a thousand feet in thickness, is by no means improbable, for in some volcanic archipelagos a period of thirty or forty centuries might well suffice for such an accumulation of matter.

We stated that at the distance of about forty miles from the base of the delta of the Ganges, there is a circular space about fifteen miles in diameter where soundings of a thousand feet sometimes fail to reach the bottom. As during the flood season the quantity of mud and sand poured by the great rivers into the Bay of Bengal, is so great that the sea only recovers its transparency at the distance of sixty miles from the coast, this depression must be gradually shoaling, especially as during the monsoons the sea, loaded with mud and sand, is beaten back in that direction towards the delta. Now if a ship or human body sink down to the bottom in such a spot, it is by no means improbable that it may become buried under a depth of three or four thousand feet of sediment in the same number of years.

Even on that part of the floor of the ocean whither no accession of drift matter is carried, (a part which we believe to constitute, at any given period, by far the larger proportion of the whole submarine area,) there are circumstances accompanying a wreck which favour the conservation of skeletons. For when the vessel fills suddenly with water, especially in the night, many persons are drowned between decks and in their cabins, so that their bodies are prevented from rising again to the surface. The vessel often strikes upon an uneven bottom and is overturned, in which case the ballast consisting of sand, shingle, and rock, or the cargo, frequently composed of heavy and durable materials, may be thrown down upon the carcasses. In the case of ships of war, cannon, shot, and other warlike stores, may press down with their weight the timbers of the vessel when they decay, and beneath these and the metallic substances the bones of man may be preserved.

When we reflect on the number of curious monuments consigned to the bed of the ocean in the course of every naval war from the earliest times, our conceptions are greatly raised respecting the multiplicity of lasting memorials which man is leaving of his labours. During our last great struggle with France, thirty-two of our ships of the line went to the bottom in the space of twenty-two years, besides seven fifty-gun ships, eighty-six frigates, and a multitude of smaller vessels. The navies of the other European powers, France, Holland, Spain, and Denmark, were almost annihilated during the same period, so that the aggregate of their losses must have many times exceeded that of Great Britain. In every one of these ships were batteries of cannon constructed of iron or brass, whereof a great number had the dates and places of their manufacture inscribed upon them in letters cast in metal. In each there were coins of copper, silver, and often many of gold, capable of serving as valuable historical monuments; in each were an infinite variety of instruments of the arts of war and peace, many formed of materials, such as glass and earthenware, capable of lasting for indefinite ages when once removed from the mechanical action of the waves, and buried under a mass of matter which may exclude the corroding action of sea-water.

But the reader must not imagine that the fury of war is more conducive than the peaceful spirit of commercial enterprise to the accumulation of wrecked vessels in the bed of the sea. From an examination of Lloyd's lists from the year 1793, to the commencement of 1829, it has appeared that the number of *British vessels* alone lost during that period amounted, on an average, to no less than one and a half *daily*, a greater number than we should have anticipated, although we learn from Moreau's tables that the number of merchant vessels employed at one time in the navigation of England and Scotland, amounts to about twenty thousand, having one with another a mean burden of one hundred and twenty tons. Out of five hundred and fifty-one ships of the royal navy lost to the country during the period above mentioned, only one hundred and sixty were taken or destroyed by the enemy, the rest having either stranded or foundered, or having been burnt by accident, a striking proof that the dangers of our naval warfare, however great, may be far exceeded by the storm, the hurricane, the shoal, and all the other perils of the deep.

Millions of dollars and other coins have been sometimes submerged in a single ship, and on these, when they happen to be enveloped in a matrix capable of protecting them from chemical changes, much information of historical interest will remain inscribed and endure for periods as indefinite as have the delicate markings of zoophytes or lapidified plants in some of the ancient secondary rocks. In almost every large ship, moreover, there are some precious stones set in seals, and other articles of use and ornament composed of the hardest substances in nature, on which letters and various images are carved – engravings which they may retain when included in subaqueous strata, as long as a crystal preserves its natural form.

It was a splendid boast, that the deeds of the English chivalry at Agincourt made Henry's chronicle

——————— as rich with praise
As is the ooze and bottom of the deep
With sunken wreck and sumless treasuries;[53]

for it is probable that a greater number of monuments of the skill and industry of man will, in the course of ages, be collected together in the bed of the ocean, than will be seen at one time on the surface of the continents.

If our species be of as recent a date as we suppose, it will be vain to seek for the remains of man and the works of his hands imbedded in submarine strata, except in those regions where violent earthquakes are frequent, and the alterations of relative level so great, that the bed of the sea may have been converted into land within the historical era. We do not despair of the discovery of such monuments whenever those regions which have been peopled by man from the earliest ages, and which are at the same time the principal theatres of volcanic action, shall be examined by the joint skill of the antiquary and the geologist.

There can be no doubt that human remains are as capable of resisting decay as are the harder parts of the inferior animals; and we have already cited the remark of Cuvier, that 'in ancient fields of battle the bones of men have suffered as little decomposition as those of horses which were buried in the same grave.'[54] In the delta of the Ganges bones of men have been found in digging a well at the depth of ninety feet; but as that river frequently shifts its course and fills up its ancient channels, we are not called upon to suppose that these bodies are of extremely high antiquity, or that they were buried when that part of the surrounding delta where they occur was first gained from the sea.

Several skeletons of men, more or less mutilated, have been found in the West Indies, on the north-west coast of the mainland of Guadaloupe, in a kind of rock which is known to be forming daily, and which consists of minute fragments of shells and corals, incrusted with a calcareous cement resembling travertin, which has also bound the different grains together. The lens shows that some of the fragments of coral composing this stone, still retain the same red colour which is seen in the reefs of living coral which surround the island. The shells belong to species of the neighbouring sea intermixed with some terrestrial kinds which now live on the island, and among them is the *Bulimus Guadaloupensis* of Férussac. The human skeletons still retain some of their animal matter, and all their phosphate of lime. One of them, of which the head is wanting, may now be seen in the British Museum, and another in the Royal Cabinet at Paris. According to Mr. König, the rock in which the former is inclosed is harder under the mason's saw and chisel, than statuary marble. It is described as forming a kind of glacis, probably an indurated beach, which slants from the steep cliffs of the island to the sea, and is nearly all submerged at high tide.

Similar formations are in progress in the whole of the West Indian archipelago, and they have greatly extended the plain of Cayes in St. Domingo, where fragments of vases and other human works have been found at a depth of twenty feet. In digging wells also near Catania, tools have been discovered in a rock somewhat similar.

When a vessel is stranded in shallow water, it usually becomes the nucleus of a sand bank, as has been exemplified in several of our harbours, and this circumstance tends greatly to its preservation. About fifty years ago, a vessel from Purbeck, laden with three hundred tons of stone, struck on a shoal off the entrance of Poole harbour and foundered; the crew were saved, but the vessel and cargo remain to this day at the bottom. Since that period the shoal at the entrance of the harbour has so extended itself in a westerly direction towards Peveril Point in Purbeck, that the navigable channel is thrown a mile nearer that Point. The cause is obvious; the tidal current deposits the sediment with which it is charged around any object which checks its velocity. Matter also drifted along the bottom is arrested by any obstacle, and accumulates round it just as the African sand-winds, before described, raise a small hillock over the carcasses of every dead camel exposed on the surface of the desert.

We alluded, in the former volume, to an ancient Dutch vessel, discovered in the deserted channel of the river Rother, in Sussex, of which the oak wood was much blackened, but its texture unchanged. The interior was filled with fluviatile silt, as was also the case in regard to a vessel discovered in a former bed of the Mersey, and another disinterred where the St. Catherine Docks are excavated in the alluvial plain of the Thames. In like manner many ships have been found preserved entire in modern strata, formed by the silting up of estuaries along the southern shores of the Baltic, especially in Pomerania. Between Bromberg and Nakel, for example, a vessel and two anchors in a very perfect state were dug up far from the sea.

At the mouth of a river in Nova Scotia, a schooner of thirty-two tons, laden with live stock, was lying with her side to the tide, when the bore, or tidal wave, which rises there about ten feet in perpendicular height, rushed into the estuary and overturned the vessel, so that it instantly disappeared. After the tide had ebbed, the schooner was so totally buried in the sand, that the taffrel or upper rail of the deck was alone visible. We are informed by Leigh, that, on draining Martin Meer, a lake eighteen miles in circumference, in Lancashire, a bed of marl was laid dry, wherein no fewer than eight canoes were found imbedded. In figure and dimensions they were not unlike those now used in America. In a morass about nine miles distant from this Meer, a whetstone and an axe of mixed metal were dug up. In Ayrshire also, three canoes were found in Loch Doon some few years ago; and during the present year (1831) four others, each hewn out of separate oak trees. They were twenty-three feet in length, two and a half in depth, and nearly four feet in breadth at the stern. In the mud which filled one of them, was found a war club of oak and a stone battle-axe.

The only examples of buried vessels to which we can obtain access, are in such situations as we have mentioned, but we are unable to examine those which have been subjected to great pressure, at the bottom of a deep ocean. It is extremely possible that the submerged wood-work of ships which have sunk where the sea is two or three miles deep, has undergone greater chemical changes in an equal space of time, for the experiments of Scoresby before mentioned show that wood may at certain depths be impregnated in a single hour with salt-water, so that its specific gravity is entirely altered.

It may often happen that hot springs charged with carbonate of lime, silex and other mineral ingredients, may issue at great depths, in which case every pore of the vegetable tissue may be injected with the lapidifying liquid, whether calcareous or siliceous, before the smallest decay commences. The conversion also of wood into lignite is probably more rapid under such enormous pressure. But the change of the timber into lignite or coal would not prevent the original form of a ship from being distinguished, for as we find in strata of the carboniferous era, the bark of the hollow reed-like trees converted into coal, and the central cavity filled with sandstone, so might we trace the outline of a ship in coal, and in the indurated mud, sandstone, or limestone filling the interior, we might discover instruments of human art, ballast consisting of rocks foreign to the rest of the stratum, and other contents of the ship.

Many of the metallic substances which fall into the waters, probably lose, in the course of ages, the forms artificially imparted to them; but under many circumstances these may be preserved for indefinite periods. The cannon inclosed in a calcareous rock, drawn up from the delta of the Rhone, which is now in the museum at Montpellier, might probably have endured as long as the calcareous matrix; but even if the metallic matter had been removed and had entered into new combinations, still a mould of its original shape would have been left, corresponding to those impressions of shells which we see in rocks, from which all the carbonate of lime has been subtracted. About the year 1776, says Mr. King, some fishermen sweeping for anchors in the Gull stream, (a part of the sea near the Downs,) drew up a very curious old swivel gun, near eight feet in length. The barrel, which was about five feet long, was of brass; but the handle by which it was traversed, was about three feet in length, and the swivel and pivot on which it turned were of iron. Around these latter were formed incrustations of sand converted into a kind of stone, of an exceeding strong texture and firmness; whereas round the barrel of the gun, except where it was near adjoining to the iron, there was no such incrustation, the greater part of it being clean and in good condition, just as if it had still continued in use. In the incrusting stone, adhering to it on the outside, were a number of shells and corallines, 'just as they are often found in a fossil state.' These were all so strongly attached, that it required as much force to separate them from the matrix, 'as to break a fragment off any hard rock.'[55]

In the year 1745, continues the same writer, the Fox man-of-war was stranded on the coast of East Lothian and went to pieces. About thirty-three years afterwards a violent storm laid bare a part of the wreck, and threw up near the place several masses 'consisting of iron, ropes and balls,' covered over with ochreous sand concreted and hardened into a kind of stone. The substance of the rope was very little altered. The consolidated sand retained perfect impressions of parts of an iron ring, 'just in the same manner as impressions of extraneous fossil bodies are found in various kinds of strata.'[55]

After a storm in the year 1824, which occasioned a considerable shifting of the sands near St. Andrew's, in Scotland, a gun barrel of ancient construction was found, which is conjectured to have belonged to one of the wrecked vessels of the Spanish armada. It is now in the museum of the Antiquarian Society of Scotland, and is encrusted over by a thin coating of sand, the grains of which are cemented by brown ferruginous matter. Attached to this coating are fragments of various shells, as of the common cardium, mya, &c.

Many other examples are recorded of iron instruments taken up from the bed of the sea near the British coasts, incased by a thick coating of conglomerate, consisting of pebbles and sand, cemented by oxide of iron.

Dr. Davy describes in the Philosophical Transactions, a bronze helmet of the antique Grecian form, taken up in 1825, from a shallow part of the sea, between the citadel of Corfu and the village of Castrades. Both the interior and exterior of the helmet were partially encrusted with shells, and a deposit of carbonate of lime. The surface generally, both under the incrustation and where freed from it, was of a variegated colour, mottled with spots of green, dirty white, and red. On minute inspection with a lens, the green and red patches proved to consist of crystals of the red oxide and carbonate of copper, and the dirty white chiefly of oxide of tin.

The mineralizing process, says Dr. Davy, which has produced these new combinations, has in general penetrated very little into the substance of the helmet. The incrustation and rust removed, the metal is found bright beneath, in some places considerably corroded, in others very slightly. It proves on analysis to be copper alloyed with 18.5 per cent. of tin. Its colour is that of our common brass, and it possesses a considerable degree of flexibility: –

'It is a curious question,' he adds, 'how the crystals were formed in the helmet, and on the adhering calcareous deposit. There being no reason to suppose deposition from solution, are we not under the necessity of inferring, that the mineralizing process depends on a small motion and separation of the particles of the original compound? This motion may have been due to the operation of electro-chemical powers which may have separated the different metals of the alloy.'[57]

Effects of the Submersion of Land by Earthquakes

We have hitherto considered the transportation of plants and animals from the land by *aqueous* agents, and their inhumation in lacustrine or submarine deposits, and we may now inquire what tendency the subsidence of tracts of land by *earthquakes* may have to produce analogous effects. Several examples of the sinking down of buildings and portions of towns near the shore to various depths beneath the level of the sea, during subterranean movements, were enumerated in a former volume, when we treated of the changes brought about by *inorganic* causes. The events alluded to were comprised within a brief portion of the historical period, and confined to a small number of the regions of active volcanos. Yet these authentic facts, relating merely to the last century and a half, gave indications of considerable change which must have taken place in the physical geography of the globe. If, during the earthquake of Jamaica in 1692, some of the houses in Port Royal subsided, together with the ground they stood upon, to the depth of twenty-four, thirty-six and forty-eight feet under water, we are not to suppose that this was the only spot throughout the whole range of the coasts of that island or the bed of the surrounding sea which suffered similar depressions. If the quay at Lisbon sank at once to the depth of six hundred feet in 1755, we must not imagine that this was the only point on the shores of the peninsula where similar phenomena might have been witnessed.

If during the short period since South America has been colonized by Europeans we have proof of alterations of level at the three principal ports on the western shores, Callao,* Valparaiso, and Conception, we cannot for a moment suspect that these cities so distant from each other have been selected as the peculiar points where the desolating power of the earthquake has expended its chief fury. 'It would be a knowing arrow that could choose out the brave men from the cowards,' retorted the young Spartan, when asked if his comrades who had fallen on the field of battle were braver than he and his fellow prisoners;[58] we might in the same manner remark that a geologist must attribute no small discrimination and malignity to the subterranean force, if he should suppose it to spare habitually a line of coast many thousand miles in length, with the exception of those few spots where populous towns have been erected. If then we consider how small is the area occupied by the sea-ports of this disturbed region, – points where alone each slight change of the relative level of sea and land can be recognized, and reflect on the proofs in our possession of the local revolutions that have happened on the site of each port, within the last century and a half, our conceptions must be greatly exalted respecting the magnitude of the alterations which the Andes may have undergone even in the course of the last six thousand years.

We cannot better illustrate the manner in which a large extent of surface may be submerged, so that the terrestrial plants and animals may become imbedded in subaqueous strata, than by referring to the earthquake of Cutch, in 1819, alluded to by us in a former volume. We shall enter

---

* It is well known that during the great earthquake of Lima, in 1746, part of the promontory south of Callao sank down, and it is a common story at Lima that its former termination became the present isle of San Lorenzo, between which and the main land there is now a navigable channel. The submerged arches of a church, and the present position of other buildings, are said to indicate that the site of Callao underwent, during the earthquakes, a change of level; an interesting fact, the evidences of which we hope will soon be examined by some of our naval officers, and other intelligent persons frequenting that port.

somewhat more fully into details concerning that catastrophe than the immediate subject of the present chapter might require, in order to lay before the reader the information obtained during the recent survey of Cutch.

The published account of Lieutenant A. Burnes, who examined that portion of the delta of the Indus in 1826 and 1829, confirms the facts before enumerated by us, and furnishes the following important particulars. The tract around Sindree, which subsided during the earthquake in June, 1819, was converted from dry land into sea in the course of a few hours, the new-formed mere extending for a distance of sixteen miles on each side of the fort, and probably exceeding in area the lake of Geneva. Neither the rush of the sea into this new depression, nor the movement of the earthquake, threw down the small fort of Sindree, the interior of which is said to have become *a tank*, the water filling the space within the walls, and the four towers continuing to stand, so that on the day after the earthquake the people in the fort who had ascended to the top of one of the towers saved themselves in boats. Immediately after the shock the inhabitants of Sindree saw, at the distance of five miles from their village, a long elevated mound, where previously there had been a low and perfectly level plain. To this uplifted tract they gave the name of 'Ullah bund,' or 'the Mound of God,' to distinguish it from an artificial barrier previously thrown across an arm of the Indus.

It is already ascertained that this newly raised country is *upwards of fifty miles* in length from east to west, running parallel to that line of subsidence before mentioned, which caused the grounds around Sindree to be flooded. The range of this elevation extends from Puchum island towards Gharee; its breadth from north to south is conjectured to be in some parts *sixteen miles*, and its greatest ascertained height above the original level of the delta is ten feet, an elevation which appears to the eye to be very uniform throughout.

For several years after the convulsion of 1819, the course of the Indus was very unsettled, and at length in 1826, the river burst its banks above Sinde, and forcing its way in a more direct course to the sea, cut right through the 'Ullah bund,' whereby a natural section was obtained. In the perpendicular cliffs thus laid open, Lieutenant Burnes found that the upraised land consisted of beds of clay filled with shells. The new channel of the river, where it intersected the 'bund,' was eighteen feet deep, and during the swells in 1826, it was two or three hundred yards in width, but in 1828 the channel was still further enlarged. The Indus, when it first opened this new passage, threw such a body of water into the new lake or salt lagoon of Sindree, that it became fresh for many months, but it had recovered its saltness in 1828, when the supply of river-water was less copious, and finally it became more salt than the sea, in consequence, as the natives suggested to Lieutenant Burnes, of the saline particles with which 'the Runn of Cutch' is impregnated.

Besides *Ullah bund*, there appears to have been another elevation south of Sindree, parallel to that before mentioned, respecting which, however, no exact information has yet been communicated. There is a tradition of an earthquake, which, about three centuries before, upheaved a large area of the bed of the sea, and converted it into land in the district now called 'the Runn,' so that numerous harbours were laid dry and ships were wrecked and engulphed; in confirmation of which account it was observed in 1819, that in the jets of black muddy water thrown out of fissures in that region, there were cast up numerous pieces of wrought iron and ship nails.

We must not conclude without alluding to a *moral* phenomenon connected with this tremendous catastrophe, which we regard as highly deserving the attention of geologists. The author above cited states that 'these wonderful events passed *unheeded* by the inhabitants of Cutch,'[59] for the region convulsed, though once fertile, had for a long period been reduced to sterility by want of irrigation, so that the natives were indifferent as to its fate. Now it is to this profound apathy, which all but highly civilized nations feel in regard to physical events, not having an immediate influence on their worldly fortunes, that we must ascribe the extraordinary dearth of historical information concerning changes of the earth's surface, which modern observations show to be by no means of rare occurrence in the ordinary course of nature.

It is stated that, for some years after the earthquake, the withered tamarisks and other shrubs protruded their tops above the waves, in parts of the submerged tract around Sindree; but after the flood of 1826 they were seen no longer. Every geologist will at once perceive that forests sunk by such subterranean movements, may become imbedded in subaqueous deposits both fluviatile and marine, and the trees may still remain erect, or sometimes the roots and part of the trunks may continue in their original position, while the current may have broken off, or levelled with the ground, their upper stems and branches.

But although a certain class of geological phenomena may be referred to the repetition of such catastrophes, we must hesitate before we call in to our aid the action of earthquakes, to explain what have been termed submarine forests, observed at various points around the shores of Great Britain. We have already hinted that the explanation of some of these may be sought in the encroachments of the sea, in estuaries, and the varying level of the tides, at distant periods on the same parts of our coast. After examining, in 1829, the so called submarine forest of Happisborough in Norfolk, I found that it was nothing more than a tertiary lignite of the 'Crag' period, which becomes exposed in the bed of the sea as soon as the waves sweep away the superincumbent strata of bluish clay. So great has been the advance of the sea upon our eastern shores within the last eight centuries, that whenever we find a mass of submerged timber near the sea side, or at the foot of the existing cliffs which we cannot suppose to be a mere accumulation of drift, vegetable matter, we should endeavour to find a solution of the problem, by reference to any cause rather than an earthquake. For we can scarcely doubt that the present outline of our coast, the shape of its estuaries, and the formation of its cliffs are of very modern date, probably within the human era, whereas we have no reason whatever to imagine that this part of Europe has been agitated by subterranean convulsions, capable of altering the relative level of land and sea, at so extremely recent a period.

Some of the buildings which have at different times subsided beneath the level of the sea, have been immediately covered up to a certain extent with strata of volcanic matter showered down upon them. Such was the case at Tomboro in Sumbawa, in the present century, and at the site of the Temple of Serapis, in the environs of Puzzuoli, probably in the 12th century. The entrance of a river charged with sediment in the vicinity, may still more frequently occasion the rapid envelopement of buildings in regularly stratified formations. But if no foreign matter be introduced, the buildings when once removed to a depth where the action of the waves is insensible, and where no great current happens to flow, may last for indefinite periods, and be as durable as the floor of the ocean itself, which may often be composed of the very same materials. There is no reason to doubt the tradition mentioned by the classic writers, that the submerged Grecian towns of Bura and Helice were seen under water; and I am informed by an eye-witness that eighty-eight years after the convulsion of 1692, the houses of Port Royal were still visible at the bottom of the sea.

We cannot conclude this chapter without recalling to the reader's mind a memorable passage written by Berkely a century ago, in which he inferred, on grounds which may be termed strictly geological, the recent date of the creation of man. 'To any one,' says he, 'who considers that on digging into the earth such quantities of shells, and in some places bones and horns of animals are found sound and entire, after having lain there in all probability some thousands of years; it should seem probable that guns, medals and implements in metal or stone might have lasted entire, buried under ground forty or fifty thousand years if the world had been so old. How comes it then to pass that no remains are found, no antiquities of those numerous ages preceding the Scripture accounts of time; that no fragments of buildings, no public monuments, no intaglias, cameos, statues, basso-relievos, medals, inscriptions, utensils, or artificial works of any kind are ever discovered, which may bear testimony to the existence of those mighty empires, those successions of monarchs, heroes, and demi-gods for so many thousand years? Let us look forward and suppose ten or twenty thousand years to come, during which time we will suppose that plagues, famine, wars and *earthquakes* shall have made great havoc in the world, is it not highly probable that at the end of such a period, pillars, vases, and statues now in being of granite, or porphyry, or jasper, (stones of such hardness as we know them to have lasted two thousand years above ground, without any considerable alteration) would bear record of these and past ages? Or that some of our current coins might then be dug up, or old walls and the foundations of buildings shew themselves, as well as the shells and stones of *the primeval world*, which are preserved down to our times?'[60]

That many signs of the agency of man would have lasted at least as long as 'the shells of the primeval world,' had our race been so ancient, we are as fully persuaded as Berkely; and we anticipate with confidence that many edifices and implements of human workmanship, and the skeletons of men, and casts of the human form, will continue to exist when a great part of the present mountains, continents, and seas have disappeared. Assuming the future duration of the planet to be indefinitely protracted, we can foresee no limit to the perpetuation of some of the memorials of man, which are continually entombed in the bowels of the earth or in the bed of the ocean, unless we carry forward our views to a period sufficient to allow the various causes of change both igneous and aqueous, to remodel more than once the entire crust of the earth. *One* complete revolution will be inadequate to efface every monument of our existence, for many works of art might enter again and again into the formation of successive eras, and escape obliteration even though the very rocks in which they had been for ages imbedded were destroyed, just as pebbles included in the conglomerates of one epoch often contain the organized remains of beings which flourished during a prior era.

Yet it is no less true, as a late distinguished philosopher has declared, 'that none of the works of a mortal being can be eternal.'[61] They are in the first place wrested from the hands of man, and lost as far as regards their subserviency to his use, by the instrumentality of those very causes which place them in situations where they are enabled to endure for indefinite periods. And even when they have been included in rocky strata, when they have been made to enter as it were into the solid framework of the globe itself, they must nevertheless eventually perish, for every year some portion of the earth's crust is shattered by earthquakes or melted by volcanic fire, or ground to dust by the moving waters on the surface. 'The river of Lethe,' as Bacon eloquently remarks, 'runneth as well above ground as below.'[62]

[*The analysis of fossilization processes continues in Chapter 17, which turns to the preservation of aquatic species in fresh and salt water environments. Chances of the remains of such creatures being preserved – especially if they have hard parts – are relatively good. This means that assemblages of molluscs become significant geological markers.*]

CHAPTER 18

Corals and Coral Reefs

The powers of the organic creation in modifying the form and structure of those parts of the earth's crust, which may be said to be undergoing repair, or where new rock-formations are continually in progress, are most conspicuously displayed in the labours of the coral animals. We may compare the operation of these zoophytes in the ocean, to the effects produced on a smaller scale upon the land, by the plants which generate peat. In the case of the Sphagnum, the upper part vegetates while the lower portion is entering into a mineral mass, where the traces of organization usually remain, but in which life has entirely ceased. In the corals, in like manner, the more durable materials of the generation that has passed away, serve as the foundation on which living animals are continuing to rear a similar structure.

The calcareous masses usually termed coral reefs, are by no means exclusively composed of zoophytes, but also a great variety of shells; some of the largest and heaviest of known species contributing to augment the mass. In the south Pacific, great beds of oysters, mussels, *pinnæ marinæ*, and other shells, cover in great profusion almost every reef; and, on the beach of coral islands, are seen the shells of echini and the broken fragments of crustaceous animals. Large shoals of fish also are discernible through the clear blue water, and their teeth and hard palates are probably preserved, although a great portion of their soft cartilaginous bones may decay.

Of the numerous species of zoophytes which are engaged in the production of coral banks, some of the most common belong to the genera Meandrina, Caryophyllia and Astrea, but especially the latter.

The reefs, which just raise themselves above the level of the sea, are usually of a circular or oval form, and are surrounded by a deep and often unfathomable ocean. In the centre of each, there is usually a comparatively shallow lagoon where there is still water, and where the smaller and more delicate kinds of zoophytes find a tranquil abode, while

334

the more strong species live on the exterior margin of the isle, where a great surf usually breaks. When the reef, says M. Chamisso, a naturalist who accompanied Kotzebue, is of such a height that it remains almost dry at low water, the corals leave off building. A continuous mass of solid stone is seen composed of the shells of molluscs and echini, with their broken off prickles and fragments of coral, united by the burning sun, through the medium of the cementing calcareous sand, which has arisen from the pulverization of shells. Fragments of coral limestone are thrown up by the waves, until the ridge becomes so high, that it is covered only during some seasons of the year by the high tides. The heat of the sun often penetrates the mass of stone when it is dry, so that it splits in many places. The force of the waves is thereby enabled to separate and lift blocks of coral, frequently six feet long and three or four in thickness, and throw them upon the reef. 'After this the calcareous sand lies undisturbed, and offers to the seeds of trees and plants cast upon it by the waves, a soil upon which they rapidly grow, to overshadow its dazzling white surface. Entire trunks of trees, which are carried by the rivers from other countries and islands, find here, at length, a resting-place after their long wanderings: with these come some small animals, such as lizards and insects, as the first inhabitants. Even before the trees form a wood, the sea-birds nestle here; strayed land-birds take refuge in the bushes; and, at a much later period, when the work has been long since completed, man also appears, builds his hut on the fruitful soil formed by the corruption of the leaves of the trees, and calls himself lord and proprietor of this new creation.'[63]

The Pacific ocean throughout, a space comprehended between the thirtieth parallel of latitude on each side of the equator, is extremely productive of coral. The Arabian gulf is rapidly filling with the same, and it is said to abound in the Persian gulf. Between the coast of Malabar and that of Madagascar, there is also a great sea of coral. Flinders describes an unbroken reef three hundred and fifty miles in length, on the east coast of New Holland; and, between that country and New Guinea, Captain P. King found the coral formations to extend throughout a distance of seven hundred miles, interrupted by no intervals exceeding thirty miles in length.

The chain of coral reefs and islets, called the Maldivas, situated in the Indian ocean to the south-west of Malabar, form a chain four hundred and eighty geographical miles in length, running due north and south. It is composed throughout of a series of circular assemblages of islets, the larger groups being from forty to fifty miles in their longest diameter. Captain Horsburgh, whose chart of these islands is subjoined, informs me that outside of each circle or atoll, as it is termed, there are coral reefs sometimes extending to the distance of two or three miles, beyond which there are no soundings at immense depths. But in the centre of each atoll there is a lagoon from fifteen to twenty fathoms deep. In the channels between the atolls, no soundings have been obtained at the depth of one hundred and fifty fathoms.

The Laccadive islands run in the same line with the Maldivas, on the north, as do the isles of the Chagos Archipelago, on the south, so that these may be continuations of the same chain of submarine mountains, crested in a similar manner by coral limestone. It would be rash to hazard the hypothesis, that they are all the summits of volcanos, yet we might imagine, that if Java and Sumatra were submerged, they would give rise to a somewhat similar shape in the bottom of the sea; for the volcanos of those islands observe a linear direction, and are often separated from each other by intervals, corresponding to the atolls of the Maldivas; and as they rise to various heights, from five to ten thousand feet above their base, they might leave an unfathomable ocean in the intermediate spaces.

In regard to the thickness of the masses of coral, MM. Quoy and Gaimard are of opinion, that the species which contribute most actively to the formation of solid masses do not grow where the water is deeper than twenty-five or thirty feet. But the branched madrepores, which live at considerable depth, may form the first foundation of a reef, and raise a platform on which other species may build, and the sand and broken fragments washed by the waves from reefs may, in time, produce calcareous rocks of great thickness.

The rapidity of the growth of coral is by no means great, according to the report of the natives to Captain Beechey. In an island west of Gambier's group, our navigators observed the *Chama gigas* (Tridacna, Lam.) while the animal was yet living, so completely overgrown by coral, that a space only of two inches was left for the extremity of the shell to open and shut. But conchologists suppose, that the chama may require thirty years or more to attain its full size, so that the fact is quite consistent with a very slow rate of increase in the calcareous reefs. In the late expedition to the Pacific no positive information could be obtained, of any channel having been filled up within a given period, and it seems established that several

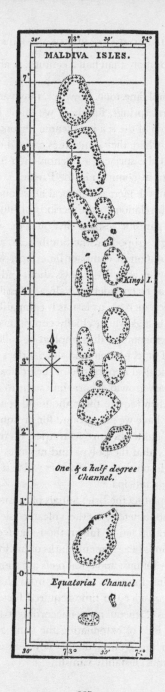

MALDIVA ISLES.

Kings I.

One & a half degree
Channel.

Equatorial Channel

4. Maldiva Isles

reefs had remained for more than half a century, at about the same depth from the surface.

The increase of coral limestone, however, may vary greatly according to the sites of mineral springs, for these we know often issue in great numbers at the bottom of the sea in volcanic regions, as in the Mediterranean, for example, where they sometimes cause the sea at great depths to be fresher than at the surface, a phenomenon also declared by the South Sea islanders to be common in the Pacific.

But when we admit the increase of coral limestone to be slow, we are merely speaking with relation to the periods of human observation. It often happens, that parasitic testacea live and die on the shells of the larger slow-moving gasteropods in the South Seas, and become entirely inclosed in an incrustation of compact limestone, while the animal, to whose habitation they are attached, crawls about and bears upon his back these shells, which may be considered as already fossilized. It is, therefore, probable, that the reefs increase as fast as is compatible with the thriving state of the organic beings which chiefly contribute to their formation; and if the rate of augmentation thus implied be called, in conformity to our ordinary ideas of time, gradual and slow, it does not diminish, in the least degree, the geological importance of such calcareous masses.

Suppose the ordinary growth of coral limestone to amount to six inches in a century, it will then require three thousand years to produce a reef fifteen feet thick; but have we any ground for presuming that, at the end of that period, or of ten times thirty centuries, there will be a failure in the supply of lime, or that the polyps and molluscs will cease to act, or that the hour of the dissolution of our planet will first arrive, as the earlier geologists were fain to anticipate?

Instead of contemplating the brief annals of human events, let us turn to some natural chronometers, to the volcanic isles of the Pacific, for example, which shoot up ten or fifteen thousand feet above the level of the ocean. These islands bear evident marks of having been produced by successive volcanic eruptions; and coral reefs are sometimes found on the volcanic soil, reaching for some distance from the sea-shore into the interior. When we consider the time required for the accumulation of such mountain masses of igneous matter according to the analogy of known volcanic agency, all idea of extenuating the comparative magnitude of coral limestones, on the ground of the slowness of the operations of lithogenous polyps, must instantly vanish.

5. View of Witsunday Island

The information collected during the late expedition to the Pacific throws much additional light on the peculiarities of form and structure of coral islands. Of thirty-two of these, examined by Captain Beechey, the largest was thirty miles in diameter, and the smallest less than a mile. They were of various shapes, all formed of living coral, except one, which, although of coral formation, was raised about eighty feet above the level of the sea, and encompassed by a reef of living coral. All were increasing their dimensions by the active operations of the lithophytes which appeared to be gradually extending and bringing the immersed parts of their structure to the surface. Twenty-nine of the number had lagoons in their centres, which had probably existed in the others, until they were filled, in the course of time, by zoophytic and other substances.

In the above-mentioned islands, the strips of dry coral encircling the lagoons when divested of loose sandy materials heaped upon them, are rarely elevated more than two feet above the level of the sea; and were it not for the abrupt descent of the external margin which causes the sea to break upon it, these strips would be wholly inundated. 'Those parts of the strip which are beyond the reach of the waves are no longer inhabited by the animals that reared them, but have their cells filled with a hard calcareous substance, and present a brown rugged appearance. The parts which are still immersed, or are dry at low water only, are intersected by small channels, and are so full of hollows that the tide, as it recedes, leaves small lakes of water upon them. The width of the plain or strip of dead coral, in the islands which fell under our observation, in no instance exceeded half a mile from the usual wash of the sea to the edge of the lagoon, and in general was only about three or four hundred yards.'[64] Beyond these limits the sides of the island descend rapidly, apparently by a succession of inclined ledges, each terminating in a precipice. The depth of the lagoons is various; in some entered by Captain Beechey, it was from twenty to thirty-eight fathoms.

In the annexed cut (No. 5), one of these circular islands is represented just rising above the waves, covered with the cocoa-nut and other trees, and inclosing within, a lagoon of tranquil water.

The accompanying section will enable the reader to comprehend the usual form of such islands. (No. 6.)

(*a a*).   Habitable part of the island, consisting of a strip of coral, inclosing the lagoon.

(*b b*).   The lagoon.

6. Section of a coral island

The subjoined cut (No. 7) exhibits a small part of the section of a coral island on a larger scale.

(*a b*).   Habitable part of the island.

(*b e*).   Slope of the side of the island, plunging at an angle of forty-five to the depth of fifteen hundred feet.

(*c c*).   Part of the lagoon.

(*d d*).   Knolls of coral in the lagoon, with over-hanging masses of coral, resembling the capitals of columns.

7. Section of part of a coral island

The circular or oval forms of the numerous coral isles of the Pacific, with the lagoons in their centre, naturally suggest the idea that they are nothing more than the crests of submarine volcanos, having the rims and bottoms of their craters overgrown by corals. This opinion is strengthened by the conical form of the submarine mountain, and the steep angle at which it plunges on all sides into the surrounding ocean. It is also well known that the Pacific is a great theatre of volcanic action, and every island yet examined in the wide region termed Eastern Oceanica, consists either of volcanic rocks or coral limestones.

It has also been observed that, although within the circular coral reefs, there is usually nothing discernible but a lagoon, the bottom of which is covered with coral, yet within some of these basins, as in Gambier's group, rocks composed of porous lava and other volcanic substances, rise up, resembling the two Kameni's, and other eminences of igneous origin, which have been thrown up within the times of history, in the midst of the Gulf of Santorin.

We mentioned that in volcanic archipelagos there is generally one large habitual vent, and many smaller volcanos formed at different points and at irregular intervals, all of which have usually a linear arrangement. Now in several of the groups of Eastern Oceanica there appears to be a similar disposition, the great islands, such as Otaheite, Owhyhee, and Terra del Spirito Santo, being habitual vents, and the lines of small circular coral isles which are dependent on them being very probably trains of minor volcanos, which may have been in eruption singly and at irregular intervals.

The absence of circular groups in the West Indian seas, and the tropical parts of the Atlantic, where corals are numerous, has been adduced as an additional argument, inasmuch as volcanic vents, though existing in those regions, are very inferior in importance to those in the Pacific and Indian seas. It may be objected that the circles formed by some coral reefs or groups of coral islets, varying as they do from ten to thirty miles and upwards in diameter, are so great as to preclude the idea of their being volcanic craters. In regard to this objection we may refer to what we have said in a former volume respecting the size of the so-called craters of elevation, many of which, we conceive, may be the ruins of truncated cones.

There is yet another phenomenon attending the circular reefs, to which we have not alluded, viz., the deep narrow passage which almost invariably leads from the sea into the lagoon, and is kept open by the efflux of the sea at low tides. It is sufficient that a reef should rise a few feet above low-water mark to cause the waters to collect in the lagoon at high tide, and, when the sea falls, to rush out violently at one or more points where the reef happens to be lowest or weakest. At first there are probably many openings; but the growth of the corals tends to obstruct all those which do not serve as the principal channels of discharge, so that their number is gradually reduced to a few, and often finally to one. This event is strictly analogous to that witnessed in our estuaries, where a body of salt-water accumulated during the flow, issues with great velocity at the ebb of the tide, and scours out or keeps open a deep passage through the bar, which is almost always formed at the mouth of a river.

When we controverted in our first volume Von Buch's theory of 'elevation craters,' we suggested that the single gorge leading from the central cavity to the sea, may have been produced by a stream of water issuing from a lake filling the original crater, and which had in process of time cut a deep channel; but we overlooked the more probable cause, the action of the tides, which affords, we think, a most satisfactory explanation. Suppose a volcanic cone, having a deep crater, to be at first submarine, and to be then *gradually* elevated by earthquakes in an ocean where tides prevail, a ravine cannot fail to be cut like that which penetrates into the Caldera of the isle of Palma. The opening would at first be made on that side where the rim of the crater was originally lowest, and it would afterwards be deepened as the island rose, so as always to descend somewhat lower than the level of the sea. Captain Beechey's observations, therefore, of the effect of the tides on the coral islands, corroborate the opinion which we offered respecting the mode of formation of islands having a configuration like Palma; whereas the theory of the *sudden* upheaving of horizontal strata into a conical form, affords no explanation whatever of the *single* ravine which intersects one side of these circular islands.

In the coral reefs surrounding those volcanic islands in the Pacific which are large enough to feed small rivers, there is generally an opening or channel opposite the point where the stream of fresh water enters the sea. The depth of these channels rarely exceeds twenty-five feet, and they may be attributed, says Captain Beechey, to the aversion of the lithophytes to fresh water, and to the probable absence of the mineral matter of which they construct their habitations.

But there is yet another peculiarity of the low coral islands, the explanation of which is by no means so obvious. They follow one general rule in having their windward side higher and more perfect than the other. 'At Gambier and Matilda islands this inequality is very conspicuous, the weather side of both being wooded, and of the former inhabited, while the other sides are from twenty to thirty feet under water, where, however, they might be perceived to be equally *narrow* and well defined. It is on the leeward side also that the entrances into the lagoons occur; and although they may sometimes be situated on a side that runs in the direction of the wind, as at Bow Island, yet there are none to windward.'[65] These observations of Captain Beechey accord perfectly with those which Captain Horsburgh and other hydrographers have made in regard to the coral islands of other seas. Thus the Chagos Isles in the Indian Ocean are chiefly of a horse-shoe form, the openings being to the north-west; whereas the prevailing wind blows regularly from the south-east. From this fortunate circumstance ships can enter and sail out again with ease, whereas, if the narrow inlets were to windward, vessels which once entered might not succeed for months in making their way out again. The well-known security of many of these harbours, depends entirely on this fortunate peculiarity in their structure.

In what manner is this singular conformation to be accounted for? The action of the waves is seen to be the cause of the superior elevation of some reefs on their windward sides, where sand and large masses of coral rock are thrown up by the breakers; but there are a variety of cases where this cause alone is inadequate to solve the problem; for reefs submerged at considerable depths, where the movements of the sea cannot exert much power, have, nevertheless, the same conformation, the leeward being much lower than the windward side.

I am informed by Captain King, that on examining the reefs called Rowley Shoals, which lie off the north-west coast of Australia, where the east and west monsoons prevail alternately, he found the open side of one crescent-shaped reef, the Impérieuse, turned to the east, and of another, the Mermaid, turned to the west; while a third oval reef, of the same group, was entirely submerged. This want of conformity is exactly what we should expect, where the winds vary periodically.

It seems impossible to refer the phenomenon now under consideration to any original uniformity in the configuration of submarine volcanos, on the summits of which we may suppose the coral reefs to grow; for although it is very common for craters to be broken down on one side only, we cannot imagine any cause that should breach them all in the same direction. But, if we mistake not, the difficulty will be removed if we call in another part of the volcanic agency – subsidence by earthquakes. Suppose the windward barrier to have been raised by the mechanical action of the waves to the height of two or three yards above the wall on the leeward side, and then the whole island to sink down a few fathoms, the appearances described would then be presented by the submerged reef. A repetition of such operations by the alternate elevation and depression of the same mass (an hypothesis strictly conformable to analogy) might produce still greater inequality in the two sides, especially as the violent efflux of the tide has probably a strong tendency to check the accumulation of the more tender corals on the leeward reef, while the action of the breakers contributes to raise the windward barrier.

* * *

The reefs in the Pacific are sometimes of great extent: thus the inhabitants of Disappointment Islands, and those of Duff's Group, pay visits to each other by passing over long lines of reefs from island to island, a distance of six hundred miles and upwards. When on their route they present the appearance of troops marching upon the surface of the ocean.

A reference to our first volume will show that a series of ordinary earthquakes might, in the course of a few centuries, convert such a tract of sea into dry land; and it is, therefore, a remarkable circumstance that there should be so immense an area in eastern Oceanica, studded with minute islands, without one single spot where there is a wider extent of land than belongs to such islands as Otaheite, Owhyhee, and a few others, which either have been or are still the seats of active volcanos. If an equilibrium only were maintained between the upheaving and depressing force of earthquakes, large islands would very soon be formed in the Pacific; for, in that case, the growth of limestone, the flowing of lava, and the ejection of volcanic ashes, would combine with the upheaving force to form new land.

Suppose the shoal which we have described as six hundred miles in length, to sink fifteen feet, and then to remain unmoved for a thousand years; during that interval the growing coral may again approach the surface. Then let the mass be re-elevated fifteen feet, so that the original reef is restored to its former position: in this case the new coral formed since the first subsidence, will constitute an island six hundred miles long. An analogous result would have occurred if a lava-current fifteen feet thick had overflowed the submerged reef. The absence, therefore, of more extensive tracts of land in the Pacific seems to show that the amount of subsidence by earthquakes exceeds in that quarter of the globe at present the elevation due to the same cause.

* * *

The calcareous masses which we have now considered, constitute, together with the associated volcanic formations, the most extensive of the groups of rocks which can be demonstrated to be now in progress. The space in the sea which they occupy is so vast, that we may safely infer that they exceed in area any group of ancient rocks which can be proved to have been of contemporaneous origin. We grant that each of the great archipelagos of the Pacific are separated by unfathomable abysses, where no zoophytes may live and no lavas flow, where not even a particle of coral sand or volcanic scoriæ may be drifted: we confine our view to the extent of reef ascertained to exist, and assume that a certain space around each volcanic or coral isle has been covered with ejections or matter from the waste of cliffs, and it will then be seen that the space occupied by these formations may equal, and perhaps exceed in area that part of our continents which has been accurately explored by the geologist.

That the increase of these calcareous masses should be principally, if not entirely, confined to the shallower parts of the ocean, or, in other words, to the summits of submarine ranges of mountains and elevated platforms, is a circumstance of the highest interest to the geologist; for, if parts of the bed of such an ocean should be upraised, so as to form large continents, mountain-chains might appear, capped and flanked by calcareous strata of great thickness, and replete with organic remains, while in the intervening lower regions no rocks of contemporary origin would ever have existed.

A modern writer has attempted to revive the theory of some of the earlier geologists, that all limestones have originated in organized substances. If we examine, he says, the quantity of limestone in the primary strata, it will be found to bear a much smaller proportion to the siliceous and argillaceous rocks than in the secondary, and this may have some connexion with the rarity of testaceous animals in the ancient ocean. He farther infers that in consequence of the operations of animals, 'the quantity of calcareous earth deposited in the form of mud or stone is always increasing; and that as the secondary series far exceeds the primary in this respect, so a third series may hereafter arise from the depths of the sea, which may exceed the last in the proportion of its calcareous strata.'[66]

If these propositions went no farther than to suggest that every particle of lime that now enters into the crust of the globe, may possibly in its turn have been subservient to the purposes of life by entering into the composition of organized bodies, we should not deem the speculation improbable; but when it is hinted that lime may be an animal product combined by the powers of vitality from some simple elements, we can discover no sufficient grounds for such an hypothesis, and many facts which militate against it.

If a large pond be made, in almost any soil, and filled with rain water, it may usually become tenanted by testacea, for carbonate of lime is almost universally diffused in small quantities. But if no calcareous matter be supplied by waters flowing from the surrounding high grounds or by springs, no tufa or shell-marl are formed. The thin shells of one generation of molluscs decompose, so that their elements afford nutriment to the succeeding races; and it is only where a stream enters a lake, which may introduce a fresh supply of calcareous matter, or where the lake is fed by springs, that shells accumulate and form marl.

All the lakes in Forfarshire which have produced deposits of shell-marl, have been the sites of springs which still evolve much carbonic acid, and a small quantity of carbonate of lime. But there is no marl in Loch Fithie, near Forfar, where there are *no springs*, although that lake is surrounded by these calcareous deposits, and although, in every other respect, the site is favourable to the accumulation of aquatic testacea.

We find those charæ which secrete the largest quantity of calcareous matter in their stems, to abound near springs impregnated with carbonate of lime. We know that if the common hen be deprived altogether of calcareous nutriment, the shells of her eggs will become of too slight a consistency to protect the contents, and some birds eat chalk greedily during the breeding season.

If on the other hand we turn to the phenomena of inorganic nature, we observe that, in volcanic countries, there is an enormous evolution of carbonic acid, mixed with water or in a gaseous form, and that the springs of such districts are usually impregnated with carbonate of lime in great abundance. No one who has travelled in Tuscany, through the region of extinct volcanos and its confines, or who has seen the map recently constructed by Targioni to show the principal sites of mineral springs, can doubt for a moment, that, if this territory was submerged beneath the sea, it might supply materials for the most extensive coral reefs. The importance of these springs is not to be estimated by the magnitude of the rocks which they have thrown down on the slanting sides of hills, although of these alone large cities might be built, nor by a coating of travertin that covers the soil in some districts for miles in length. The greater part of the calcareous matter passes down in a state of solution to the sea; and a geologist might as well assume the mass of alluvium formed in a few years in the bed of the Po, or the Ganges, to be the measure of the quantity deposited in the course of centuries in the deltas of those rivers, as conceive that the influence of the carbonated springs in Italy can be estimated by the mass of tufa precipitated by them near their sources.

It is generally admitted that the abundance of carbonate of lime given out by springs, in regions where volcanic eruptions or earthquakes prevail, is referrible to the solvent power of carbonic acid. For, as the acidulous waters percolate calcareous strata, they take up a certain portion of lime and carry it up to the surface where, under diminished pressure in the atmosphere, it may be deposited, or, being absorbed by animals and vegetables, may be secreted by them. In Auvergne, springs charged with carbonate of lime rise through granite, in which case we must suppose the calcareous matter to be derived from some primary rock, unless we imagine it to rise up from the volcanic foci themselves.

We see no reason for supposing that the lime now on the surface, or in the crust of the earth, may not, as well as the silex, alumine, or any other mineral substance, have existed before the first organic beings were created, if it be assumed that the arrangement of the inorganic materials of our planet preceded in the order of time the introduction of the first organic inhabitants.

But if the carbonate of lime secreted by the testacea and corals of the Pacific, be chiefly derived *from below*, and if it be a very general effect of the action of subterranean heat to subtract calcareous matter from the *inferior* rocks, and to cause it to ascend to the surface, no argument can be derived in favour of the progressive increase of limestone from the magnitude of coral reefs, or the greater proportion of calcareous strata, in the more modern formations. A constant transfer of carbonate of lime from the inferior parts of the earth's crust to its surface, would cause throughout all future time, and for an indefinite succession of geological epochs, a preponderance of calcareous matter in the newer, as contrasted with the older formations.

View of the volcanoes around Olot in Catalonia

VOLUME III

CHAPTER I

Methods of Theorizing
in Geology

Having considered, in the preceding volumes, the actual operation of the causes of change which affect the earth's surface and its inhabitants, we are now about to enter upon a new division of our inquiry, and shall therefore offer a few preliminary observations, to fix in the reader's mind the connexion between two distinct parts of our work, and to explain in what manner the plan pursued by us differs from that more usually followed by preceding writers on Geology.[1]

All naturalists, who have carefully examined the arrangement of the mineral masses composing the earth's crust, and who have studied their internal structure and fossil contents, have recognized therein the signs of a great succession of former changes; and the causes of these changes have been the object of anxious inquiry. As the first theorists possessed but a scanty acquaintance with the present economy of the animate and inanimate world, and the vicissitudes to which these are subject, we find them in the situation of novices, who attempt to read a history written in a foreign language, doubting about the meaning of the most ordinary terms; disputing, for example, whether a shell was really a shell, – whether sand and pebbles were the result of aqueous trituration, – whether stratification was the effect of successive deposition from water; and a thousand other elementary questions which now appear to us so easy and simple, that we can hardly conceive them to have once afforded matter for warm and tedious controversy.

In the first volume we enumerated many prepossessions which biassed the minds of the earlier inquirers, and checked an impartial desire of arriving at truth. But of all the causes to which we alluded, no one contributed so powerfully to give rise to a false method of philosophizing as the entire unconsciousness of the first geologists of the extent of their own ignorance respecting the operations of the existing agents of change.

They imagined themselves sufficiently acquainted with the mutations now in progress in the animate and inanimate world, to entitle them at once to affirm, whether the solution of certain problems in geology could ever be derived from the observation of the actual economy of nature, and having decided that they could not, they felt themselves at liberty to indulge their imaginations, in guessing at what *might be*, rather than in inquiring *what is*; in other words, they employed themselves in conjecturing what might have been the course of nature at a remote period, rather than in the investigation of what was the course of nature in their own times.

It appeared to them more philosophical to speculate on the possibilities of the past, than patiently to explore the realities of the present, and having invented theories under the influence of such maxims, they were consistently unwilling to test their validity by the criterion of their accordance with the ordinary operations of nature. On the contrary, the claims of each new hypothesis to credibility appeared enhanced by the great contrast of the causes or forces introduced to those now developed in our terrestrial system during a period, as it has been termed, of *repose*.

Never was there a dogma more calculated to foster indolence, and to blunt the keen edge of curiosity, than this assumption of the discordance between the former and the existing causes of change. It produced a state of mind unfavourable in the highest conceivable degree to the candid reception of the evidence of those minute, but incessant mutations, which every part of the earth's surface is undergoing, and by which the condition of its living inhabitants is continually made to vary. The student, instead of being encouraged with the hope of interpreting the enigmas presented to him in the earth's structure, – instead of being prompted to undertake laborious inquiries into the natural history of the organic world, and the complicated effects of the igneous and aqueous causes now in operation, was taught to despond from the first. Geology, it was affirmed, could never rise to the rank of an exact science, – the greater number of phenomena must for ever remain inexplicable, or only be partially elucidated by ingenious conjectures. Even the mystery which invested the subject was said to constitute one of its principal charms, affording, as it did, full scope to the fancy to indulge in a boundless field of speculation.

The course directly opposed to these theoretical views consists in an earnest and patient endeavour to reconcile the former indications of change with the evidence of gradual mutations now in progress; restricting us, in the first instance, to known causes, and then speculating on those which may be in activity in regions inaccessible to us. It seeks an interpretation of geological monuments by comparing the changes of which they give evidence with the vicissitudes now in progress, or *which may be in progress*.

We shall give a few examples in illustration of the practical results already derived from the two distinct methods of theorizing, for we have now the advantage of being enabled to judge by experience of their respective merits, and by the relative value of the fruits which they have produced.

In our historical sketch of the progress of geology, the reader has seen that a controversy was maintained for more than a century, respecting the origin of fossil shells and bones – were they organic or inorganic substances? That the latter opinion should for a long time have prevailed, and that these bodies should have been supposed to be fashioned into their present form by a plastic virtue, or some other mysterious agency, may appear absurd; but it was, perhaps, as reasonable a conjecture as could be expected from those who did not appeal, in the first instance, to the analogy of the living creation, as affording the only source of authentic information. It was only by an accurate examination of living testacea, and by a comparison of the osteology of the existing vertebrated animals with the remains found entombed in ancient strata, that this favourite dogma was exploded, and all were, at length, persuaded that these substances were exclusively of organic origin.

In like manner, when a discussion had arisen as to the nature of basalt and other mineral masses, evidently constituting a particular class of rocks, the popular opinion inclined to a belief that they were of aqueous, not of igneous origin. These rocks, it was said, might have been precipitated from an aqueous solution, from a chaotic fluid, or an ocean which rose over the continents, charged with the requisite mineral ingredients. All are now agreed that it would have been impossible for human ingenuity to invent a theory more distant from the truth; yet we must cease to wonder, on that account, that it gained so many proselytes, when we remember that its claims to probability arose partly from its confirming the assumed want of all analogy between geological causes and those now in action.

By what train of investigation were all theorists brought round at length to an opposite opinion, and induced to assent to the igneous origin of these formations? By an examination of the structure of active volcanos, the mineral composition of their lavas and ejections, and by comparing the undoubted products of fire with the ancient rocks in question.

We shall conclude with one more example. When the organic origin of fossil shells had been conceded, their occurrence in strata forming some of the loftiest mountains in the world, was admitted as a proof of a great alteration of the relative level of sea and land, and doubts were then entertained whether this change might be accounted for by the partial drying up of the ocean, or by the elevation of the solid land. The former hypothesis, although afterwards abandoned by general consent, was at first embraced by a vast majority. A multitude of ingenious speculations were hazarded to show how the level of the ocean might have been depressed, and when these theories had all failed, the inquiry, as to what vicissitudes of this nature might now be taking place, was, as usual, resorted to in the last instance. The question was agitated, whether any changes in the level of sea and land had occurred during the historical period, and, by patient research, it was soon discovered that considerable tracts of land had been permanently elevated and depressed, while the level of the ocean remained unaltered. It was therefore necessary to reverse the doctrine which had acquired so much popularity, and the unexpected solution of a problem at first regarded as so enigmatical, gave perhaps the strongest stimulus ever yet afforded to investigate the ordinary operations of nature. For it must have appeared almost as improbable to the earlier geologists, that the laws of earthquakes should one day throw light on the origin of mountains, as it must to the first astronomers, that the fall of an apple should assist in explaining the motions of the moon.

Of late years the points of discussion in geology have been transferred to new questions, and those, for the most part, of a higher and more general nature; but, notwithstanding the repeated warnings of experience, the ancient method of philosophising has not been materially modified.

We are now, for the most part, agreed as to what rocks are of igneous, and what of aqueous origin, – in what manner fossil shells, whether of the sea or of lakes, have been imbedded in strata, – how sand may have been converted into sandstone, – and are unanimous as to other propositions which are not of a complicated nature; but when we ascend to those of a higher order, we find as little disposition, as formerly, to make a strenuous effort, in the first instance, to search out an explanation in the ordinary economy of Nature. If, for example, we seek for the causes why mineral masses are associated together in certain groups; why they are arranged in a certain order which is never inverted; why there are many breaks in the continuity of the series; why different organic remains are found in distinct sets of strata; why there is often an abrupt passage from an assemblage of species contained in one formation to that in another immediately superimposed, – when these and other topics of an equally extensive kind are discussed, we find the habit of indulging conjectures, respecting irregular and extraordinary causes, to be still in full force.

We hear of sudden and violent revolutions of the globe, of the instantaneous elevation of mountain chains, of paroxysms of volcanic energy, declining according to some, and according to others increasing in violence, from the earliest to the latest ages. We are also told of general catastrophes and a succession of deluges, of the alternation of periods of repose and disorder, of the refrigeration of the globe, of the sudden annihilation of whole races of animals and plants, and other hypotheses, in which we see the ancient spirit of speculation revived, and a desire manifested to cut, rather than patiently to untie, the Gordian knot.

In our attempt to unravel these difficult questions, we shall adopt a different course, restricting ourselves to the known or possible operations of existing causes; feeling assured that we have not yet exhausted the resources which the study of the present course of nature may provide, and therefore that we are not authorized, in the infancy of our science, to recur to extraordinary agents. We shall adhere to this plan, not only on the grounds explained in the first volume, but because, as we have above stated, history informs us that this method has always put geologists on the road that leads to truth, – suggesting views which, although imperfect at first, have been found capable of improvement, until at last adopted by universal consent. On the other hand, the opposite method, that of speculating on a former distinct state of things, has led invariably to a multitude of contradictory systems, which have been overthrown one after the other, – which have been found quite incapable of modification, – and which are often required to be precisely reversed.

In regard to the subjects treated of in our first two volumes, if systematic treatises had been written on these topics, we should willingly have entered at once upon the description of geological monuments properly so called, referring to other authors for the elucidation of elementary and collateral questions, just as we shall appeal to the best authorities in conchology and comparative anatomy, in proof of many positions which, but for the labours of naturalists devoted to these departments, would have demanded long digressions. When we find it asserted, for example, that the bones of a fossil animal at Œningen were those of man, and the fact adduced as a proof of the deluge, we are now able at once to dismiss the argument as nugatory, and to affirm the skeleton to be that of a reptile, on the authority of an able anatomist; and when we find among ancient writers the opinion of the gigantic stature of the human race in times of old, grounded on the magnitude of certain fossil teeth and bones, we are able to affirm these remains to belong to the elephant and rhinoceros, on the same authority.

But since in our attempt to solve geological problems, we shall be called upon to refer to the operation of aqueous and igneous causes, the geographical distribution of animals and plants, the real existence of species, their successive extinction, and so forth, we were under the necessity of collecting together a variety of facts, and of entering into long trains of reasoning, which could only be accomplished in preliminary treatises.

These topics we regard as constituting the alphabet and grammar of geology; not that we expect from such studies to obtain a key to the interpretation of all geological phenomena, but because they form the groundwork from which we must rise to the contemplation of more general questions relating to the complicated results to which, in an indefinite lapse of ages, the existing causes of change may give rise.

CHAPTER 2

General Arrangement of the Materials Composing the Earth's Crust

When we examine into the structure of the earth's crust (by which we mean the small portion of the exterior of our planet accessible to human observation), whether we pursue our investigations by aid of mining operations, or by observing the sections laid open in the sea cliffs, or in the deep ravines of mountainous countries, we discover everywhere a series of mineral masses, which are not thrown together in a confused heap, but arranged with considerable order; and even where their original position has undergone great subsequent disturbance, there still remain proofs of the order that once reigned.

We have already observed, that if we drain a lake, we frequently find at the bottom a series of recent deposits disposed with considerable regularity one above the other; the uppermost, perhaps, may be a stratum of peat, next below a more compact variety of the same, still lower a bed of laminated shell marl, alternating with peat, and then other beds of marl, divided by layers of clay. Now if a second pit be sunk through the same continuous lacustrine deposit, at some distance from the first, we often meet with nearly the same series of beds, yet with slight variations; some, for example, of the layers of sand, clay, or marl may be wanting, one or more of them having thinned out and given place to others, or sometimes one of the masses, first examined, is observed to increase in thickness to the exclusion of other beds. Besides this limited continuity of particular strata, it is obvious that the whole assemblage must terminate somewhere; as, for example, where they reach the boundary of the original lake-basin, and where they will come in contact with the rocks which form the boundary of, and, at the same time, pass under all the recent accumulations.

In almost every estuary we may see, at low water, analogous phenomena where the current has cut away part of some newly-formed bank, consisting of a series of horizontal strata of peat, sand, clay, and, sometimes, interposed beds of shells. Each of these may often be traced over a considerable area, some extending farther than others, but all of necessity confined within the basin of the estuary. Similar remarks are applicable, on a much more extended scale, to the recent delta of a great river, like the Ganges, after the periodical inundations have subsided, and when sections are exposed of the river-banks and the cliffs of numerous islands, in which horizontal beds of clay and sand may be traced over an area many hundred miles in length, and more than a hundred in breadth.

Subaqueous deposits. – The greater part of our continents are evidently composed of subaqueous deposits; and in the manner of their arrangement we discover many characters precisely similar to those above described; but the different groups of strata are, for the most part, on a greater scale, both in regard to depth and area, than any observable in the new formations of lakes, deltas, or estuaries. We find, for example, beds of limestone several hundred feet in thickness, containing imbedded corals and shells, stretching from one country to another, yet always giving place, at length, to a distinct set of strata, which either rise up from under it like the rocks before alluded to as forming the borders of a lake, or cover and conceal it. In other places, we find beds of pebbles, and sand, or of clay of great thickness. The different formations composed of these materials usually contain some peculiar organic remains; as, for example, certain species of shells and corals, or certain plants.

Volcanic rocks. – Besides these strata of aqueous origin, we find other rocks which are immediately recognized to be the products of fire, from their exact resemblance to those which have been produced in modern times by volcanos, and thus we immediately establish two distinct orders of mineral masses composing the crust of the globe – the sedimentary and the volcanic.

Primary rocks. – But if we investigate a large portion of a continent which contains within it a lofty mountain range, we rarely fail to discover another class, very distinct from either of those above alluded to, and which we can neither assimilate to deposits such as are now accumulated in lakes or seas, nor to those generated by ordinary volcanic action. The class alluded to, consists of granite, granitic schist, roofing slate, and many other rocks, of a much more compact and crystalline texture than the sedimentary and volcanic divisions before mentioned. In the unstratified portion of these crystalline rocks, as in the granite for example, no organic fossil remains have ever been discovered, and only a few faint traces of them in some of the *stratified* masses of the same class; for we should state, that a considerable portion of these rocks are divided, not only into strata, but into laminæ, so closely imitating the internal arrangement of well-known aqueous deposits, as to leave scarcely any reasonable doubt that they owe this part of their texture to similar causes.

These remarkable formations have been called *primitive*, from being supposed to constitute the most ancient mineral productions known to us, and from a notion that they originated before the earth was inhabited by living beings, and while yet the planet was in a nascent state. Their high relative antiquity is indisputable; for in the oldest sedimentary strata, containing organic remains, we often meet with rounded pebbles of the older crystalline rocks, which must therefore have been consolidated before the derivative strata were formed out of their ruins. They rise up from beneath the rocks of mechanical origin, entering into the structure of lofty mountains, so as to constitute, at the same time, the lowest and the most elevated portions of the crust of the globe.

Origin of primary rocks. – Nothing strictly analogous to these ancient formations can now be seen in the progress of formation on the habitable surface of the earth, nothing, at least, within the range of human observation. The first speculators, however, in Geology, found no difficulty in explaining their origin, by supposing a former condition of the planet perfectly distinct from the present, when certain chemical processes were developed on a great scale, and whereby crystalline precipitates were formed, some more suddenly, in huge amorphous masses, such as granite; others by successive deposition and with a foliated and stratified structure, as in the rocks termed gneiss and mica-schist. A great part of these views have since been entirely abandoned, more especially with regard to the origin of granite, but it is interesting to trace the train of reasoning by which they were suggested. First, the stratified primitive rocks exhibited, as we before mentioned, well-defined marks of successive accumulation, analogous to those so common in ordinary subaqueous deposits. As the latter formations were found divisible into natural groups, characterized by certain peculiarities of mineral composition, so also were the primitive. In the next place, there were discovered, in many districts, certain members of the so-called primitive series, either alternating with, or passing by intermediate gradations into rocks of a decidedly mechanical origin, containing traces of organic remains. From such gradual passage the aqueous origin of the stratified crystalline rocks was fairly inferred; and as we find in the different strata of subaqueous origin every gradation between a mechanical and a purely crystalline texture; between sand, for example, and saccharoid gypsum, the latter having, probably, been precipitated originally in a crystalline form, from water containing sulphate of lime in solution, so it was imagined that, in a former condition of the planet, the different degrees of crystallization in the older rocks might have been dependent on the varying state of the menstruum from which they were precipitated.

The presence of certain crystalline ingredients in the composition of many of the primary rocks, rendered it necessary to resort to many arbitrary hypotheses, in order to explain their precipitation from aqueous solution, and for this reason a difference in the condition of the planet, and of the pristine energy of chemical causes, was assumed. A train of speculation originally suggested by the observed effects of aqueous agents, was thus pushed beyond the limits of analogy, and it was not until a different and almost opposite course of induction was pursued, beginning with an examination of volcanic products, that more sound theoretical views were established.

Granite of igneous origin. – As we are merely desirous, in this chapter, of fixing in the reader's mind the leading divisions of the rocks composing the earth's crust, we cannot enter, at present, into a detailed account of these researches, but shall only observe, that a passage was first traced from lava into other more crystalline igneous rocks, and from these again to granite, which last was found to send forth dikes and veins into the contiguous strata in a manner strictly analogous to that observed in volcanic rocks, and producing at the point of contact such changes as might be expected to result from the influence of a heated mass cooling down slowly under great pressure from a state of fusion. The want of stratification in granite supplied another point of analogy in confirmation of its igneous origin; and as some masses were found to send out veins through others, it was evident that there were granites of different ages, and that instead of forming in all cases the oldest part of the earth's crust, as had at first been supposed, the granites were often of comparatively recent origin, sometimes newer than the stratified rocks which covered them.

Stratified primary rocks. – The theory of the origin of the other crystalline rocks was soon modified by these new views respecting the nature of granite. First it was shown, by numerous examples, that ordinary volcanic dikes might produce great alterations in the sedimentary strata which they traversed, causing them to assume a more crystalline texture, and obliterating all traces of organic remains, without, at the same time, destroying either the lines of stratification, or even those which mark the division into laminæ. It was also found, that granite dikes and veins produced analogous, though somewhat different changes; and hence it was suggested as highly probable, that the effects to which small veins gave rise, to the distance of a few yards, might be superinduced on a much grander scale where immense masses of fused rock, intensely heated for ages, came in contact at great depths from the surface with sedimentary formations. The slow action of heat in such cases, it was thought, might occasion a state of semi-fusion, so that, on the cooling down of the masses, the different materials might be re-arranged in new forms, according to their chemical affinities, and all traces of organic remains might disappear, while the stratiform and lamellar texture remained.

May be of different ages. – According to these views, the primary strata may have assumed their crystalline structure at as many successive periods as there have been distinct eras of the formation of granite, and their difference of mineral composition may be attributed, not to an original difference of the conditions under which they were deposited at the surface, but to subsequent modifications superinduced by heat at great depths below the surface.

The strict propriety of the term primitive, as applied to granite and to the granitiform and associated rocks, thus became questionable, and the term primary was very generally substituted, as simply expressing the fact, that the crystalline rocks, as a mass, were older than the *secondary*, or those which are unequivocally of a mechanical origin and contain organic remains.[2]

Transition formations. – The reader may readily conceive, even from the hasty sketch which we have thus given of the supposed origin of the stratified primary rocks, that they may occasionally graduate into the secondary; accordingly, an attempt was made, when the classification of rocks was chiefly derived from mineral structure, to institute an order called *transition*, the characters of which were intermediate between those of the primary and secondary formations. Some of the shales, for example, associated with these strata, often passed insensibly into clay slates, undistinguishable from those of the granitic series; and it was often difficult to determine whether some of the compound rocks of this transition series, called greywacke, were of mechanical or chemical origin. The imbedded organic remains were rare, and sometimes nearly obliterated; but by their aid the groups first called transition were at length identified with rocks, in other countries, which had undergone much less alteration, and wherein shells and zoophytes were abundant.

The term transition, however, was still retained, although no longer applicable in its original signification. It was now made to depend on the identity of certain species of organized fossils; yet reliance on mineral peculiarities was not fairly abandoned, as constituting part of the characters of the group. This circumstance became a fertile source of ambiguity and confusion; for although the species of the transition strata denoted a certain epoch, the intermediate state of mineral character gave no such indications, and ought never to have been made the basis of a chronological division of rocks.

Order of succession of stratified masses. – All the subaqueous strata which we before alluded to as overlying the primary, were at first called secondary; and when they had been found divisible into different groups, characterised by certain organic remains and mineral peculiarities, the relative position of these groups became a matter of high interest. It was soon found that the order of succession was never inverted, although the different formations were not coextensively distributed; so that, if there be four different formations, as *a*, *b*, *c*, *d*, in the annexed diagram (No. 1), which, in certain localities, may be seen in vertical superposition, the uppermost or newest of them, *a*, will in other places be in contact with *c*, or with the lowest of the whole series, *d*, all the intermediate formations being absent.

1. Diagram showing the order of succession of stratified masses

Tertiary formations. – After some progress had been made in classifying the secondary rocks, and in assigning to each its relative place in a chronological series, another division of sedimentary formations was established, called *tertiary*, as being of newer origin than the secondary. The fossil contents of the deposits belonging to this newly-instituted order are,

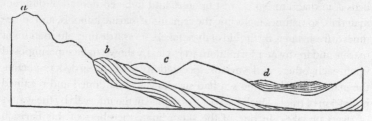

a. Primary rocks.
b. Older secondary formations.
c. Chalk.
d. Tertiary formation.

2. *Diagram showing the relative position of the Primary, Secondary, and Tertiary strata*

upon the whole, very dissimilar from those of the secondary rocks, not only all the species, but many of the most remarkable animal and vegetable forms, being distinct. The tertiary formations were also found to consist very generally of detached and isolated masses, surrounded on all sides by primary and secondary rocks, and occupying a position, in reference to the latter, very like that of the waters of lakes, inland seas, and gulfs, in relation to a continent, and, like such waters, being often of great depth, though of limited area. The imbedded organic remains were chiefly those of marine animals, but with frequent intermixtures of terrestrial and freshwater species so rarely found among the secondary fossils. Frequently there was evidence of the deposits having been purely lacustrine, a circumstance which has never yet been clearly ascertained in regard to any secondary group.

We shall consider more particularly, in the next chapter, how far this distinction of rocks into secondary and tertiary is founded in nature, and in what relation these two orders of mineral masses may be supposed to stand to each other. But before we offer any general views of this kind, it may be useful to present the reader with a succinct sketch of the principal points in the history of the discovery and classification of the tertiary strata.

Paris Basin. – The first series of deposits belonging to this class, of which the characters were accurately determined, were those which occur in the neighbourhood of Paris, first described by MM. Cuvier and

Brongniart. They were ascertained to fill a depression in the chalk (as the beds *d*, in diagram No. 2, rest upon *c*), and to be composed of different materials, sometimes including the remains of marine animals, and sometimes of freshwater. By the aid of these fossils, several distinct alternations of marine and freshwater formations were clearly shown to lie superimposed upon each other, and various speculations were hazarded respecting the manner in which the sea had successively abandoned and regained possession of tracts which had been occupied in the intervals by the waters of rivers or lakes. In one of the subordinate members of this Parisian series, a great number of scattered bones and skeletons of land animals were found entombed, the species being perfectly dissimilar from any known to exist, as indeed were those of almost all the animals and plants of which any portions were discovered in the associated deposits.

We shall defer, to another part of this work, a more detailed account of this interesting formation, and shall merely observe in this place, that the investigation of the fossil contents of these beds forms an era in the progress of the science. The French naturalists brought to bear upon their geological researches so much skill and proficiency in comparative anatomy and conchology, as to place in a strong light the importance of the study of organic remains, and the comparatively subordinate interest attached to the mere investigation of the structure and mineral ingredients of rocks.

A variety of tertiary formations were soon afterwards found in other parts of Europe, as in the south-east of England, in Italy, Austria, and different parts of France, especially in the basins of the Loire and Gironde, all strongly contrasted to the secondary rocks. As in the latter class many different divisions had been observed to preserve the same mineral characters and organic remains over wide areas, it was natural that an attempt should first be made to trace the different subdivisions of the Parisian tertiary strata throughout Europe, for some of these were not inferior in thickness to several of the secondary formations that had a wide range.

But in this case the analogy, however probable, was not found to hold good, and the error, though almost unavoidable, retarded seriously the progress of geology. For as often as a new tertiary group was discovered, as that of Italy, for example, an attempt was invariably made, in the first instance, to discover in what characters it agreed with some one or more subordinate members of the Parisian type. Every fancied point of correspondence was magnified into undue importance, and such trifling circumstances, as the colour of a bed of sand or clay, were dwelt upon as proofs of identification, while the difference in the mineral character and organic contents of the group from the whole Parisian series was slurred over and thrown into the shade.

By the influence of this illusion, the succession and chronological relations of different tertiary groups were kept out of sight. The difficulty of clearly discerning these, arose from the frequent isolation of the position of the tertiary formations before described, since, in proportion as the areas occupied by them are limited, it is rare to discover a place where one set of strata overlap another, in such a manner that the geologist might be enabled to determine the difference of age by direct superposition.

Origin of the European Tertiary Strata at Successive Periods

We shall now very briefly enumerate some of the principal steps which eventually led to a conviction of the necessity of referring the European tertiary formations to distinct periods, and the leading data by which such a chronological series may be established.

London and Hampshire Basins. – Very soon after the investigation, before alluded to, of the Parisian strata, those of Hampshire and of the Basin of the Thames were examined in our own country. Mr. Webster found these English tertiary deposits to repose, like those in France, upon the chalk or newest rock of the secondary series. He identified a great variety of the shells occurring in the British and Parisian strata, and ascertained that, in the Isle of Wight, an alternation of marine and fresh-water beds occurred, very analogous to that observed in the basin of the Seine. But no two sets of strata could well be more dissimilar in mineral composition, and they were only recognized to belong to the same era, by aid of the specific identity of their organic remains. The discordance, in other respects, was as complete as could well be imagined, for the principal marine formation in the one country consisted of blue clay, in the other of white limestone, and a variety of curious rocks in the neighbourhood of Paris had no representatives whatever in the south of England.

Subapennine Beds. – The next important discovery of tertiary strata was in Italy, where Brocchi traced them along the flanks of the Apennines, from one extremity of the peninsula to the other, usually forming a lower range of hills, called by him the Subapennines. These formations, it is true, had been pointed out by the older Italian writers, and some correct ideas, as we have seen, had been entertained respecting their recent origin, as compared to the inclined secondary rocks on which they rested. But accurate data were now for the first time collected, for instituting a comparison between them and other members of the great European series of tertiary formations.

Brocchi came to the conclusion that nearly one-half of several hundred species of fossil shells procured by him from these Subapennine beds were identical with those now living in existing seas, an observation which did not hold true in respect to the organic remains of the Paris basin. It might have been supposed that this important point of discrepancy would at once have engendered great doubt as to the identity, in age, of any part of the Subapennine beds to any one member of the Parisian series; but, for reasons above alluded to, this objection was not thought of much weight, and it was supposed that a group of strata, called 'the upper marine formation,' in the basin of the Seine, might be represented by all the Subapennine clays and yellow sand.

English Crag. – Several years before, an English naturalist, Mr. Parkinson, had observed, that certain shelly strata, in Suffolk, which overlaid the blue clay of London, contained distinct fossil species of testacea, and that a considerable portion of these might be identified with species now inhabiting the neighbouring sea. These overlying beds, which were provincially termed 'Crag,' were of small thickness, and were not regarded as of much geological importance. But when duly considered, they presented a fact worthy of great attention, viz., the superposition of a tertiary group, inclosing, like the Subapennine beds, a great intermixture of recent species of shells, upon beds wherein a very few remains of recent or living species were entombed.

Mr. Conybeare, in his excellent classification of the English strata, placed the crag as the uppermost of the British series, and several geologists began soon to entertain an opinion that this newest of our tertiary formations might correspond in age to the Italian strata described by Brocchi.

Tertiary Strata of Touraine. – The next step towards establishing a succession of tertiary periods was the evidence adduced to prove that certain formations, more recent than the uppermost members of the Parisian series, were also older than the Subapennine beds, so that they constituted deposits of an age intermediate between the two types above alluded to. Mr. Desnoyers, for example, ascertained that a group of marine strata in

Touraine, in the basin of the Loire (*e*, diagram No. 3), rest upon the uppermost subdivision of the Parisian group *d*, which consists of a lacustrine formation, extending continuously throughout a platform which intervenes between the basin of the Seine and that of the Loire. These overlaying marine strata, M. Desnoyers assimilated to the English crag, to which they bear some analogy, although their organic remains differ considerably, as will be afterwards shown.

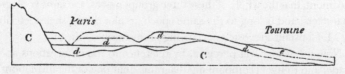

C. Chalk and other secondary formations.
d. Tertiary formation of Paris basin.
e. Superimposed marine tertiary beds of the Loire.

3. Diagram showing the relative age of the strata of the Paris basin, and those of the basin of the Loire, in Touraine

A large tertiary deposit had already been observed in the south-west of France, around Bordeaux and Dax, and a description of its fossils had been published by M. de Basterot. Many of the species were peculiar, and differed from those of the strata now called Subapennine; yet these same peculiar and characteristic fossils reappeared in Piedmont, in a series of strata inferior in position to the Subapennines (as *e* underlies *f*, diagram No. 4).

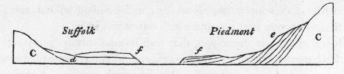

C. Chalk and older formations.
d. London clay (old tertiary).
e. Tertiary strata of same age as beds of the Loire.
f. Crag and Subapennine tertiary deposits.

4. Diagram showing the relative age of the strata of Suffolk and Piedmont

This inferior group, *e*, composed principally of green sand, occurs in the hills of Mont Ferrat, and beds of the same age are seen in the valley of the Bormida. They also form the hill of the Superga, near Turin, where M. Bonelli formed a large collection of their fossils, and identified them with those discovered near Bordeaux and in the basin of the Gironde.

But we are indebted to M. Deshayes for having proved, by a careful comparison of the entire assemblage of shells found in the above-mentioned localities, in Touraine, in the south-east of France, and in Piedmont, that the whole of these three groups possess the same zoological characters, and belong to the same epoch, as also do the shells described by M. Constant Prevost, as occurring in the basin of Vienna.

Now the reader will perceive, by reference to the observations above made, and to the accompanying diagrams, that one of the formations of this intervening period, *e*, has been found superimposed upon the highest member of the Parisian series, *d*; while another of the same set has been observed to underlie the Subapennine beds, *f*. Thus the chronological series, *d*, *e*, *f*, is made out, in which the deposits, originally called tertiary, those of the Paris and London basins, for example, occupy the lowest position, and the beds called 'the Crag,' and 'the Subapennines,' the highest.

Tertiary Strata newer than the Subapennine. – The fossil remains which characterize each of the three successive periods above alluded to, approximate more nearly to the assemblage of *species* now existing, in proportion as their origin is less remote from our own era, or, in other words, the recent species are always more numerous, and the extinct more rare, in proportion to the low antiquity of the formations. But the discordance between the state of the organic world indicated by the fossils of the Subapennine beds and the actual state of things is still considerable, and we naturally ask, are there no monuments of an intervening period? – no evidences of a gradual passage from one condition of the animate creation to that which now prevails, and which differs so widely?

It will appear, in the sequel, that such monuments are not wanting, and that there are marine strata entering into the composition of extensive districts, and of hills of no trifling height, which contain the exuviæ of testacea and zoophytes, hardly distinguishable, as a group, from those now peopling the neighbouring seas. Thus the line of demarcation between the actual period and that immediately antecedent, is quite evanescent, and the newest members of the tertiary series will be often found to blend with the formations of the historical era.

In Europe, these modern strata have been found in the district around Naples, in the territory of Otranto and Calabria, and more particularly in the Island of Sicily; and the bare enumeration of these localities cannot fail to remind the reader, that they belong to regions where the volcano and the earthquake are now active, and where we might have anticipated the discovery of emphatic proofs, that the conversion of sea into land had been of frequent occurrence at very modern periods.

Different Circumstances under which the Secondary and Tertiary Formations may have Originated

We have already glanced at the origin of some of the principal points of difference in the characters of the primary and secondary rocks, and may now briefly consider the relation in which the secondary stand to the tertiary, and the causes of that succession of tertiary formations described in the last chapter.

It is evident that large parts of Europe were simultaneously submerged beneath the sea when different portions of the secondary series were formed, because we find homogeneous mineral masses, including the remains of marine animals, referrible to the secondary period, extending over great areas; whereas the detached and isolated position of tertiary groups, in basins or depressions bounded by secondary and primary rocks, favours the hypothesis of a sea interrupted by extensive tracts of dry land.

State of the Surface when the Secondary Strata were formed

Let us consider the changes that must be expected to accompany the gradual conversion of part of the bed of an ocean into a continent, and the different characters that might be imparted to subaqueous deposits formed during the period when the sea prevailed, as contrasted with those that might belong to the subsequent epoch when the land should predominate. First, we may suppose a vast submarine region, such as the bed of the western Atlantic, to receive for ages the turbid waters of several great rivers, like the Amazon, Orinoco, or Mississippi, each draining a considerable continent. The sediment thus introduced might be characterized by a peculiar colour and composition, and the same homogeneous mixture might be spread out over an immense area by the action of a powerful current, like the Gulf-stream. First one submarine basin, and then another, might be filled, or rendered shallow, by the influx of transported matter, the same species of animals and plants still continuing to inhabit the sea, so that the organic, as well as the mineral characters, might be constant throughout the whole series of deposits.

In another part of the same ocean, let us suppose masses of coralline and shelly limestone to grow, like those of the Pacific, simultaneously over a space several thousand miles in length, and thirty or forty degrees of latitude in breadth, while volcanic eruptions give rise, at different intervals, to igneous rocks, having a common subaqueous character in different parts of the vast area.

It is evident that, during such a state of a certain quarter of the globe, beds of limestone and other rocks might be formed, and retain a common character over spaces equal to a large portion of Europe.

State of the Surface when the Tertiary Groups were formed

But when the area under consideration began to be converted into land, a very different condition of things would succeed. A series of subterranean movements might first give rise to small rocks and isles, and then, by subsequent elevations, to larger islands, by the junction of the former. These lands would consist partly of the mineral masses before described, whether coralline, sedimentary, or volcanic, and partly of the subjacent rocks, whatever they may have been, which constituted the original bed of the ocean. Now the degradation of these lands would commence immediately upon their emergence, the waves of the sea undermining the cliffs, and torrents flowing from the surface, so that new strata would begin to form in different places; and in proportion as the lands increased, these deposits would augment.

At length by the continued rising and sinking of different parts of the bed of the ocean, a number of distinct basins would be formed, wherein different kinds of sediment, each distinguished by some local character, might accumulate. Some of the groups of isles that had first risen would, in the course of ages, become the central mountain ranges of continents, and different lofty chains might thus be characterized by similar rocks of contemporaneous origin, the component strata having originated under analogous circumstances in the ocean before described.

Finally, when large tracts of land existed, there would be a variety of disconnected gulfs, inland seas, and lakes, each receiving the drainage of distinct hydrographical basins, and becoming the receptacles of strata distinguished by marked peculiarities of mineral composition. The organic remains would also be more varied, for in one locality fresh-water species would be imbedded, as in deposits now forming in the lakes of Switzerland and the north of Italy; in another, marine species, as in the Aral and Caspian; in a third region, gulfs of brackish water would be converted into land, like those of Bothnia and Finland in the Baltic; in a fourth, there might be great fluviatile and marine formations along the borders of a chain of inland seas, like the deltas now growing at the mouths of the Don, Danube, Nile, Po, and Rhone, along the shores of the Azof, Euxine, and Mediterranean. These deposits would each partake more or less of the peculiar mineral character of adjoining lands, the degradation of which would supply sediment to the different rivers.

Now if such be, in a great measure, the distinction between the circumstances under which the secondary and tertiary series originated, it is quite natural that particular tertiary groups should occupy areas of comparatively small extent, – that they should frequently consist of littoral and lacustrine deposits, and that they should often contain those admixtures of terrestrial, fresh-water, and marine remains, which are so rare in secondary rocks. It might also be expected, that the tertiary volcanic formations should be much less exclusively submarine, and this we accordingly find to be the case.

Causes of the Superposition of Successive Formations having Distinct Mineral and Organic Characters

But we have still to account for those remarkable breaks in the series of superimposed formations, which are common both to the secondary and tertiary rocks, but are more particularly frequent in the latter.

The elucidation of this curious point is the more important, because geologists of a certain school appeal to phenomena of this kind in support of their doctrine of great catastrophes, out of the ordinary course of nature, and sudden revolutions of the globe.

It is only by carefully considering the combined action of all the causes of change now in operation, whether in the animate or inanimate world, that we can hope to explain such complicated appearances as are exhibited in the general arrangement of mineral masses. In attempting, therefore, to trace the origin of these violations of continuity, we must re-consider many of the topics treated of in our two former volumes, such as the effects of the various agents of decay and reproduction, the imbedding of organic remains, and the extinction of species.

Shifting of the Areas of Sedimentary Deposition. – By reverting to our survey of the destroying and renovating agents, it will be seen that the surface of the terraqueous globe may be divided into two parts, one of which is undergoing repair, while the other, constituting, at any one period, by far the largest portion of the whole, is either suffering degradation, or remaining stationary without loss or increment. The reader will assent at once to this proposition, when he reflects that the dry land is, for the most part, wasting by the action of rain, rivers, and torrents, while the effects of vegetation have, as we have shown, only a conservative tendency, being very rarely instrumental in adding new masses of mineral matter to the surface of emerged lands; and when he also reflects that part of the bed of the sea is exposed to the excavating action of currents, while the greater part, remote from continents and islands, probably receives no new deposits whatever, being covered for ages with the clear blue waters uncharged with sediment. Here the relics of organic beings, lying in the ooze of the deep, may decompose like the leaves of the forest in autumn, and leave no wreck behind, but merely supply nourishment, by their decomposition, to succeeding races of marine animals and plants.

The other part of the terraqueous surface is the receptacle of new deposits, and in this portion alone, as we pointed out in the last volume, the remains of animals and plants become fossilized. Now the position of this area, where new formations are in progress, and where alone any memorials of the state of organic life are preserved, is always varying, and must for ever continue to vary; and, for the same reason, that portion of the terraqueous globe which is undergoing waste, also shifts its position, and these fluctuations depend partly on the action of aqueous, and partly of igneous causes.

In illustration of these positions we may observe, that the sediment of the Rhone, which is thrown into the lake of Geneva, is now conveyed to a spot a mile and a half distant from that where it accumulated in the tenth century, and six miles from the point where the delta began originally to form. We may look forward to the period when the lake will be filled up, and then a sudden change will take place in the distribution of the transported matter; for the mud and sand brought down from the Alps will thenceforth, instead of being deposited near Geneva, be carried nearly two hundred miles southwards, where the Rhone enters the Mediterranean.

The additional matter thus borne down to the lower delta of the Rhone would not only accelerate its increase, but might affect the mineral character of the strata there deposited, and thus give rise to an upper group, or subdivision of beds, having a distinct character. But the filling up of a lake, and the consequent transfer of the sediment to a new place, may sometimes give rise to a more abrupt transition from one group to another; as, for example, in a gulf like that of the St. Lawrence, where no deposits are now accumulated the river being purged of all its impurities in its previous course through the Canadian lakes. Should the lowermost of these lakes be at any time filled up with sediment, or laid dry by earthquakes, the waters of the river would thenceforth become turbid, and strata would begin to be deposited in the gulf, where a new formation would immediately overlie the ancient rocks now constituting the bottom. In this case there would be an abrupt passage from the inferior and more ancient, to the newer superimposed formation.

The same sudden coming on of new sedimentary deposits, or the suspension of those which were in progress, must frequently occur in different submarine basins where there are currents which are always liable, in the course of ages, to change their direction. Suppose, for instance, a sea to be filling up in the same manner as the Adriatic, by the influx of the Po, Adige, and other rivers. The deltas, after advancing and converging, may at last come within the action of a transverse current, which may arrest the further deposition of matter, and sweep it away to a distant point. Such a current now appears to prey upon the delta of the Nile, and to carry eastward the annual accessions of sediment that once added rapidly to the plains of Egypt.

On the other hand, if a current charged with sediment vary its course, a circumstance which, as we have shown, must happen to all of them in the lapse of ages, the accumulation of transported matter will at once cease in one region, and commence in another.

Although the causes which occasion the transference of the places of sedimentary deposition are continually in action in every region, yet they are most frequent where subterranean movements alter, from time to time, the levels of land, and they must be immense during the successive elevations and depressions which must be supposed to accompany the rise of a great continent from the deep. A trifling change of level may sometimes throw a current into a new direction, or alter the course of a considerable river. Some tracts will be alternately submerged and laid dry by subterranean movements; in one place a shoal will be formed, whereby the waters will drift matter over spaces where they once threw down their burden, and new cavities will elsewhere be produced, both marine and lacustrine, which will intercept the waters bearing sediment, and thereby stop the supply once carried to some distant basin.

We have before stated, that a few earthquakes of moderate power might cause a subsidence which would connect the sea of Azof with a large part of Asia now below the level of the ocean. This vast depression, recently shown by Humboldt to extend over an area of eighteen thousand square leagues, surrounds Lake Aral and the Caspian, on the shores of which seas it sinks in some parts to the depth of three hundred feet below the level of the ocean. The whole area might thus suddenly become the receptacle of new beds of sand and shells, probably differing in mineral character from the masses previously existing in that country, for an exact correspondence could only arise from a precise identity in the whole combination of circumstances which should give rise to formations produced at different periods in the same place.

Without entering into more detailed explanations, the reader will perceive that, according to the laws now governing the aqueous and igneous causes, distinct deposits must, at different periods, be thrown down on various parts of the earth's surface, and that, in the course of ages, the same area may become, again and again, the receptacle of such dissimilar sets of strata. During intervening periods, the space may either remain unaltered, or suffer what is termed *denudation*, in which case a superior set of strata are removed by the power of running water, and subjacent beds are laid bare, as happens wherever a sea encroaches upon a line of coast. By such means, it is obvious that the discordance in age of rocks in contact must often be greatly increased.

The frequent unconformability in the stratification of the inferior and overlying formation is another phenomenon in their arrangement, which may be considered as a natural consequence of those movements that accompany the gradual conversion of part of an ocean into land; for by such convulsions the older set of strata may become rent, shattered, inclined, and contorted to any amount. If the movement entirely cease before a new deposit is formed in the same tract, the superior strata may repose horizontally upon the dislocated series. But even if the subterranean convulsions continue with increasing violence, the more recent formations must remain comparatively undisturbed, because they cannot share in the immense derangement previously produced in the older beds, while the latter, on the contrary, cannot fail to participate in all the movements subsequently communicated to the newer.

Change of Species everywhere in progress. – If, then, it be conceded, that the combined action of the volcanic and the aqueous forces would give rise to a succession of distinct formations, and that these would be sometimes unconformable, let us next inquire in what manner these groups might become characterized by different assemblages of fossil remains.

We endeavoured to show, in the last volume, that the hypothesis of the gradual extinction of certain animals and plants, and the successive introduction of new species, was quite consistent with all that is known of the existing economy of the animate world; and if it be found the only hypothesis which is reconcilable with geological phenomena, we shall have strong grounds for conceiving that such is the order of nature.

Fossilization of Plants and Animals partial. – We have seen that the causes which limit the duration of species are not confined, at any one time, to a particular part of the globe; and, for the same reason, if we suppose that their place is supplied, from time to time, by new species, we may suppose their introduction to be no less generally in progress. Hence, from all the foregoing premises, it would follow, that the change of species would be in simultaneous operation everywhere throughout the habitable surface of sea and land; whereas the fossilization of plants and animals must always be confined to those areas where new strata are produced. These areas, as we have proved, are always shifting their position, so that the fossilizing process, whereby the commemoration of the particular state of the organic world, at any given time, is effected, may be said to move about, visiting and revisiting different tracts in succession.

In order more distinctly to elucidate our idea of the working of this machinery, let us compare it to a somewhat analogous case that might easily be imagined to occur in the history of human affairs. Let the mortality of the population of a large country represent the successive extinction of species, and the births of new individuals the introduction of new species. While these fluctuations are gradually taking place everywhere, suppose commissioners to be appointed to visit each province of the country in succession, taking an exact account of the number, names, and individual peculiarities of all the inhabitants, and leaving in each district a register containing a record of this information. If, after the completion of one census, another is immediately made after the same plan, and then another, there will, at last, be a series of statistical documents in each province. When these are arranged in chronological order, the contents of those which stand next to each other will differ according to the length of the intervals of time between the taking of each census. If, for example, all the registers are made in a single year, the proportion of deaths and births will be so small during the interval between the compiling of two consecutive documents, that the individuals described in each will be nearly identical; whereas, if there are sixty provinces, and the survey of each requires a year, there will be an almost entire discordance between the persons enumerated in two consecutive registers.

There are undoubtedly some other causes besides the mere quantity of time which may augment or diminish the amount of discrepancy. Thus, for example, at some periods a pestilential disease may lessen the average duration of human life, or a variety of circumstances may cause the births to be unusually numerous, and the population to multiply, or, a province may be suddenly colonized by persons migrating from surrounding districts.

We must also remind the reader, that we do not propose the above case as an exact parallel to those geological phenomena which we desire to illustrate; for the commissioners are supposed to visit the different provinces in rotation, whereas the commemorating processes by which organic remains become fossilized, although they are always shifting from one area to another, are yet very irregular in their movements. They may abandon and revisit many spaces again and again, before they once approach another district; and besides this source of irregularity, it may often happen, that while the depositing process is suspended, denudation may take place, which may be compared to the occasional destruction of some of the statistical documents before mentioned. It is evident, that where such accidents occur, the want of continuity in the series may become indefinitely great, and that the monuments which follow next in succession will by no means be equidistant from each other in point of time.

If this train of reasoning be admitted, the frequent distinctness of the fossil remains, in formations immediately in contact, would be a necessary consequence of the existing laws of sedimentary deposition, accompanied by the gradual birth and death of species.

We have already stated, that we should naturally look for a change in the mineral character in strata thrown down at distant intervals in the same place; and, in like manner, we must also expect, for the reason last set forth, to meet occasionally with sudden transitions from one set of organic remains to another. But the causes which have given rise to such differences in mineral characters have no necessary connexion with those which have produced a change in the species of imbedded plants and animals.

When the lowest of two sets of strata are much dislocated over a wide area, the upper being undisturbed, there is usually a considerable discordance in the organic remains of the two groups; but this coincidence must not be ascribed to the agency of the disturbing forces, as if they had exterminated the living inhabitants of the surface. The immense *lapse of time* required for the development of so great a series of subterranean movements, has in these cases allowed the species also throughout the globe to vary, and hence the two phenomena are usually concomitant.

Although these inferences appear to us very obvious, we are aware that they are directly opposed to many popular theories respecting catastrophes; we shall, therefore, endeavour to place our views in a still clearer light before the reader. Suppose we had discovered two buried cities at the foot of Vesuvius, immediately superimposed upon each other, with a great mass of tuff and lava intervening, just as Portici and Resina, if now covered with ashes, would overlie Herculaneum. An antiquary might possibly be entitled to infer, from the inscriptions on public edifices, that the inhabitants of the inferior and older town were Greeks, and those of the modern, Italians. But he would reason very hastily, if he also concluded from these data, that there had been a sudden change from the Greek to the Italian language in Campania. Suppose he afterwards found *three* buried cities, one above the other, the intermediate one being Roman, while, as in the former example, the lowest was Greek, and the uppermost Italian, he would then perceive the fallacy of his former opinion, and would begin to suspect that the catastrophes, whereby the cities were inhumed, might have no relation whatever to the fluctuations in the language of the inhabitants; and that, as the Roman tongue had evidently intervened between the Greek and Italian, so many other dialects may have been spoken in succession, and the passage from the Greek to the Italian may have been very gradual, some terms growing obsolete, while others were introduced from time to time.

If this antiquary could have shown that the volcanic paroxysms of Vesuvius were so governed as that cities should be buried one above the other, just as often as any variation occurred in the language of the inhabitants, then, indeed, the abrupt passage from a Greek to a Roman, and from a Roman to an Italian city, would afford proof of fluctuations no less sudden in the language of the people.

So in Geology, if we could assume that it is part of the plan of nature to preserve, in every region of the globe, an unbroken series of monuments to commemorate the vicissitudes of the organic creation, we might infer the sudden extirpation of species, and the simultaneous introduction of others, as often as two formations in contact include dissimilar organic fossils. But we must shut our eyes to the whole economy of the existing causes, aqueous, igneous, and organic, if we fail to perceive *that such is not the plan of Nature.*

CHAPTER 4

Determination of the Relative
Ages of Rocks

In attempting to classify the mineral masses which compose the crust of the earth, the principal object which the geologist must keep in view, is to determine with accuracy their chronological relations, for it is abundantly clear, that different rocks have been formed in succession; and in order thoroughly to comprehend the manner in which they enter into the structure of our continents, we should study them with reference to the time and mode of their formation.

We shall now, therefore, consider by what characters the relative ages of different rocks may be established, whereby we may be supplied at once with sound information of the greatest practical utility, and which may throw, at the same time, the fullest light on the ancient history of the globe.

Proofs of Relative Age by Superposition

It is evident that where we find a series of horizontal strata, of sedimentary origin, the uppermost bed must be newer than those which it overlies, and that when we observe one distinct set of strata reposing upon another, the inferior is the older of the two. In countries where the original position of mineral masses has been disturbed, at different periods, by convulsions of extraordinary violence, as in the Alps and other mountainous districts, there are instances where the original position of strata has been reversed; but such exceptions are rare, and are usually on a small scale, and an experienced observer can generally ascertain the true relations of the rocks in question, by examining some adjoining districts where the derangement has been less extensive.

In regard to volcanic formations, if we find a stratum of tuff or ejected matter, or a stream of lava covering sedimentary strata, we may infer, with confidence, that the igneous rock is the more recent; but, on the other hand, the superposition of aqueous deposits to a volcanic mass does not always prove the former to be of newer origin. If, indeed, we discover strata of tuff with imbedded shells, or, as in the Vicentine and other places, rolled blocks of lava with adhering shells and corals, we may then be sure that these masses of volcanic origin covered the bottom of the sea, before the superincumbent strata were thrown down. But as lava rises from below, and does not always reach the surface, it may sometimes penetrate a certain number of strata, and then cool down, so as to constitute a solid mass of newer origin, although inferior in position. It is, for the most part, by the passage of veins proceeding from such igneous rocks through contiguous sedimentary strata, or by such hardening and other alteration of the overlying bed, as might be expected to result from contact with a heated mass, that we are enabled to decide whether the volcanic matter was previously consolidated, or subsequently introduced.

Proofs by Included Fragments of Older Rocks

A Geologist is sometimes at a loss, after investigating a district composed of two distinct formations, to determine the relative ages of each, from want of sections exhibiting their superposition. In such cases, another kind of evidence, of a character no less conclusive, can sometimes be obtained. One group of strata has frequently been derived from the degradation of another in the immediate neighbourhood, and may be observed to include within it fragments of such older rocks. Thus, for example, we may find chalk with flints, and in another part of the same country, a distinct series, consisting of alternations of clay, sand, and pebbles. If some of these pebbles consist of flints, with silicified fossil-shells of the same species as those in the chalk, we may confidently infer, that the chalk is the oldest of the two formations.

We remarked in the second chapter, that some granite must have existed before the most ancient of our secondary rocks, because some of the latter contain rounded pebbles of granite. But for the existence of such evidence, we might not have felt assured that all the granite which we see had not been protruded from below in a state of fusion, subsequently to the origin of the secondary strata.

Proofs of Contemporaneous Origin derived from Mineral Characters

When we have established the relative age of two formations in a given place, by direct superposition, or by other evidence, a far more difficult task remains, to trace the continuity of the same formation, or, in other cases, to find means of referring detached groups of rocks to a contemporaneous origin. Such identifications in age are chiefly derivable from two sources – mineral character and organic contents; but the utmost skill and caution are required in the application of such tests, for scarcely any general rules can be laid down respecting either, that do not admit of important exceptions.

If, at certain periods of the past, rocks of peculiar mineral composition had been precipitated simultaneously upon the floor of an 'universal ocean,' so as to invest the whole earth in a succession of concentric coats, the determination of relative dates in geology might have been a matter of the greatest simplicity. To explain, indeed, the phenomenon would have been difficult, or rather impossible, as such appearances would have implied a former state of the globe, without any analogy to that now prevailing. Suppose, for example, there were three masses extending over every continent, – the upper of chalk and chloritic sand; the next below, of blue argillaceous limestone; and the third and lowest, of red marl and sandstone; we must imagine that all the rivers and currents of the world had been charged, at the first period, with red mud and sand; at the second, with blue calcareo-argillaceous mud; and at a subsequent epoch, with chalky sediment and chloritic sand.

But if the ocean were universal, there could have been no land to waste away by the action of the sea and rivers, and, therefore, no known source whence the homogeneous sedimentary matter could have been derived. Few, perhaps, of the earlier geologists went so far as to believe implicitly in such universality of formations, but they inclined to an opinion, that they were continuous over areas almost indefinite; and since such a disposition of mineral masses would, if true, have been the least complex and most convenient for the purposes of classification, it is probable that a belief in its reality was often promoted by the hope that it might prove true. As to the objection, that such an arrangement of mineral masses could never result from any combination of causes now in action, it never weighed with the earlier cultivators of the science, since they indulged no expectation of being ever able to account for geological phenomena by reference to the known economy of nature. On the contrary, they set out, as we have already seen, with the assumption that the past and present conditions of the planet were too dissimilar to admit of exact comparison.

But if we inquire into the true composition of any stratum, or set of strata, and endeavour to pursue these continuously through a country, we often find that the character of the mass changes gradually, and becomes at length so different, that we should never have suspected its identity, if we had not been enabled to trace its passage from one form to another.

We soon discover that rocks dissimilar in mineral composition have originated simultaneously; we find, moreover, evidence in certain districts, of the recurrence of rocks of precisely the same mineral character at very different periods; as, for example, two formations of red sandstone, with a great series of other strata intervening between them. Such repetitions might have been anticipated, since these red sandstones are produced by the decomposition of granite, gneiss, and mica-schist; and districts composed exclusively of these, must again and again be exposed to decomposition, and to the erosive action of running water.

But notwithstanding the variations before alluded to in the composition of one continuous set of strata, many rocks retain the same homogeneous structure and composition, throughout considerable areas, and frequently, after a change of mineral character, preserve their new peculiarities throughout another tract of great extent. Thus, for example, we may trace a limestone for a hundred miles, and then observe that it becomes more arenaceous, until it finally passes into sand or sandstone. We may then follow the last-mentioned formation throughout another district as extensive as that occupied by the limestone first examined.

Proofs of Contemporaneous Origin derived from Organic Remains

We devoted several chapters, in the last volume, to show that the habitable surface of the sea and land may be divided into a considerable number of distinct provinces, each peopled by a peculiar assemblage of animals and plants, and we endeavoured to point out the origin of these separate divisions. It was shown that climate is only one of many causes on which they depend, and that difference of longitude, as well as latitude, is generally accompanied by a dissimilarity of indigenous species of organic beings.

As different seas, therefore, and lakes are inhabited at the same period, by different species of aquatic animals and plants, and as the lands adjoining these may be peopled by distinct terrestrial species, it follows that distinct organic remains are imbedded in contemporaneous deposits. If it were otherwise – if the same species abounded in every climate, or even in every part of the globe where a corresponding temperature, and other conditions favourable to their existence were found, the identification of mineral masses of the same age, by means of their included organic contents, would be a matter of much greater facility.

But, fortunately, the extent of the same zoological provinces, especially those of marine animals, is very great, so that we are entitled to expect, from analogy, that the identity of fossil species, throughout large areas, will often enable us to connect together a great variety of detached and dissimilar formations.

Thus, for example, it will be seen, by reference to our first volume, that deposits now forming in different parts of the Mediterranean, as in the deltas of the Rhone and the Nile, are distinct in mineral composition; for calcareous rocks are precipitated from the waters of the former river, while pebbles are carried into its delta, and there cemented, by carbonate of lime, into a conglomerate; whereas strata of soft mud and fine sand are formed exclusively in the Nilotic delta. The Po, again, carries down fine sand and mud into the Adriatic; but since this sediment is derived from the degradation of a different assemblage of mountains from those drained by the Rhone or the Nile, we may safely assume that there will never be an exact identity in their respective deposits.

If we pass to another quarter of the Mediterranean, as, for example, to the sea on the coast of Campania, or near the base of Etna in Sicily, or to the Grecian archipelago, we find in all these localities that distinct combinations of rocks are in progress. Occasional showers of volcanic ashes are falling into the sea, and streams of lava are flowing along its bottom; and in the intervals between volcanic eruptions, beds of sand and clay are frequently derived from the waste of cliffs, or the turbid waters of rivers. Limestones, moreover, such as the Italian travertins, are here and there precipitated from the waters of mineral springs, while shells and corals accumulate in various localities. Yet the entire Mediterranean, where the above-mentioned formations are simultaneously in progress, may be considered as one zoological province; for, although certain species of testacea and zoophytes may be very local, and each region may probably have some species peculiar to it, still a considerable number are common to the whole sea. If, therefore, at some future period, the bed of this inland sea should be converted into land, the geologist might be enabled, by reference to organic remains, to prove the contemporaneous origin of various mineral masses throughout a space equal in area to a great portion of Europe. The Black Sea, moreover, is inhabited by so many identical species, that the delta of the Danube and the Don might, by the same evidence, be shown to have originated simultaneously.

Such identity of fossils, we may remark, not only enables us to refer to the same era, distinct rocks widely separated from each other in the horizontal plane, but also others which may be considerably distant in the vertical series. Thus, for example, we may find alternating beds of clay, sand, and lava, two thousand feet in thickness, the whole of which may be proved to belong to the same epoch, by the specific identity of the fossil shells dispersed throughout the whole series. It may be objected, that different species would, during the same zoological period, inhabit the sea at different depths, and that the case above supposed could never occur; but, for reasons explained in the last volume, we believe that rivers and tidal currents often act upon the banks of littoral shells, so that a sea of great depth may be filled with strata, containing throughout a considerable number of the same fossils.

The reader, however, will perceive, by referring to what we have said of zoological provinces, that they are sometimes separated from each other by very narrow barriers, and for this reason contiguous rocks may be formed at the same time, differing widely both in mineral contents and organic remains. Thus, for example, the testacea, zoophytes, and fish of the Red Sea, may be considered, as a group, to be very distinct from those inhabiting the adjoining parts of the Mediterranean, although the two seas are only separated by the narrow isthmus of Suez. We shall show, in a subsequent chapter, that calcareous formations have accumulated, on a great scale, in the Red Sea, in modern times, and that fossil shells of existing species are well preserved therein; while we know that, at the mouth of the Nile, large deposits of mud are amassed, including the remains of Mediterranean species. Hence it follows, that if, at some future period, the bed of the Red Sea should be laid dry, the geologist might experience great difficulties in endeavouring to ascertain the relative age of these formations, which, although dissimilar both in organic and mineral characters, were of synchronous origin.

There might, perhaps, be no means of clearing up the obscurity of such a question, yet we must not forget that the north-western shores of the Arabian Gulf, the plains of Egypt, and the isthmus of Suez, are all parts of one province of *terrestrial* species. Small streams, therefore, occasional land-floods, and those winds which drift clouds of sand along the deserts, might carry down into the Red Sea the same shells of fluviatile and land testacea, which the Nile is sweeping into its delta, together with some remains of terrestrial plants, whereby the groups of strata, before alluded to, might, notwithstanding the discrepancy of their mineral composition, and *marine* organic fossils, be shown to have belonged to the same epoch.

In like manner, the rivers which descend into the Caribbean Sea and Gulf of Mexico on one side, and into the Pacific on the other, carry down the same fluviatile and terrestrial spoils into seas which are inhabited by different groups of marine species.

But it will much more frequently happen, that the coexistence of *terrestrial* species, of distinct zoological and botanical provinces, will be proved by the specific identity of the *marine* organic remains which inhabited the intervening space. Thus, for example, the distinct terrestrial species of the south of Europe, north of Africa, and north-west of Asia, might all be shown to have been contemporaneous, if we suppose the rivers flowing from those three countries to carry the remains of different species of the animal and vegetable kingdoms into the Mediterranean.

In like manner, the sea intervening between the northern shores of Australia and the islands of the Indian ocean, contains a great proportion of the same species of corallines and testacea, yet the *land animals and plants* of the two regions are very dissimilar, even the islands nearest to Australia, as Java, New Guinea, and others, being inhabited by a distinct assemblage of terrestrial species. It is well known that there are calcareous rocks, volcanic tuff, and other strata in progress, in different parts of these intermediate seas, wherein marine organic remains might be preserved and associated with the terrestrial fossils above alluded to.

As it frequently happens that the barriers between different provinces of animals and plants are not very strongly marked, especially where they are determined by differences of temperature, there will usually be a passage from one set of species to another, as in a sea extending from the temperate to the tropical zone. In such cases, we may be enabled to prove, by the fossils of intermediate deposits, the connexion between the distinct provinces, since these intervening spaces will be inhabited by many species, common both to the temperate and equatorial seas.

On the other hand, we may be sometimes able, by aid of a peculiar homogeneous deposit, to prove the former coexistence of distinct animals and plants in distant regions. Suppose, for example, that in the course of ages the sediment of a river, like that of the Red River in Louisiana, is dispersed over an area several hundred leagues in length, so as to pass from the tropics into the temperate zone, the fossil remains imbedded in red mud might indicate the different forms which inhabited, at the same period, those remote regions of the earth.

It appears, then, that mineral and organic characters, although often inconstant, may, nevertheless, enable us to establish the contemporaneous origin of formations in distant countries. As the same species of organic beings usually extend over wider areas than deposits of a homogeneous composition, they are more valuable in geological classification than mineral peculiarities; but it fortunately happens, that where the one criterion fails, we can often avail ourselves of the other. Thus, for example, sedimentary strata are as likely to preserve the same colour and composition in a part of the ocean reaching from the borders of the tropics to the temperate zone, as in any other quarter of the globe; but in such spaces the variation of species is always most considerable.

In regard to the habitations of species, the marine tribes are of more importance than the terrestrial, not only because they are liable to be fossilized in subaqueous deposits in the greatest abundance, but because they have, for the most part, a wider geographical range. Sometimes, however, it may happen, as we have shown, that the remains of species of some one province of terrestrial plants and animals may be carried down into two seas inhabited by distinct marine species; and here again we have an illustration of the principle, that when one means of identification fails, another is often at hand to assist us.

In conclusion, we may observe, that in endeavouring to prove the contemporaneous origin of strata in remote countries by organic remains, we must form our conclusions from a great number of species, since a single species may be enabled to survive vicissitudes in the earth's surface, whereby thousands of others are exterminated. When a change of climate takes place, some may migrate and become denizens of other latitudes, and so abound there, as to characterize strata of a subsequent era. In the last volume we have stated our reasons for inferring that such migrations are never sufficiently general to interfere seriously with geological conclusions, provided we do not found our theories on the occurrence of a small number of fossil species.

CHAPTER 5

Classification of Tertiary Formations in Chronological Order

We explained in the last chapter the principles on which the relative ages of different formations may be ascertained, and we found the character to be chiefly derivable from superposition, mineral structure, and organic remains. It is by combining the evidence deducible from all these sources, that we determine the chronological succession of distinct formations, and this principle is well illustrated by the investigation of those European tertiary strata to the discovery of which we have already alluded.

It will be seen, that in proportion as we have extended our inquiries over a larger area, it has become necessary to intercalate new groups of an age intermediate between those first examined, and we have every reason to expect that, as the science advances, new links in the chain will be supplied, and that the passage from one period to another will become less abrupt. We may even hope, without travelling to distant regions, – without even transgressing the limits of western Europe, to render the series far more complete. The fossil shells, for example, of many of the Subalpine formations, on the northern limits of the plain of the Po, have not yet been carefully collected and compared with those of other countries, and we are almost entirely ignorant of many deposits known to exist in Spain and Portugal.

The theoretical views developed in the last chapter, respecting breaks in the sequence of geological monuments, will explain our reasons for anticipating the discovery of intermediate gradations as often as new regions of great extent are explored.

Comparative Value of Different Classes of Organic Remains

In the mean time, we must endeavour to make the most systematic arrangement in our power of those formations which are already known, and in attempting to classify these in chronological order, we have already stated that we must chiefly depend on the evidence afforded by their fossil organic contents. In the execution of this task, we have first to consider what class of remains are most useful, for although every kind of fossil animal and plant is interesting, and cannot fail to throw light on the former history of the globe at a certain period, yet those classes of remains which are of rare and casual occurrence, are absolutely of no use for the purposes of general classification. If we have nothing but plants in one assemblage of strata, and the bones of mammalia in another, we can obviously draw no conclusion respecting the number of species of organic beings common to two epochs; or if we have a great variety, both of vertebrated animals and plants, in one series, and only shells in another, we can form no opinion respecting the remoteness or proximity of the two eras. We might, perhaps, draw some conclusions as to relative antiquity, if we could compare each of these monuments to a third; as, for example, if the species of shells should be almost all identical with those now living, while the plants and vertebrated animals were all extinct; for we might then infer that the shelly deposit was the most recent of the two. But in this case it will be seen that the information flows from a direct comparison of the species of corresponding orders of the animal and vegetable kingdoms, – of plants with plants, and shells with shells; the only mode of making a systematic arrangement by reference to organic remains.

Although the bones of mammalia in the tertiary strata, and those of reptiles in the secondary, afford us instruction of the most interesting kind, yet the species are too few, and confined to too small a number of localities, to be of great importance in characterizing the minor subdivisions of geological formations. Skeletons of fish are by no means frequent in a good state of preservation, and the science of ichthyology must be farther advanced, before we can hope to determine their specific character with sufficient precision. The same may be said of fossil botany, notwithstanding the great progress that has recently been made in that department; and even in regard to zoophytes, which are so much more abundant in a fossil state than any of the classes above enumerated, we are still greatly impeded in our endeavour to classify strata by their aid, in consequence of the smallness of the number of recent species which have been examined in those tropical seas where they occur in the greatest profusion.

Fossil remains of testacea of chief importance. – The testacea are by far the most important of all classes of organic beings which have left their spoils in the subaqueous deposits; they are the medals which nature has chiefly selected to record the history of the former changes of the globe. There is scarcely any great series of strata that does not contain some marine or fresh-water shells, and these fossils are often found so entire, especially in the tertiary formations, that when disengaged from the matrix, they have all the appearance of having been just procured from the sea. Their colour, indeed, is usually wanting, but the parts whereon specific characters are founded remain unimpaired; and although the animals themselves are gone, yet their form and habits can generally be inferred from the shell which covered them.

The utility of the testacea, in geological classification, is greatly enhanced by the circumstance, that some forms are proper to the sea, others to the land, and others to fresh-water. Rivers scarcely ever fail to carry down into their deltas some land shells, together with species which are at once fluviatile and lacustrine. The Rhone, for example, receives annually, from the Durance, many shells which are drifted down in an entire state from the higher Alps of Dauphiny, and these species, such as *Bulimus montanus*, are carried down into the delta of the Rhone to a climate far different from that of their native habitation. The young hermit crabs may often be seen on the shores of the Mediterranean, near the mouth of the Rhone, inhabiting these univalves, brought down to them from so great a distance. At the same time that some fresh-water and land species are carried into the sea, other individuals of the same become fossil in inland lakes, and by this means we learn what species of fresh-water and marine testacea coexisted at particular eras; and from this again we are able to make out the connexion between various plants and mammifers imbedded in those lacustrine deposits, and the testacea which lived in the ocean at the same time.

There are two other characters of the molluscous animals which render them extremely valuable in settling chronological questions in geology. The first of these is a wide geographical range, and the second (probably a consequence of the former), is the superior duration of species in this class. It is evident that if the habitation of a species be very local, it cannot aid us greatly in establishing the contemporaneous origin of distant groups of strata, in the manner pointed out in the last chapter; and if a wide geographical range be useful in connecting formations far separated in space, the longevity of species is no less serviceable in establishing the relations of strata considerably distant from each other in point of time.

We shall revert in the sequel to the curious fact, that in tracing back these series of tertiary deposits, many of the existing species of testacea accompany us after the disappearance of all the recent mammalia, as well as the fossil remains of living species of several other classes. We even find the skeletons of extinct quadrupeds in deposits wherein all the land and fresh-water shells are of recent species.

Necessity of accurately determining species. – The reader will already perceive that the systematic arrangement of strata, so far as it rests on organic remains, must depend essentially on the accurate determination of *species*, and the geologist must therefore have recourse to the ablest naturalists, who have devoted their lives to the study of certain departments of organic nature. It is scarcely possible that they who are continually employed in laborious investigations in the field, and in ascertaining the relative position and characters of mineral masses, should have leisure to acquire a profound knowledge of fossil osteology, conchology, and other branches; but it is desirable that, in the latter science at least, they should become acquainted with the principles on which the specific characters are determined, and on which the habits of species are inferred from their peculiar forms. When the specimens are in an imperfect state of preservation, or the shells happen to belong to genera in which it is difficult to decide on the species, except when the inhabitant itself is present, or when any other grounds of ambiguity arise, we must reject, or lay small stress upon, the evidence, lest we vitiate our general results by false identifications and analogies. We cannot do better than consider the steps by which the science of botanical geography has reached its present stage of advancement, and endeavour to introduce the same severe comparison of the specific characters, in drawing all our geological inferences.

Tables of shells by M. Deshayes. – In the Appendix[3] the reader will find a tabular view of the results obtained by the comparison of more than three thousand tertiary shells, with nearly five thousand living species, all of which, with few exceptions, are contained in the rich collection of M. Deshayes. Having enjoyed an opportunity of examining, again and again, the specimens on which this eminent conchologist has founded his identifications, and having been witness to the great time and labour devoted by him to this arduous work, I feel confidence in the results, so far as the data given in his list will carry us. It was necessary to compare nearly forty thousand specimens, in order to construct these tables, since not only the varieties of every species required examination, but the

392

different individuals, also, belonging to each which had been found fossil in various localities. The correctness of the localities themselves was ascertained with scrupulous exactness, together with the relative position of the strata; and if any doubts existed on these questions, the specimens were discarded as of no geological value. A large proportion of the shells were procured, by M. Deshayes himself, from the Paris basin, many were contributed by different French geologists, and some were collected by myself from different parts of Europe.

It would have been impossible to give lists of more than three thousand fossil-shells in a work not devoted exclusively to conchology; but we were desirous of presenting the reader with a catalogue of those fossils which M. Deshayes has been able to identify with living species, as also of those which are common to two distinct tertiary eras. By this means a comparison may be made of the testacea of each geological epoch, with the actual state of the organic creation, and, at the same time, the relations of different tertiary deposits to each other exhibited. The number of shells mentioned by name in the tables, in order to convey this information, is seven hundred and eighty-two, of which four hundred and twenty-six have been found both living and fossil, and three hundred and fifty-six fossil only, but in the deposits of more than one era. An exception, however, to the strictness of this rule has been made in regard to the fossil-shells common to the London and Paris basins, fifty-one of which have been enumerated by name, though these formations do not belong to different eras.

It has been more usual for geologists to give tables of characteristic shells; that is to say, of those found in the strata of one period and not common to any other. These typical species are certainly of the first importance, and some of them will be seen figured in the plates illustrative of the different tertiary eras; but we were more anxious, in this work, to place in a clear light a point of the greatest theoretical interest, which has been often overlooked or controverted, viz., the identity of many living and fossil species, as also the connexion of the zoological remains of deposits formed at successive periods.

The value of such extensive comparisons, as those of which the annexed tables of M. Deshayes give the results, depends greatly on the circumstance, that all the identifications have been made by the same naturalist. The amount of variation which ought to determine a species is, in cases where they approach near to each other, a question of the nicest discrimination, and requires a degree of judgment and tact that can hardly be possessed by different zoologists in exactly the same degree. The standard, therefore, by which differences are to be measured, can scarcely ever be perfectly invariable, and one great object to be sought for is, that, at least, it should be uniform. If the distinctions are all made by the same naturalist, and his knowledge and skill be considerable, the results may be relied on with sufficient confidence, as far as regards our geological conclusions.

If one conchologist should inform us that out of 1122 species of fossil testacea, discovered in the Paris basin, he has only been enabled to identify thirty-eight with recent species, while another should declare, that out of two hundred and twenty-six Sicilian fossil shells, no less than two hundred and sixteen belonged to living species, we might suspect that one of these observers allowed a greater degree of latitude to the variability of the specific character than the other; but when, in both instances, the conclusions are drawn by the same eminent conchologist, we are immediately satisfied that the relations of these two groups, to the existing state of the animate creation, are as distinct as are indicated by the numerical results.

It is not pretended that the tables, to which we refer, comprise all the known tertiary shells. In the museums of Italy there are magnificent collections, to which M. Deshayes had no access, and the additions to the recent species in the cabinets of conchologists in London have been so great of late years, that in many extensive genera the number of species has been more than doubled. But as the greater part of these newly-discovered shells have been brought from the Pacific and other distant seas, it is probable that these accessions would not materially alter the results given in the tables, and it must, at all events, be remembered, that the only effect of such additional information would be, to increase the number of identifications of recent with fossil species, while the proportional number of analogues in the different periods might probably remain nearly the same.

Subdivisions of the Tertiary Epoch

Recent formations. – We shall now proceed to consider the subdivisions of tertiary strata which may be founded on the results of a comparison of their respective fossils, and to give names to the periods to which they each belong. The tertiary epoch has been divided into three periods in the tables; we shall, however, endeavour to establish *four*, all distinct from the actual period, or that which has elapsed since the earth has been tenanted by man. To the events of this latter era, which we shall term the *recent*, we have exclusively confined ourselves in the two preceding volumes. All sedimentary deposits, all volcanic rocks, in a word, every geological monument, whether belonging to the animate or inanimate world, which appertains to this epoch, may be termed *recent*. Some *recent* species, therefore, are found *fossil* in various tertiary periods, and, on the other hand, others, like the Dodo, may be *extinct*, for it is sufficient that they should once have coexisted with man, to make them referrible to this era.

Some authors apply the term *contemporaneous* to all the formations which have originated during the human epoch; but as the word is so frequently in use to express the synchronous origin of distinct formations, it would be a source of great inconvenience and ambiguity, if we were to attach to it a technical sense.

We may sometimes prove, that certain strata belong to the recent period by aid of historical evidence, as parts of the delta of the Po, Rhone, and Nile, for example; at other times, by discovering imbedded remains of man or his works; but when we have no evidence of this kind, and we hesitate whether to ascribe a particular deposit to the recent era, or that immediately preceding, we must generally incline to refer it to the latter, for it will appear in the sequel, that the changes of the historical era are quite insignificant when contrasted with those even of the newest tertiary period.

Newer Pliocene period. – This most modern of the four subdivisions of the whole tertiary epoch, we propose to call the *Newer Pliocene*, which, together with the *Older* Pliocene, constitute one group in the annexed tables of M. Deshayes.

We derive the term Pliocene from πλειων, major, and καινος, recens, as the major part of the fossil testacea of this epoch are referrible to recent species.* Whether in all cases there may hereafter prove to be an absolute

---

* In the terms Pliocene, Miocene, and Eocene, the Greek diphthongs *ei* and *ai* are changed into the vowels *i* and *e*, in conformity with the idiom of our language. Thus we have Encenia, an inaugural ceremony, derived from εν and καινος, recens; and as examples of the conversion of *ei* into *i*, we have icosahedron.

I have been much indebted to my friend, the Rev. W. Whewell, for assisting me in inventing and anglicizing these terms, and I sincerely wish that the numerous foreign diphthongs, barbarous terminations, and Latin plurals, which have been so plentifully introduced of late years into our scientific language, had been avoided as successfully as they are by French naturalists, and as they were by the earlier English writers, when our language was more

preponderance of recent species, in every group of strata assigned to this period in the tables, is very doubtful; but the proportion of living species, where least considerable, usually approaches to one-half of the total number, and appears always to exceed a third; and as our acquaintance with the testacea of the Mediterranean, and some other seas, increases, it is probable that a greater proportion will be identified.

The newer Pliocene formations, before alluded to, pass insensibly into those of the *Recent* epoch, and contain an immense preponderance of recent species. It will be seen that of two hundred and twenty-six species, found in the Sicilian beds, only ten are of extinct or unknown species, although the antiquity of these tertiary deposits, as contrasted with our most remote historical eras, is immensely great. In the volcanic and sedimentary strata of the district round Naples, the proportion appears to be even still smaller.

Older Pliocene period. – These formations, therefore, and others wherein the plurality of living species is so very decided, we shall term the *Newer* Pliocene, while those of the tertiary period immediately preceding may be called the *Older* Pliocene. To the latter belong the formations of Tuscany, and of the Subapennine hills in the north of Italy, as also the English Crag.

It appears that in the period last mentioned, the proportion of recent species varies from upwards of a third to somewhat more than half of the entire number; but it must be recollected, that this relation to the recent epoch is only *one* of its zoological characters, and that certain *peculiar species* of testacea also distinguish its deposits from all other strata. The relative position of the beds referrible to this era has been explained in diagrams Nos. 3 and 4, letter *f*, chapter 2 [p. 367].

Miocene period. – The next antecedent tertiary epoch we shall name Miocene, from μειων, minor, and καινος, recens, a minority only of fossil shells imbedded in the formations of this period being of recent species. The total number of Miocene shells, referred to in the annexed tables, amounts to 1021, of which one hundred and seventy-six only are recent, being in the proportion of rather less than eighteen in one hundred.

_____

flexible than it is now. But while I commend the French for accommodating foreign terms to the structure of their own language, I must confess that no naturalists have been more unscholar-like in their mode of fabricating Greek derivatives and compounds, many of the latter being a bastard offspring of Greek and Latin.

Of species common to this period, and to the two divisions of the Pliocene epoch before alluded to, there are one hundred and ninety-six, whereof one hundred and fourteen are living, and the remaining eighty-two extinct, or only known as fossil.

As there are a certain number of fossil species which are characteristic of the Pliocene strata before described, so also there are many shells exclusively confined to the Miocene period. We have already stated, that in Touraine and in the South of France near Bordeaux, in Piedmont, in the basin of Vienna, and other localities, these Miocene formations are largely developed, and their relative position has been shown in diagrams Nos. 3 and 4, letter *e*, chapter 2.

Eocene period. – The period next antecedent we shall call Eocene, from ἠώς, aurora, and καινός, recens, because the extremely small proportion of living species contained in these strata, indicates what may be considered the first commencement, or *dawn*, of the existing state of the animate creation. To this era the formations first called tertiary, of the Paris and London basins, are referrible. Their position is shown in the diagrams Nos. 3 and 4, letter *d*, in the second chapter.

The total number of fossil shells of this period already known, is one thousand two hundred and thirty-eight, of which number forty-two only are living species, being nearly in the proportion of three and a half in one hundred. Of fossil species, not known as recent, forty-two are common to the Eocene and Miocene epochs. In the Paris basin alone, 1122 species have been found fossil, of which thirty-eight only are still living.

The geographical distribution of those recent species which are found fossil in formations of such high antiquity as those of the Paris and London basins, is a subject of the highest interest.

It will be seen by reference to the tables, that in the more modern formations, where so large a proportion of the fossil shells belong to species still living, they also belong, for the most part, to species now inhabiting the seas immediately adjoining the countries where they occur fossil; whereas the recent species, found in the older tertiary strata, are frequently inhabitants of distant latitudes, and usually of warmer climates. Of the forty-two Eocene species, which occur fossil in England, France, and Belgium, and which are still living, about half now inhabit within, or near the tropics, and almost all the rest are denizens of the more southern parts of Europe. If some Eocene species still flourish in the same latitudes where they are found fossil, they are species which, like *Lucina divaricata*, are now found in many seas, even those of different quarters of the globe, and this wide geographical range indicates a capacity of enduring a variety of external circumstances, which may enable a species to survive considerable changes of climate and other revolutions of the earth's surface. One fluviatile species (*Melania inquinata*), fossil in the Paris basin, is now only known in the Philippine islands, and during the lowering of the temperature of the earth's surface, may perhaps have escaped destruction by transportation to the south. We have pointed out in the second volume (chap. 7), how rapidly the eggs of fresh-water species might, by the instrumentality of water-fowl, be transported from one region to another. Other Eocene species, which still survive and range from the temperate zone to the equator, may formerly have extended from the pole to the temperate zone, and what was once the southern limit of their range may now be the most northern.

Even if we had not established several remarkable facts in attestation of the longevity of certain tertiary species, we might still have anticipated that the duration of the living species of aquatic and terrestrial testacea would be very unequal. For it is clear that those which now inhabit many different regions and climates, may survive the influence of destroying causes, which might extirpate the greater part of the species now living. We might expect, therefore, some species to survive several successive states of the organic world, just as Nestor was said to have outlived three generations of men.

The distinctness of periods may indicate our imperfect information. – In regard to distinct zoological periods, the reader will understand, from our observations in the third chapter, that we consider the wide lines of demarcation that sometimes separate different tertiary epochs, as quite unconnected with extraordinary revolutions of the surface of the globe, and as arising, partly, like chasms in the history of nations, out of the present imperfect state of our information, and partly from the irregular manner in which geological memorials are preserved, as already explained. We have little doubt that it will be necessary hereafter to intercalate other periods, and that many of the deposits, now referred to a single era, will be found to have been formed at very distinct periods of time, so that, notwithstanding our separation of tertiary strata into four groups, we shall continue to use the term *contemporaneous* with a great deal of latitude.

We throw out these hints, because we are apprehensive lest zoological periods in geology, like artificial divisions in other branches of natural history, should acquire too much importance, from being supposed to be founded on some great interruptions in the regular series of events in the organic world, whereas, like the genera and orders in zoology and botany, we ought to regard them as invented for the convenience of systematic arrangement, always expecting to discover intermediate gradations between the boundary lines that we have first drawn.

In natural history we select a certain species as a generic type, and then arrange all its congeners in a series, according to the degrees of their deviation from that type, or according as they approach to the characters of the genus which precedes or follows. In like manner, we may select certain geological formations as typical of particular epochs; and having accomplished this step, we may then arrange the groups referred to the same period in chronological order, according as they deviate in their organic contents from the *normal* groups, or according as they approximate to the type of an antecedent or subsequent epoch.

If intermediate formations shall hereafter be found between the Eocene and Miocene, and between those of the last period and the Pliocene, we may still find an appropriate place for all, by forming subdivisions on the same principle as that which has determined us to separate the lower from the upper Pliocene groups. Thus, for example, we might have three divisions of the Eocene epoch, – the older, middle, and newer; and three similar subdivisions, both of the Miocene and Pliocene epochs. In that case, the formations of the middle period must be considered as the types from which the assemblage of organic remains in the groups immediately antecedent or subsequent will diverge.

The recent strata form a common point of departure in all countries. – We derive one great advantage from beginning our classification of formations by a comparison of the fossils of the more recent strata with the species now living, namely, the acquisition of a common point of departure in every region of the globe. Thus, for example, if strata should be discovered in India or South America, containing the same small proportion of recent shells as are found in the Paris basin, *they* also might be termed Eocene, and, on analogous data, an approximation might be made to the relative dates of strata placed in the arctic and tropical regions, or the comparative age ascertained of European deposits, and those which are trodden by our antipodes.

There might be no species common to the two groups; yet we might infer their synchronous origin from the common relation which they bear to the existing state of the animate creation. We may afterwards avail ourselves of the dates thus established, as eras to which the monuments of preceding periods may be referred.

Numerical proportion of recent shells in the different Tertiary periods. – There are seventeen species of shells discovered, which are common to all the tertiary periods, thirteen of which are still living, while four are extinct, or only known as fossil. These seventeen species show a connexion between all these geological epochs, whilst we have seen that a much greater number are common to the Eocene and Miocene periods, and a still greater to the Miocene and Pliocene.

We have already stated, that in the older tertiary formations, we find a very small proportion of fossil species identical with those now living, and that, as we approach the superior and newer sets of strata, we find the remains of existing animals and plants in greater abundance. It is almost as difficult to find an unknown species in some of the newer Pliocene deposits, although very ancient, and elevated at great heights above the level of the sea, as to meet with recent species in the Eocene strata.

This increase of existing species, and gradual disappearance of the extinct, as we trace the series of formations from the older to the newer, is strictly analogous, as we before observed, to the fluctuations of a population such as might be recorded at successive periods, from the time when the oldest of the individuals now living was born to the present moment. The disappearance of persons who never were contemporaries of the greater part of the present generation, would be seen to have kept pace with the birth of those who now rank amongst the oldest men living, just as the Eocene and Miocene species are observed to have given place to those Pliocene testacea which are now contemporary with man.

In reference to the organic remains of the different groups which we have named, we may say that about a thirtieth part of the Eocene shells are of recent species, about one-fifth of the Miocene, more than a third, and often more than half, of the older Pliocene, and nine-tenths of the newer Pliocene.

Mammiferous remains of the successive tertiary eras. – But although a thirtieth part of the Eocene testacea have been identified with species now living, none of the associated mammiferous remains belong to species which now exist, either in Europe or elsewhere. Some of these equalled the horse, and others the rhinoceros, in size, and they could not possibly have escaped observation, had they survived down to our time. More than forty of these Eocene mammifers are referrible to a division of the order Pachydermata, which has now only four living representatives on the globe. Of these, not only the species but the genera are distinct from any of those which have been established for the classification of living animals.

In the Miocene mammalia we find a few of the generic forms most frequent in the Eocene strata associated with some of those now existing, and in the Pliocene we find an intermixture of extinct and recent species of quadrupeds. There is, therefore, a considerable degree of accordance between the results deducible from an examination of the fossil testacea, and those derived from the mammiferous fossils. But although the latter are more important in respect to the unequivocal evidence afforded by them of the extinction of species, yet, for reasons before explained, they are of comparatively small value in the general classification of strata in geology.

It will appear evident, from what we have said in the last volume respecting the fossilization of terrestrial species, that the imbedding of their remains depends on rare casualties, and that they are, for the most part, preserved in detached alluvions covering the emerged land, or in osseous breccias and stalagmites formed in caverns and fissures, or in isolated lacustrine formations. These fissures and caves may sometimes remain open during successive geological periods, and the alluvions, spread over the surface, may be disturbed, again and again, until the mammalia of successive epochs are mingled and confounded together. Hence we must be careful, when we endeavour to refer the remains of mammalia to certain tertiary periods, that we ascertain, not only their association with testacea of which the date is known, but also that the remains were intermixed in such a manner as to leave no doubt of the former coexistence of the species.

In the next page will be found a Synoptical Table of the Recent and Tertiary formations alluded to in this chapter.

Synoptical Table of Recent and Tertiary Formations

PERIODS.		Character of Formations.	Localities of the different Formations.
I. RECENT.		Marine.	Coral Formations of Pacific. Delta of Po, Ganges, &c.
		Freshwater.	Modern deposits in Lake Superior – Lake of Geneva – Marl lakes of Scotland – Italian travertin, &c.
		Volcanic.	Jorullo – Monte Nuovo – Modern lavas of Iceland, Etna, Vesuvius, &c.
II. TERTIARY.	1. Newer Pliocene.	Marine.	Strata of the Val di Noto in Sicily, Ischia, Morea? Uddevalla.
		Freshwater.	Valley of the Elsa around Colle in Tuscany.
		Volcanic.	Older parts of Vesuvius, Etna, and Ischia – Volcanic rocks of the Val di Noto in Sicily.
	2. Older Pliocene.	Marine.	Northern Subapennine formations, as at Parma, Asti, Sienna, Perpignan, Nice – English Crag.
		Freshwater.	Alternating with marine beds near the town of Sienna.
		Volcanic.	Volcanos of Tuscany and Campagna di Roma.
	3. Miocene.	Marine.	Strata of Touraine, Bordeaux, Valley of the Bormida, and the Superga near Turin – Basin of Vienna.
		Freshwater.	Alternating with marine at Saucats, twelve miles south of Bordeaux.
		Volcanic.	Hungarian and Transylvanian volcanic rocks. Part of the volcanos of Auvergne, Cantal, and Velay?
	4. Eocene.	Marine.	Paris and London Basins.
		Freshwater.	Alternating with marine in Paris basin – Isle of Wight – purely lacustrine in Auvergne, Cantal, and Velay.
		Volcanic.	Oldest part of volcanic rocks of Auvergne.

Newer Pliocene Formations – Sicily

Having endeavoured, in the last chapter, to explain the principles on which the different tertiary formations may be arranged in chronological order, we shall now proceed to consider the newest division of formations, or that which we have named the newer Pliocene.

It may appear to some of our readers, that we reverse the natural order of historical research by thus describing, in the first place, the monuments of a period which immediately preceded our own era, and passing afterwards to the events of antecedent ages. But, in the present state of our science, this retrospective order of inquiry is the only one which can conduct us gradually from the known to the unknown, from the simple to the more complex phenomena. We have already explained our reasons for beginning this work with an examination, in the first two volumes, of the events of the *recent* epoch, from which the greater number of rules of interpretation in geology may be derived. The formations of the newer Pliocene period will be considered next in order, because these have undergone the least degree of alteration, both in position and internal structure, subsequently to their origin. They are monuments of which the characters are more easily deciphered than those belonging to more remote periods, for they have been less mutilated by the hand of time. The organic remains, more especially of this era, are most important, not only as being in a more perfect state of preservation, but also as being chiefly referrible to species now living; so that their habits are known to us by direct comparison, and not merely by inference from analogy, as in the case of extinct species.

Geological structure of Sicily. – We shall first describe an extensive district in Sicily, where the newer Pliocene strata are largely developed, and where they are raised to considerable heights above the level of the sea. After presenting the reader with a view of these formations, we shall endeavour to explain the manner in which they originated, and speculate on the subterranean changes of which their present position affords evidence.

The island of Sicily consists partly of primary and secondary rocks, which occupy, perhaps, about two-thirds of its superficial area, and the remaining part is covered by tertiary formations, which are of great extent in the southern and central parts of the island, while portions are found bordering nearly the whole of the coasts.

* * *

[*The remainder of Chapter 6 describes the newer Pliocene of Sicily in detail. The discussion centres on a group of limestones, sandstones and marls in the Val di Noto, a district south of Etna. Although these beds are elevated several thousand feet above sea level, the molluscan fossils in them are similar to those found in the surrounding waters of the Mediterranean; almost none are extinct. Lyell shows that the deposits accumulated gradually and over a long period of time. For example, he describes a limestone containing volcanic pebbles covered with the calcareous tubes of marine worms, 'a beautiful proof of a considerable interval of time having elapsed between the rounding of these pebbles and their inclosure in a solid stratum.' At another place, a twenty-foot thick bed of oysters – 'perfectly identifiable with our common eatable species' – is sandwiched between two flows of basaltic lava.*

The critical significance of these strata is revealed in Chapter 7, where Lyell infers that analogous formations, of the same age as those in the Val di Noto, underlie the volcanic cone of Etna. This is an extraordinary conclusion, for it suggests that the largest accumulation of volcanic deposits in Europe was formed in a very recent geological period. But as Lyell says, if there was time for the Val di Noto limestones to be uplifted several thousand feet, then there 'may also have been sufficient time for the growth of a volcanic pile like Etna, since the newer Pliocene strata now seen at the base of the volcano originated.'

In the second half of Chapter 7, Lyell turns to Etna itself:]

* * *

Internal Structure of the Cone of Etna

In our first volume we merely described that part of Etna which has been formed during the historical era; an insignificant portion of the whole mass. Nearly all the remainder may be referred to the tertiary period immediately antecedent to the *recent* epoch. We before stated, that the great cone is, in general, of a very symmetrical form, but is broken, on its eastern side, by a deep valley, called the Val del Bove,* which, commencing near the summit of the mountain, descends into the woody region, and is then continued, on one side, by a second and narrower valley, called the Val di Calanna. Below the latter another, named the Val di St. Giacomo, begins, – a long narrow ravine, which is prolonged to the neighbourhood of Zaffarana (*e*, No. 17), on the confines of the fertile region. These natural incisions, into the side of the volcano, are of such depth, that they expose to view a great part of the structure of the entire mass, which, in the Val del Bove, is laid open to the depth of from four thousand to five thousand feet from the summit of Etna. The geologist thus enjoys an opportunity of ascertaining how far the internal conformation of the cone corresponds with what he might have anticipated as the result of that mode of increase which has been witnessed during the historical era.

a. highest cone.   *b.* Montagnuola.   *c.* Head of Val del Bove.   *d, d.* Serre del Solfizio.   *e.* Zaffarana.   *f.* One of the lateral cones.   *g.* Monti Rossi.

17. Great valley on the east side of Etna

It is clear, from what we before said of the gradual manner in which the principal cone increases, partly by streams of lava and showers of volcanic ashes ejected from the summit, partly by the throwing up of

* In the provincial dialect of the peasants called 'Val del Bué,' for here the herdsman

> . . . in the Vale beneath he views
> His wandring Sheep, and grazing Cows.

Dr. Buckland was, I believe, the first English geologist who examined this valley with attention, and I am indebted to him for having described it to me, before my visit to Sicily, as more worthy of attention than any single spot in that island, or perhaps in Europe. I have already stated, that the view of this valley, which I have given in the frontispiece of the second volume, does not pretend to convey any idea of the grandeur of the scene.[4]

minor hills and the issuing of lava-currents on the flanks of the mountain, that the whole cone must consist of a series of cones enveloping others, the regularity of each being only interrupted by the interference of the lateral volcanos.

We might, therefore, have anticipated that a section of Etna, as exposed in a ravine which should begin near the summit and extend nearly to the sea, would correspond very closely to the section of the ancient Vesuvius, commencing with the escarpment of Somma, and ending with the Fossa Grande; but with this difference, that where the ravine intersects the woody region of Etna, indications must appear of changes brought about by lateral eruptions. Now the section before alluded to, which can be traced from the head of the Val del Bove to the inferior borders of the woody region, fully answers such expectations. We find, almost everywhere, a series of layers of tuff and breccia interstratified with lavas, which slope gently to the sea, at an angle of from twenty to thirty degrees; and as we rise to the parallel of the zone of lateral eruptions, and still more as we approach the summit, we discover indications of disturbances, occasioned by the passage of lava from below, and the successive inhumation of lateral cones.

Val di Calanna. – On leaving Zaffarana, on the borders of the fertile region, we enter the ravine-like valley of St. Giacomo, and see on the north side, or on our right as we ascend, rising ground composed of the modern lavas of Etna. On our left, a lofty cliff, wherein a regular series of beds is exhibited, composed of tuffs and lavas, descending with a gentle inclination towards the sea. In this lower part of the section there are no intersecting dikes, nor any signs of minor cones interfering with the regular slope of the alternating volcanic products. If we then pass upwards through a defile, called the 'Portello di Calanna,' we enter a second valley, that of Calanna, resembling the ravine before mentioned, but wider and much deeper. Here again we find, on our right, many currents of modern lava, piled one upon the other, and on our left a continuation of our former section, in a perpendicular cliff from four hundred to five hundred feet high. As this lofty wall sweeps in a curve, it has very much the appearance of the escarpment which Somma presents towards Vesuvius, and this resemblance is increased by the occurrence of two or three vertical dikes which traverse the gently-inclined volcanic beds. When I first beheld this precipice, I fancied that I had entered a lateral crater, but was soon undeceived, by discovering that on all sides, both at the head of the valley, in the hill of Zocolaro, and at its side and lower extremity, the dip of the beds was always in the same direction, all slanting to the east, or towards the sea, instead of sloping to the north, east, and south, as would have been the case had they constituted three walls of an ancient crater.

* * *

Val del Bove. – After passing up through the defile, called the 'Rocca di Calanna,' we enter a third valley of truly magnificent dimensions – the Val del Bove – a vast amphitheatre four or five miles in diameter, surrounded by nearly vertical precipices, varying from one thousand to above three thousand feet in height, the loftiest being at the upper end, and the height gradually diminishing on both sides. The feature which first strikes the geologist as distinguishing this valley from those before mentioned, is the prodigious multitudes of vertical dikes, which are seen in all directions traversing the volcanic beds. The circular form of this great chasm, and the occurrence of these countless dikes, amounting perhaps to several thousands in number, so forcibly recalled to my mind the phenomena of the Atrio del Cavallo, on Vesuvius, that I imagined once more that I had entered a vast crater, on a scale as far exceeding that of Somma, as Etna surpasses Vesuvius in magnitude.

But having already been deceived in regard to the crescent-shaped precipice of the valley of Calanna, I began attentively to explore the different sides of the great amphitheatre, in order to satisfy myself whether the semicircular wall of the Val del Bove had ever formed the boundary of a crater, and whether the beds had the same quâquâ-versal dip which is so beautifully exhibited in the escarpment of Somma. If the supposed analogy between Somma and the Val del Bove should hold true, the tuffs and lavas, at the head of the valley, would dip to the west, those on the north side towards the north, and those on the southern side to the south. But such I did not find to be the inclination of the beds; they all dip towards the sea, or nearly east, as was before seen to be the case in the Valley of Calanna.

There are undoubtedly exceptions to this general rule, which might deceive a geologist who was strongly prepossessed with a belief that he had discovered the hollow of an ancient crater. It is evident that, wherever lateral cones are intersected in the precipices, a series of tuffs and lavas, very similar to those which enter into the structure of the great cone, will be seen dipping at a much more rapid angle.

The lavas and tuffs, which have conformed to the sides of Etna, dip at angles of from fifteen to twenty-five degrees, while the slope of the lateral cones is from thirty-five to fifty degrees. Now, wherever we meet with sections of these buried cones in the precipices bordering the Val del Bove, (and they are frequent in the cliffs called the Serre del Solfizio, and in those near the head of the valley not far from the rock of Musara,) we find the beds dipping at high angles and inclined in various directions.*

Scenery of the Val del Bove. – Without entering at present into any further discussions respecting the origin of the Val del Bove, we shall proceed to describe some of its most remarkable features. Let the reader picture to himself a large amphitheatre, five miles in diameter, and surrounded on three sides by precipices from two thousand to three thousand feet in height. If he has beheld that most picturesque scene in the chain of the Pyrenees, the celebrated 'cirque of Gavarnie,' he may form some conception of the magnificent circle of precipitous rocks which inclose, on three sides, the great plain of the Val del Bove. This plain has been deluged by repeated streams of lava, and although it appears almost level when viewed from a distance, it is, in fact, more uneven than the surface of the most tempestuous sea. Besides the minor irregularities of the lava, the valley is in one part interrupted by a ridge of rocks, two of which, Musara and Capra, are very prominent. It can hardly be said that they

———— like giants stand
To sentinel enchanted land;[5]

for although, like the Trosachs, they are of gigantic dimensions, and appear almost isolated as seen from many points, yet the stern and severe grandeur of the scenery which they adorn is not such as would be selected by a poet for a vale of enchantment. The character of the scene would accord far better with Milton's picture of the infernal world; and if we imagine ourselves to behold in motion, in the darkness of the night, one of those fiery currents, which have so often traversed the great valley, we may well recall

* I perceive that Professor Hoffmann, who visited the Val del Bove after me (in January, 1831), has speculated on its structure as corresponding to that of the so-called elevation craters, which hypothesis would require that there should be a quâquâ-versal dip, such as I have above alluded to. I can only account for this difference of opinion, by supposing the Professor to have overlooked the phenomena of the buried cones. – Archiv. für Mineralogie, &c. Berlin, 1831.

———— yon dreary plain, forlorn and wild,
The seat of desolation, void of light
Save what the glimmering of these livid flames
Cast pale and dreadful.[6]

The face of the precipices already mentioned is broken in the most picturesque manner by the vertical walls of lava which traverse them. These
masses usually stand out in relief, are exceedingly diversified in form, and
often of immense altitude. In the autumn, their black outline may often be
seen relieved by clouds of fleecy vapour which settle behind them, and do
not disperse until midday, continuing to fill the valley while the sun is shining
on every other part of Sicily, and on the higher regions of Etna.

As soon as the vapours begin to rise, the changes of scene are varied in
the highest degree, different rocks being unveiled and hidden by turns, and
the summit of Etna often breaking through the clouds for a moment with
its dazzling snows, and being then as suddenly withdrawn from the view.

An unusual silence prevails, for there are no torrents dashing from the
rocks, nor any movement of running water in this valley, such as may
almost invariably be heard in mountainous regions. Every drop of water
that falls from the heavens, or flows from the melting ice and snow, is
instantly absorbed by the porous lava; and such is the dearth of springs,
that the herdsman is compelled to supply his flocks, during the hot season,
from stores of snow laid up in hollows of the mountain during winter.

The strips of green herbage and forest-land, which have here and there
escaped the burning lavas, serve, by contrast, to heighten the desolation
of the scene. When I visited the valley, nine years after the eruption of
1819, I saw hundreds of trees, or rather the white skeletons of trees, on
the borders of the black lava, the trunks and branches being all leafless,
and deprived of their bark by the scorching heat emitted from the melted
rock; an image recalling those beautiful lines –

———— As when heaven's fire
Hath scath'd the forest oaks, or mountain pines,
With singed top their stately growth, though bare,
Stands on the blasted heath.[7]

Form, composition, and origin of the Dikes. – But without indulging the
imagination any longer in descriptions of scenery, we may observe, that
the dikes before mentioned form unquestionably the most interesting
geological phenomenon in the Val del Bove.

19. Dikes at the base of the Serre del Solfizio, Etna

Some of these are composed of trachyte, others of compact blue basalt with olivine. They vary in breadth from two to twenty feet and upwards, and usually project from the face of the cliffs, as represented in the annexed drawing (No. 19). They consist of harder materials than the strata which they traverse, and therefore waste away less rapidly under the influence of that repeated congelation and thawing to which the rocks in this zone of Etna are exposed. The dikes are, for the most part, vertical, but sometimes they run in a tortuous course through the tuffs and breccias, as represented in diagram, No. 20. In the escarpment of Somma where, as we before observed, similar walls of lava cut through alternating beds of sand and scoriæ, a coating of coal-black rock, approaching in its nature and appearance to pitch-stone, is seen at the contact of the dike with the intersected beds. I did not observe such parting layers at the junction of the Etnean dikes which I examined, but they may perhaps be discoverable.

The geographical position of these dikes is most interesting, as they occur in that zone of the mountain where lateral eruptions are frequent;

20. *Veins of lava, Punto di Guimento*

whereas, in the valley of Calanna, which is below that parallel, and in a region where lateral eruptions are extremely rare, scarcely any dikes are seen, and none whatever still lower in the valley of St. Giacomo. This is precisely what we should have expected, if we consider the vertical fissures now filled with rock to have been the feeders of lateral cones, or, in other words, the channels which gave passage to the lava-currents and scoriæ that have issued from vents in the forest-zone.

Some fissures may have been filled from above, but I did not see any which, by terminating downwards, gave proof of such an origin. Almost all the isolated masses in the Val del Bove, such as Capra, Musara, and others, are traversed by dikes, and may, perhaps, have partly owed their preservation to that circumstance, if at least the action of occasional floods has been one of the destroying causes in the Val del Bove; for there is nothing which affords so much protection to a mass of strata against the undermining action of running water, as a perpendicular dike of hard rock.

In the accompanying drawing (No. 21) the flowing of the lavas of 1811 and 1819, between the rocks Finochio, Capra, and Musara, is represented. The height of the two last-mentioned isolated masses has been much diminished by the elevation of their base, caused by these currents. They may, perhaps, be the remnants of cones, which existed before the Val del Bove was formed, and may hereafter be once more buried by the lavas that are now accumulating in the valley.

From no point of view are the dikes more conspicuous than from the

21. *View of the rocks Finochio, Capra, and Musara, Val del Bove*

summit of the highest cone of Etna; a view of some of them is given in the annexed drawing.*

Lavas and breccias. – In regard to the volcanic masses which are intersected by dikes in the Val del Bove, they consist, in great part, of graystone lavas, of an intermediate character between basalt and trachyte, and partly of the trachytic varieties of lava. Beds of scoriæ and sand, also, are very numerous, alternating with breccias formed of angular blocks of igneous rock. It is possible that some of the breccias may be referred to aqueous causes, as we have before seen that great floods do occasionally sweep down the flanks of Etna when eruptions take place in winter, and when the snows are melted by lava.

Many of the angular fragments may have been thrown out by volcanic explosions, which, falling on the hardened surface of moving lava-currents, may have been carried to a considerable distance. It may also happen, that when lava advances very slowly, in the manner of the flow of 1819, described in the first volume, the angular masses resulting from the frequent breaking of the mass as it rolls over upon itself, may produce these breccias. It is at least certain, that the upper portion of the lava-currents of

* This drawing is part of a panoramic sketch which I made from the summit of the cone, December 1st, 1828, when every part of Etna was free from clouds except the Val del Bove.

22. *View from the summit of Etna into the Val del Bove. The small cone and crater immediately below were among those formed during the eruptions of 1810 and 1811*

1811 and 1819, now consist of angular masses, to the depth of many yards.

D'Aubuisson has compared the surface of one of the ancient lavas of Auvergne to that of a river suddenly frozen over by the stoppage of immense fragments of drift-ice, a description perfectly applicable to these modern Etnean flows.

CHAPTER 8

Rocks of the Same Age in Etna

Origin of the Val del Bove

Before concluding our observations on the cone of Etna, the structure of which was considered in the last chapter, we desire to call the reader's attention to several questions: – first, in regard to the probable origin of the great valley already described; secondly, whether any estimate can be made of the length of the period required for the accumulation of the great cone; and, thirdly, whether there are any signs on the surface of the older parts of the mountain, of those devastating waves which, according to the theories of some geologists, have swept again and again over our continents.

Origin of the Val del Bove. – We explained our reasons in the last chapter for not assenting to the opinion, that the great cavity on the eastern side of Etna was the hollow of an immense crater, from which the volcanic masses of the surrounding walls were produced. On the other hand, we think it impossible to ascribe the valley to the action of running water alone; for if it had been excavated exclusively by that power, its depth would have increased in the descent; whereas, on the contrary, the precipices are most lofty at the upper extremity, and diminish gradually on approaching the lower region of the volcano.

The structure of the surrounding walls is such as we should expect to see exhibited on any other side of Etna, if a cavity of equal depth should be caused, whether by subsidence, or by the blowing up of part of the flanks of the volcano, or by either of these causes co-operating with the removing action of running water.

It is recorded, as we have already seen in our history of earthquakes, that in the year 1772 an immense subsidence took place on Papandayang, the largest volcano in the island of Java, and that, during the catastrophe, an extent of ground, *fifteen miles in length and six in breadth*, gave way, so that no less than forty villages were engulphed, and the cone lost no less than four thousand feet of its height.

Now we might imagine a similar event, or a series of subsidences to have formerly occurred on the eastern side of Etna, although such catastrophes have not been witnessed in modern times, or only on a very trifling scale. A narrow ravine, about a mile long, twenty feet wide, and from twenty to thirty-six in depth, has been formed, within the historical era, on the flanks of the volcano, near the town of Mascalucia; and a small circular tract, called the Cisterna, near the summit, sank down in the year 1792, to the depth of about forty feet, and left on all sides of the chasm a vertical section of the beds, exactly resembling those which are seen in the precipices of the Val del Bove. At some remote periods, therefore, we might suppose more extensive portions of the mountain to have fallen in during great earthquakes.

But some geologists will, perhaps, incline to the opinion, that the removed mass was blown up by paroxysmal explosions, such as that which, in the year 79, destroyed the ancient cone of Vesuvius, and gave rise to the escarpment of Somma. The Val del Bove, it will be remembered, lies within the zone of lateral eruptions, so that a repetition of volcanic explosions might have taken place, after which the action of running water may have contributed powerfully to degrade the rocks, and to transport the materials to the sea. We have before alluded to the effects of a violent flood, which swept through the Val del Bove in the year 1755, when a fiery torrent of lava had suddenly overflowed a great depth of snow in winter.

In the present imperfect state of our knowledge of the history of volcanos, we have some difficulty in deciding on the relative probability of these hypotheses; but if we embrace the theory of explosions from below, the cavity would not constitute a *crater* in the ordinary acceptation of that term, still less would it accord with the notion of the so-called 'elevation craters.'

Antiquity of the Cone of Etna

We have stated in a former volume, that confined notions in regard to the quantity of past time, have tended, more than any other prepossessions, to retard the progress of sound theoretical views in Geology; the inadequacy of our conceptions of the earth's antiquity having cramped the freedom of our speculations in this science, very much in the same way as a belief in the existence of a vaulted firmament once retarded the progress of astronomy. It was not until Descartes assumed the indefinite extent of the celestial spaces, and removed the supposed boundaries of the universe, that just opinions began to be entertained of the relative distances of the heavenly bodies; and until we habituate ourselves to contemplate the possibility of an indefinite lapse of ages having been comprised within each of the more modern periods of the earth's history, we shall be in danger of forming most erroneous and partial views in Geology.

Mode of computing the age of volcanos. – If history had bequeathed to us a faithful record of the eruptions of Etna, and a hundred other of the principal active volcanos of the globe, during the last three thousand years, – if we had an exact account of the volume of lava and matter ejected during that period, and the times of their production, – we might, perhaps, be able to form a correct estimate of the average rate of the growth of a volcanic cone. For we might obtain a mean result from the comparison of the eruptions of so great a number of vents, however irregular might be the development of the igneous action in any one of them, if contemplated singly during a brief period.

It would be necessary to balance protracted periods of inaction against the occasional outburst of paroxysmal explosions. Sometimes we should have evidence of a repose of seventeen centuries, like that which was interposed in Ischia, between the end of the fourth century, B.C., and the beginning of the fourteenth century of our era. Occasionally a tremendous eruption, like that of Jorullo, would be recorded, giving rise, at once, to a considerable mountain.

If we desire to approximate to the age of a cone such as Etna, we ought first to obtain some data in regard to the thickness of matter which has been added during the historical era, and then endeavour to estimate the time required for the accumulation of such alternating lavas and beds of sand and scoriæ as are superimposed upon each other in the Val del Bove; afterwards we should try to deduce, from observations on other volcanos, the more or less rapid increase of burning mountains in all the different stages of their growth.

Mode of increase of volcanos analogous to that of exogenous trees. – There is a considerable analogy between the mode of increase of a volcanic cone and that of trees of *exogenous* growth. These trees augment, both in height and diameter, by the successive application externally of cone upon cone of new ligneous matter, so that if we make a transverse section near the base of the trunk, we intersect a much greater number of layers than nearer to the summit. When branches occasionally shoot out from the trunk they first pierce the bark, and then, after growing to a certain size, if they chance to be broken off, they may become inclosed in the body of the tree, as it augments in size, forming knots in the wood, which are themselves composed of layers of ligneous matter, cone within cone.

In like manner a volcanic mountain, as we have seen, consists of a succession of conical masses enveloping others, while lateral cones, having a similar internal structure, often project, in the first instance, like branches from the surface of the main cone, and then becoming buried again, are hidden like the knots of a tree.

We can ascertain the age of an oak or pine, by counting the number of concentric rings of annual growth, seen in a transverse section near the base, so that we may know the date at which the seedling began to vegetate. The Baobab-tree of Senegal (*Adansonia digitata*) is supposed to exceed almost any other in longevity; Adanson inferred that one which he measured, and found to be thirty feet in diameter, had attained the age of 5150 years. Having made an incision to a certain depth, he first counted three hundred rings of annual growth, and observed what thickness the tree had gained in that period. The average rate of growth of younger trees, of the same species, was then ascertained, and the calculation made according to a supposed mean rate of increase. De Candolle considers it not improbable, that the celebrated Taxodium of Chapultepec, in Mexico (*Cupressus disticha*, Linn.), which is one hundred and seventeen feet in circumference, may be still more aged.

It is, however, impossible, until more data are collected respecting the average intensity of the volcanic action, to make anything like an approximation to the age of a cone like Etna, because, in this case, the successive envelopes of lava and scoriæ are not continuous, like the layers of wood in a tree, and afford us no definite measure of time. Each conical envelope is made up of a great number of distinct lava-currents and showers of sand and scoriæ, differing in quantity, and which may have been accumulated in unequal periods of time. Yet we cannot fail to form the most exalted conception of the antiquity of this mountain, when we consider that its base is about ninety miles in circumference; so that it would require ninety flows of lava, each a mile in breadth at their termination, to raise the present foot of the volcano as much as the average height of one lava-current.

There are no records within the historical era which lead to the opinion, that the altitude of Etna has materially varied within the last two thousand years. Of the eighty most conspicuous minor cones which adorn its flanks, only one of the largest, Monti Rossi, has been produced within the times of authentic history. Even this hill, thrown up in the year 1669, although 450 feet in height, only ranks as a cone of second magnitude. Monte Minardo, near Bronte, rises, even now, to the height of 750 feet, although its base has been elevated by more modern lavas and ejections. The dimensions of these larger cones appear to bear testimony to *paroxysms* of volcanic activity, after which we may conclude, from analogy, that the fires of Etna remained dormant for many years – since nearly a century of rest has sometimes followed a violent eruption in the historical era. It must also be remembered, that of the small number of eruptions which occur in a century, one only is estimated to issue from the summit of Etna for every two that proceed from the sides. Nor do all the lateral eruptions give rise to such cones as would be enumerated amongst the smallest of the eighty hills above enumerated; some produce merely insignificant monticules, soon destined to be buried, as we before explained.

How many years then must we not suppose to have been expended in the formation of the eighty cones? It is difficult to imagine that a fourth part of them have originated during the last thirty centuries. But if we conjecture the whole of them to have been formed in twelve thousand years, how inconsiderable an era would this portion of time constitute in the history of the volcano! If we could strip off from Etna all the lateral monticules now visible, together with the lavas and scoriæ that have been poured out from them, and from the highest crater, during the period of their growth, the diminution of the entire mass would be extremely slight! Etna might lose, perhaps, several miles in diameter at its base, and some hundreds of feet in elevation, but it would still be the loftiest of Sicilian mountains, studded with other cones, which would be recalled, as it were, into existence by the removal of the rocks under which they are now buried.

There seems nothing in the deep sections of the Val del Bove, to indicate that the lava currents of remote periods were greater in volume than those of modern times; and there are abundant proofs that the countless beds of solid rock and scoriæ were accumulated, as now, in succession. On the grounds, therefore, already explained, we must infer that a mass, eight thousand or nine thousand feet in thickness, must have required an immense series of ages anterior to our historical periods, for its growth; yet the whole must be regarded as the product of a modern portion of the newer Pliocene epoch. Such, at least, is the conclusion that we draw from the geological data already detailed, which show that the oldest parts of the mountain, if not of posterior date to the marine strata which are visible around its base, were at least of coeval origin.

Whether signs of Diluvial Waves are observable on Etna. – Some geologists contend, that the sudden elevation of large continents from beneath the waters of the sea, have again and again produced waves which have swept over vast regions of the earth, and left enormous rolled blocks strewed over the surface.[8] That there are signs of local floods of extreme violence, on various parts of the surface of the dry land, is incontrovertible, and in the former volumes we have pointed out causes which must for ever continue to give rise to such phenomena; but for the proofs of these general cataclysms we have searched in vain. It is clear that no devastating wave has passed over the forest zone of Etna, since any of the lateral cones before mentioned were thrown up; for none of these heaps of loose sand and scoriæ could have resisted for a moment the denuding action of a violent flood.

To some, perhaps, it may appear that hills of such incoherent materials cannot be of immense antiquity, because the mere action of the atmosphere must, in the course of several thousand years, have obliterated their original forms. But there is no weight in this objection, for the older hills are covered with trees and herbage, which protect them from waste; and in regard to the newer ones, such is the porosity of their component materials, that the rain which falls upon them is instantly absorbed, and, for the same reason that the rivers on Etna have a subterranean course, there are none descending the sides of the minor cones.

No sensible alteration has been observed in the form of these cones since the earliest periods of which there are memorials; and we see no reason for anticipating, that in the course of the next ten thousand or twenty thousand years they will undergo any great alteration in their appearance, unless they should be shattered by earthquakes, or covered by volcanic ejections.

We shall afterwards point out, that, in other parts of Europe, similar loose cones of scoriæ, which we believe to be of higher antiquity than the whole mass of Etna, stand uninjured at inferior elevations above the level of the sea.

CHAPTER 9

Origin of the Newer Pliocene
Strata of Sicily

Having in the last two chapters described the tertiary formations of the Val di Noto and Valdemone, both igneous and aqueous, we shall now proceed more fully to consider their origin, and the manner in which they may be supposed to have assumed their present position. The consideration of this subject may be naturally divided into three parts: first, we shall inquire in what manner the submarine formations were accumulated beneath the waters; secondly, whether they emerged slowly or suddenly, and what modifications in the earth's crust, at considerable depths below the surface, may be indicated by their rise; thirdly, the mutations which the surface and its inhabitants have undergone during and since the period of emergence.

Growth of Submarine formations. – First, then, we are to inquire in what manner the subaqueous masses, whether volcanic or sedimentary, may have been formed. On this subject we have but few observations to make, for by reference to our former volumes, the reader will learn how a single stratum, whether of sand, clay, or limestone, may be thrown down at the bottom of the sea, and how shells and other organic remains may become imbedded therein. He will also understand how one sheet of lava, or bed of scoriæ and volcanic sand, may be spread out over a wide area, and how, at a subsequent period, a second bed of sand, clay, or limestone, or a second lava-stream may be superimposed, so that in the lapse of ages a mountain mass may be produced.

It is enough that we should behold a single course of bricks or stones laid by the mason upon another, in order to comprehend how a massive edifice, such as the Coliseum at Rome, was erected; and we can have no difficulty in conceiving that a sea, three hundred or four hundred fathoms deep, might be filled up by sediment and lava, provided we admit an indefinite lapse of ages for the accumulation of the materials.

The sedimentary and volcanic masses of the newer Pliocene era, which, in the Val di Noto, attain the thickness of two thousand feet, are subdivided into a vast number of strata and lava-streams, each of which were originally formed on the subaqueous surface, just as the tuffs and lavas whereof sections are laid open in the Val del Bove, were each in their turn external additions to the Etnean cone.

It is also clear, that before any part of the mass of submarine origin began to rise above the waters, the uppermost stratum of the whole must have been deposited; so that if the date of the origin of these masses be comparatively recent, still more so is the period of their rise above the level of the sea.

Subaqueous formations how raised. – In what manner, then, and by what agency, did this rise of the subaqueous formations take place? We have seen that since the commencement of the present century, an immense tract of country in Cutch, more than fifty miles long and sixteen broad, was permanently upraised to the height of ten feet above its former position, and the earthquake which accompanied this wonderful variation of level, is reported to have terminated by a volcanic eruption at Bhooi. We have also seen, that when the Monte Nuovo was thrown up, in the year 1538, a large fissure approached the small town of Tripergola, emitting a vivid light, and throwing out ignited sand and scoriæ. At length this opening reached a shallow part of the sea close to the shore, and then widened into a large chasm, out of which were discharged blocks of lava, pumice, and ashes. But no current of melted matter flowed from the orifice, although it is perfectly evident that lava existed below in a fluid state, since so many portions of it were cast up in the form of scoriæ into the air. We have shown that the coast near Puzzuoli rose, at that time, to the height of more than twenty feet above its former level, and that it has remained permanently upheaved to this day.

On a review of the whole phenomena, it appears most probable that the elevated country was forced upwards by lava which did not escape, but which, after causing violent earthquakes, during several preceding months, produced at length a fissure from whence it discharged gaseous fluids, together with sand and scoriæ. The intruded mass then cooled down at a certain distance below the uplifted surface, and constituted a solid and permanent foundation.

If an habitual vent had previously existed near Puzzuoli, such as we may suppose to remain always open in the principal ducts of Vesuvius or Etna, the lava might, perhaps, have flowed over upon the surface, instead of heaving upwards the superficial strata. In that case, there might have been the same conversion of sea into land, the only difference being, that the lava would have been uppermost, instead of the tufaceous strata containing shells, now seen in the plain of La Starza, and on the site of the Temple of Serapis.

Subterranean lava the upheaving cause. – The only feasible theory, indeed, that has yet been proposed, respecting the causes of the permanent rise of the bed of the sea, is that which refers the phenomenon to the generation of subterranean lava. We have stated, in the first volume, that the regions now habitually convulsed by earthquakes, include within them the site of all the active volcanos. We know that the expansive force of volcanic heat is sufficiently great to overcome the resistance of columns of lava, several miles or leagues in height, forcing them up from great depths, and causing the fluid matter to flow out upon the surface. To imagine, therefore, that this same power, which is so frequently exerted in different parts of the globe, should occasionally propel a column of lava to a considerable height, yet be unable to force it through the superincumbent rocks, is quite natural.

Whenever the superimposed masses happen to be of a yielding and elastic nature, they will bend, and instead of breaking, so as to afford an escape to the melted matter through a fissure, they will allow it to accumulate in large quantities beneath the surface, sometimes in amorphous masses, and sometimes in horizontal sheets. So long as such sheets of matter retain their fluidity, and communicate with the column of lava which is still urged upwards, they must exert an enormous hydrostatic pressure on the overlying mass, tending to elevate it, and an equal force on the subjacent beds pressing them down, and probably rendering them more compact. If we consider how great is the volume of lava that sometimes flows out on the surface from volcanic vents, we must expect that it will produce great changes of level so often as its escape is impeded.

Let us only reflect on the magnitude of Iceland, an island two hundred and sixty miles long by two hundred in breadth, and which rises, at some points, to the height of six thousand feet above the level of the sea. Nearly the entire mass is represented to be of volcanic origin; but even if we suppose some parts to consist of aqueous deposits, still that portion may be more than compensated by the great volume of lava which must have been poured out upon the bottom of the surrounding sea during the growth of the entire island; for we know that submarine eruptions have been considerable near the coast during the historical era. Now if the whole of this lava had been prevented from reaching the surface, by the weight and tenacity of certain overlying rocks, it might have given rise to the gradual elevation of a tract of land nearly as large as Iceland. We say *nearly*, because the lava which cooled down beneath the surface, and under considerable pressure, would be more compact than the same when poured out in the open air, or in a sea of moderate depth, or shot up into the atmosphere by the explosive force of elastic vapours, and thus converted into sand and scoriæ.

According to this theory, we must suppose the action of the upheaving power to be intermittent, and, like ordinary volcanic eruptions, to be reiterated again and again in the same region, at unequal intervals of time and with unequal degrees of force.

If we follow this train of induction, which appears so easy and natural, to what important conclusions are we led! The reader will bear in mind that the tertiary strata have attained in the central parts of Sicily, as at Castrogiovanni, for example, an elevation of about three thousand feet above the level of the sea, and a height of from fifty to two thousand feet in different parts of the Val di Noto. In this country, therefore, we must suppose a solid support of igneous rock to have been successively introduced into part of the earth's crust immediately subjacent, equal in volume to the upraised tract, and this generation of subterranean rock must have taken place during the latter part of the newer Pliocene period. The dimensions of the Etnean cone shrink into insignificance, in comparison to the volume of this subterranean lava; and, however staggering the inference might at first appear, that the oldest foundations of Etna were laid subsequently to the period when the Mediterranean became inhabited by the living species of testacea and zoophytes, yet we may be reconciled to such conclusions, when we find incontestable proofs of still greater revolutions beneath the surface within the same modern period.

Probable structure of the recent subterranean rocks of fusion. – Let us now inquire what form these unerupted newer Pliocene lavas of Sicily have assumed? For reasons already explained, we may infer that they cannot have been converted into tuffs and peperinos, nor can we imagine that, under enormous pressure, they could have become porous, since we observe, that the lava which has cooled down under a moderate degree of pressure, in the dikes of Etna and Vesuvius, has a compact and porphyritic texture, and is very rarely porous or cellular. No signs of volcanic sand, scoriæ, breccia, or conglomerate are to be looked for, nor any of stratification, for all these imply formation in the atmosphere, or by the agency of water. The only proofs that we can expect to find of the *successive* origin of different parts of the fused mass, will be confined to the occasional passage of veins through portions previously consolidated. This consolidation would take place with extreme slowness, when nearer the source of volcanic heat and under enormous pressure, so that we must anticipate a perfectly crystalline and compact texture in all these subterranean products.

Now geologists have discovered, as we before stated, great abundance of crystalline and unstratified rocks in various parts of the globe, and these masses are particularly laid open to our view in those mountainous districts where the crust of the earth has undergone the greatest derangement. These rocks vary considerably in composition, and have received many names, such as granite, syenite, porphyry, and others. That they must have been formed by igneous fusion, and at many distinct eras, is now admitted; and their highly crystalline texture is such as might result from cooling down slowly from an intensely-heated state. They answer, therefore, admirably to the conditions required by the above hypothesis, and we therefore deem it probable that similar rocks have originated in the nether regions below the island of Sicily, and have attained a thickness of from one thousand to three thousand feet, since the newer Pliocene strata were deposited.

It is, moreover, very probable, that these fused masses have come into contact with subaqueous deposits far below the surface, in which case they may, in the course of ages, have greatly altered their structure, just as dikes of lava render more crystalline the stratified masses which they traverse, and obliterate all traces of their organic remains.

Suppose some of these changes to have been superinduced upon subaqueous deposits underlying the tertiary formations of Sicily, it is important to reflect that in that case no geological proofs would remain of the era when the alterations had taken place; and if, at some future period, the whole island should be uplifted, and these rocks of fusion, together with the altered strata, should be brought up to the surface, it would not be apparent that they had assumed their crystalline texture in the newer Pliocene period. For aught that would then appear, they might have acquired their peculiar mineral texture at epochs long anterior, and might be supposed to have been formed before the planet was inhabited by living beings; instead of having originated at an era long subsequent to the introduction of the *existing species*.

Changes of the Surface during and since the Emergence of the Newer Pliocene Strata

Valleys. – Geologists who are accustomed to attribute a great portion of the inequalities of the earth's surface to the excavating power of running water during a long series of ages, will probably look for the signs of remarkable freshness in the aspect of countries so recently elevated as the parts of Sicily already described. There is, however, nothing in the external configuration of that country which would strike the eye of the most practised observer, as peculiar and distinct in character from many other districts in Europe which are of much higher antiquity. The general outline of the hills and valleys would accord perfectly well with what may often be observed in regard to other regions of equal altitude above the level of the sea.

It is true that, towards the central parts of the island where the argillaceous deposits are of great thickness, as around Castrogiovanni, Caltanisetta, and Piazza, the torrents are observed annually to deepen the ravines in which they flow, and the traveller occasionally finds that the narrow mule-path, instead of winding round the head of a ravine, terminates abruptly in a deep trench which has been hollowed out, during the preceding winter, through soft clay. But throughout a great part of Italy, where the marls and sands of the Subapennine hills are elevated to considerable heights, the same rapid degradation is often perceived.

* * *

We have stated, in the first volume, that the waves washed the base of the inland cliff near Puzzuoli, in the Bay of Baiæ, within the historical era, and that the retiring of the sea was caused, in the sixteenth century, by an upheaving of the land to an elevation of twenty feet above its original level. At that period, a terrace twenty feet high in some parts, was laid dry between the sea and the cliff, but the Mediterranean is hastening to resume its former position, when the terrace will be destroyed, and every trace of the *successive* rise of the land will be obliterated.

We have been led into these observations, in order to show that the principal features in the physical geography of Sicily are by no means inconsistent with the hypothesis of the successive elevation of the country by the intermittent action of ordinary earthquakes. On the other hand, we consider the magnitude of the valleys, and their correspondence in form with those of other parts of the globe, to lend countenance to the theory of the slow and gradual rise of subaqueous strata.

We have remarked in the first volume, that the excavation of valleys must always proceed with the greatest rapidity when the levels of a country are undergoing alteration from time to time by earthquakes, and that it is principally when a country is rising or sinking by successive movements, that the power of aqueous causes, such as tides, currents, rivers, and land-floods, is exerted with the fullest energy.

In order to explain the present appearance of the surface, we must first go back to the time when the Sicilian formations were mere shoals at the bottom of the sea, in which the currents may have scooped out channels here and there. We must next suppose these shoals to have become small islands of which the cliffs were thrown down from time to time, as were those of Gian Greco, in Calabria, during the earthquake of 1783. The waves and currents would then continue their denuding action during the emergence of these islands, until at length, when the intervening channels were laid dry, and rivers began to flow, the deepening and widening of the valleys by rivers and land-floods would proceed in the same manner as in modern times in Calabria, according to our former description.

Before a tract could be upraised to the height of several thousand feet above the level of the sea, the joint operation of running water and subterranean movements must greatly modify the physical geography; but when the action of the volcanic forces has been suspended, when a period of tranquillity succeeds, and the levels of the land remain fixed and stationary, the erosive power of water must soon be reduced to a state of comparative equilibrium. For this reason, a country that has been raised at a very remote period to a considerable height above the level of the sea, may present nearly the same external configuration as one that has been more recently uplifted to the same height.

In other words, the time required for the raising of a mass of land to the height of several hundred yards must usually be so enormous (assuming as we do that the operation is effected by ordinary volcanic forces), that the aqueous and igneous agents will have time before the elevation is completed to modify the surface, and imprint thereon the ordinary forms of hill and valley, by which our continents are diversified. But after the cessation of earthquakes these causes of change will remain dormant, or nearly so. The greater part, therefore, of the earth's surface will at each period be at rest, simply retaining the features already imparted to it, while smaller tracts will assume, as they rise successively from the deep, a configuration perfectly analogous to that by which the more ancient lands were previously distinguished.

Migration of animals and plants. – The changes which, according to the views already explained, have been brought about in the earth's crust by the agency of volcanic heat, cannot fail to strike the imagination, when we consider how recent in the calendar of nature is the epoch to which we refer them. But if we turn our thoughts to the organic world, we shall feel, perhaps, no less surprise at the great vicissitude which it has undergone during the same period.

We have seen that a large portion of Sicily has been converted from sea to land since the Mediterranean was peopled with the living species of testacea and zoophytes. The newly emerged surface, therefore, must, during this modern zoological epoch, have been inhabited for the first time with the terrestrial plants and animals which now abound in Sicily. It is fair to infer, that the existing terrestrial species are, for the most part, of as high antiquity as the marine, and if this be the case, a large proportion of the plants and animals, now found in the tertiary districts in Sicily, must have inhabited the earth before the newer Pliocene strata were raised above the waters. The plants of the Flora of Sicily are common, almost without exception, to Italy or Africa, or some of the countries surrounding the Mediterranean, so that we may suppose the greater part of them to have migrated from pre-existing lands, just as the plants and animals of the Phlegræan fields have colonized Monte Nuovo, since that mountain was thrown up in the sixteenth century.

We are brought, therefore, to admit the curious result, that the flora and fauna of the Val di Noto, and some other mountainous regions of Sicily, are of higher antiquity than the country itself, having not only flourished before the lands were raised from the deep, but even before they were deposited beneath the waters. Such conclusions throw a new light on the adaptation of the attributes and migratory habits of animals and plants, to the changes which are unceasingly in progress in the inanimate world. It is clear that the duration of species is so great, that they are destined to outlive many important revolutions in the physical geography of the earth, and hence those innumerable contrivances for enabling the subjects of the animal and vegetable creation to extend their range, the inhabitants of the land being often carried across the ocean, and the aquatic tribes over great continental spaces. It is obviously expedient that the terrestrial and fluviatile species should not only be fitted for the rivers, valleys, plains, and mountains which exist at the era of their creation, but for others that are destined to be formed before the species shall become extinct; and, in like manner, the marine species are not only made for the deep or shallow regions of the ocean at the time when they are called into being, but for tracts that may be submerged or variously altered in depth during the time that is allotted for their continuance on the globe.

Recapitulation. – We may now briefly recapitulate some of the most striking results which we have deduced from our investigation of a single district where the newer Pliocene strata are largely developed.

In the first place, we have seen that a stratified mass of solid limestone, attaining sometimes a thickness of eight hundred feet and upwards, has been gradually deposited at the bottom of the sea, the imbedded fossil shells and corallines being almost all of recent species. Yet these fossils are frequently in the state of mere casts, so that in appearance they correspond very closely to organic remains found in limestones of very ancient date.

2dly. In some localities the limestone above-mentioned alternates with volcanic rocks such as have been formed by submarine eruptions, recurring again and again at distant intervals of time.

3dly. Argillaceous and sandy deposits have also been produced during the same period, and their accumulation has also been accompanied by submarine eruptions. Masses of mixed sedimentary and igneous origin, at least two thousand feet in thickness, can thus be shown to have accumulated since the sea was peopled with the greater number of the aquatic species now living.

4thly. These masses of submarine origin have, since their formation, been raised to the height of two thousand or three thousand feet above the level of the sea, and this elevation implies an extraordinary modification in the state of the earth's crust at some unknown depth beneath the tract so upheaved.

5thly. The most probable hypothesis in regard to the nature of this change, is the successive generation and forcible intrusion into the inferior parts of the earth's crust of lava which, after cooling down, may have assumed the form of crystalline unstratified rock, such as is frequently exhibited in those mountainous parts of the globe where the greatest alterations of level have taken place.

6thly. Great inequalities must have been caused on the surface of the new-raised lands during the emergence of the newer Pliocene strata, by the action of tides, currents, and rivers, combined with the disturbing and dislocating force of the elevatory movements.

7thly. There are no features in the forms of the valleys and sea-cliffs thus recently produced, which indicate the sudden rise of the strata to the whole or the greater part of their present altitude, while there are some proofs of distinct elevations at successive periods.

8thly. We may infer that the species of terrestrial and fluviatile animals and plants which now inhabit extensive districts, formed during the newer Pliocene era, were in existence not only before the new strata were raised, but before their materials were brought together at the bottom of the sea.

Former Changes of the Earth's Surface

[*The remaining chapters of Volume III continue the pattern of analysis so strikingly applied to the case of Etna. The records of past ages are interpreted with combinations of aqueous and igneous causes which can be observed acting at the present day.*

Chapters 10 to 22 focus on the Tertiary strata of Europe. There are discussions of the origin of the Newer Pliocene strata outside Sicily (10–11); the Older Pliocene (12–14); the Miocene (15–16); and the Eocene (17–22). For the Eocene strata of the Paris Basin, discussed in Chapter 18, the use of a geological chronometer based on marine molluscs proves especially important. The French naturalist Georges Cuvier had argued that the complete extinction of the vertebrate fauna of this epoch made it necessary to invoke a major catastrophe or 'revolution'. But the continuing presence of a proportion – albeit only 3 per cent – of existing species of molluscs allows Lyell to draw a less dramatic conclusion: no sudden catastrophe need be invoked. Far from the world being in a chaotic or unstable state, it was 'already fitted for the habitation of man' (p. 255 of Vol. III in the first edition).

Lyell's most telling attack on distinctions between past and present conditions occurs in Chapter 19. The immediate subject is the Eocene volcanic rocks of the Auvergne district in Central France. Many of the volcanic cones, known locally as 'puys', appear as fresh as those at Etna; but they were actually formed at many different periods, and probably none is younger than Miocene. The aim of Lyell's detailed report is to undermine the interpretative force of any distinction between an antediluvian and a postdiluvian period. The final pages of the chapter, included here, bring out the significance of this issue, particularly for the Biblical Deluge and the authority of geology in relation to Scripture. The wider consequences of Lyell's gradualist views, already implicit in his account of Etna, are made explicit 'with great reluctance' and in a 'digression' deep within the most empirical part of the Principles:]

Attempt to divide Volcanos into ante-diluvian and post-diluvian. – The opinions above expressed are entirely at variance with the doctrines of those writers who have endeavoured to arrange all the volcanic cones of Europe under two divisions, those of ante-diluvian and those of post-diluvian origin. To the former they attribute such hills of sand and scoriæ as exhibit on their surface evident signs of aqueous denudation; to the latter, such as betray no marks of having been exposed to such aqueous action. According to this classification almost all the minor cones of Central France must be called post-diluvian; although, if we receive this term in its ordinary acceptation as denoting posteriority of date to the Noachian deluge, we are forced to suppose that all the volcanic eruptions occurred within a period of little more than twenty centuries, or between the era of the flood, which happened about 4000 years ago, and the earliest historical records handed down to us respecting the former state of Central France. Dr. Daubeny has justly observed, that had any of these French volcanos been in a state of activity in the age of Julius Cæsar, that general, who encamped upon the plains of Auvergne, and laid siege to its principal city, (Gergovia, near Clermont,) could hardly have failed to notice them. Had there been even any record of their existence in the time of Pliny or Sidonius Apollinaris, the one would scarcely have omitted to make mention of it in his Natural History, nor the other to introduce some allusion to it among the descriptions of this his native province. This poet's residence was on the borders of the Lake Aidat, which owed its very existence to the damming up of a river by one of the most modern lava currents.

The ruins of several Roman bridges and of the Roman baths at Royat confirm the conclusion that no sensible alteration has taken place in the physical geography of the district, not even in the chasms excavated through the newest lavas since ages historically remote. We have no data at present for presuming that any one of the Auvergne cones has been produced within the last 4000 or 5000 years; and the same may be said of those of Velay. Until the bones of men or articles of human workmanship are found buried under some of their lavas, instead of the remains of extinct animals, which alone have hitherto been met with, we shall consider it probable, as we before hinted, that the latest of the volcanic eruptions may have occurred during the Miocene period.

Supposed Effects of the Flood

They who have used the terms ante-diluvian and post-diluvian in the manner above adverted to, proceed on the assumption that there are clear and unequivocal marks of the passage of a general flood over all parts of the surface of the globe. It had long been a question among the learned, even before the commencement of geological researches, whether the deluge of the Scriptures was universal in reference to the whole surface of the globe, or only so with respect to that portion of it which was then inhabited by man. If the latter interpretation be admissible, the reader will have seen, in former parts of this work, that there are two classes of phenomena in the configuration of the earth's surface, which might enable us to account for such an event. First, extensive lakes elevated above the level of the ocean; secondly, large tracts of dry land depressed below that level. When there is an immense lake, having its surface, like Lake Superior, raised 600 feet above the level of the sea, the waters may be suddenly let loose by the rending or sinking down of the barrier during earthquakes, and hereby a region as extensive as the valley of the Mississippi, inhabited by a population of several millions, might be deluged. On the other hand, there may be a country placed beneath the mean level of the ocean, as we have shown to be the case with part of Asia, and such a region must be entirely laid under water should the tract which separates it from the ocean be fissured or depressed to a certain depth. The great cavity of western Asia is 18,000 square leagues in area, and is occupied by a considerable population. The lowest parts, surrounding the Caspian Sea, are 300 feet below the level of the Euxine, – here, therefore, the diluvial waters might overflow the summits of hills rising 300 feet above the level of the plain; and if depressions still more profound existed at any former time in Asia, the tops of still loftier mountains may have been covered by a flood.

But it is undeniable, that the great majority of the older commentators have held the deluge, according to the brief account of the event given by Moses, to have consisted of a rise of waters over *the whole earth,* by which the summits of the loftiest mountains on the globe were submerged. Many have indulged in speculations concerning the instruments employed to bring about the grand cataclysm; and there has been a great division of opinion as to the effects which it might be expected to have produced on the surface of the earth. According to one school, of which De Luc in former times, and more recently Dr. Buckland, have been zealous and eloquent supporters, the passage of the flood worked a considerable alteration in the external configuration of our continents. By the last-mentioned writer the deluge is represented as a violent and transient rush of waters which tore up the soil to a great depth, excavated valleys, gave rise to immense beds of shingle, carried fragments of rock and gravel from one point to another, and, during its advance and retreat, strewed the valleys, and even the tops of many hills, with alluvium.

But we agree with Dr. Fleming, that in the narrative of Moses there are no terms employed that indicate the impetuous rushing of the waters, either as they rose or when they retreated, upon the restraining of the rain and the passing of a wind over the earth. On the contrary, the olive-branch, brought back by the dove, seems as clear an indication to us that the vegetation was not destroyed, as it was then to Noah that the dry land was about to appear.

We have been led with great reluctance into this digression, in the hope of relieving the minds of some of our readers from groundless apprehension respecting the bearing of many of the views advocated in this work. They have been in the habit of regarding the diluvial theory above controverted as alone capable of affording an explanation of geological phenomena in accordance with Scripture, and they may have felt disapprobation at our attempt to prove, in a former chapter, that the minor volcanos on the flanks of Etna may, some of them, be more than 10,000 years old. How, they would immediately ask, could they have escaped the denuding force of a diluvial rush of waters? The same objection may have presented itself when we quoted, with so much respect, the opinion of a distinguished botanist, that some living specimens of the Baobab tree of Africa, or the Taxodium of Mexico, may be five thousand years old. Our readers may also have been astonished at the high antiquity assigned by us to the greater part of the European alluviums, and the many different ages to which we refer them, as they may have been taught to consider the whole as the result of one *recent* and *simultaneous* inundation. Lastly, they may have felt some disappointment at observing, that we attach no value whatever to the hypothesis of M. Elie de Beaumont, adopted by Professor Sedgwick, that the sudden elevation of mountain-chains 'has been followed again and again by mighty waves desolating whole regions of the earth,' a phenomenon which, according to the last-mentioned of these writers, has 'taken away all anterior incredibility from the fact of a recent deluge.'[9]

For our own part, we have always considered the flood, if we are required to admit its universality in the strictest sense of the term, as a preternatural event far beyond the reach of philosophical inquiry, whether as to the secondary causes employed to produce it, or the effects most likely to result from it. At the same time, it is evident that they who are desirous of pointing out the coincidence of geological phenomena with the occurrence of such a general catastrophe, must neglect no one of the circumstances enumerated in the Mosaic history, least of all so remarkable a fact as that the olive remained standing while the waters were abating.

Recapitulation. – We shall now briefly recapitulate some of the principal conclusions to which we have been led by an examination of the volcanic districts of Central France.

1st. Some of the volcanic eruptions of Auvergne took place during the Eocene period, others at an era long subsequent, probably during the Miocene period.

2ndly. There are no proofs as yet discovered that the most recent of the volcanos of Auvergne and Velay are subsequent to the Miocene period, the integrity of many cones and craters not opposing any sound objection to the opinion that they may be of indefinite antiquity.

3rdly. There are alluviums in Auvergne of very different ages, some of them belonging to the Miocene period. Many of these have been covered by lava-currents which have been poured out in succession while the excavation of valleys was in progress.

4thly. There are a multitude of cones in Auvergne, Velay, and the Vivarais, which have never been subjected to the action of a violent rush of waters capable of modifying considerably the surface of the earth.

5thly. If, therefore, the Mosaic deluge be represented as universal, and as having exercised a violent denuding force, all these cones, several hundred in number, must be post-diluvian.

6thly. But since the beginning of the historical era, or the invasion of Gaul by Julius Cæsar, the volcanic action in Auvergne has been dormant, and there is nothing to countenance the idea that, between the date usually assigned to the Mosaic deluge and the earliest traditional and historical records of Central France (a period of little more than twenty centuries), all or any one of the more entire cones of loose scoriæ were thrown up.

Lastly, it is the opinion of some writers, that the earth's surface underwent no great modification at the era of the Mosaic deluge, and that the strictest interpretation of the scriptural narrative does not warrant us in expecting to find any geological monuments of the catastrophe, an opinion which is consistent with the preservation of the volcanic cones, however high their antiquity.

[*The campaign against diluvialist interpretations continues in Chapters 20 to 22, which examine the Eocene strata of England, especially their uplift and erosion. Chapter 23 then indicates, very briefly, that the principles of interpretation applied to the Tertiary could also account for the older sedimentary rocks of the Secondary period. A full analysis would require another volume, and the work has already exceeded its original limit.*

The final three chapters (24–26) show that mountain chains and the so-called 'Primary' rocks often found in them cannot be uniquely associated with any geological era. These issues had already been raised in the fourth chapter of Volume III. Just as in the organic world, where all classes of animals and plants are likely to be represented in all geological periods, so too in the inorganic world: all the different kinds of rocks could – at least potentially – be formed at any point in earth history. Chapter 24 argues against the French geologist Léonce Elie de Beaumont, who had recently suggested that mountain chains with similar orientations had been formed at similar times. Elie de Beaumont's influential theory had received prestigious support in England from Adam Sedgwick and William Daniel Conybeare. Chapters 25 and 26 contend (following James Hutton, John Playfair and John MacCulloch) that granite and other altered rocks have been formed at many different periods, and are not remnants of any original molten state of the planet.

Some reviewers of earlier volumes of the Principles *had taken Lyell's advocacy of constant geological forces to imply an eternal earth. Chapter 26 (and the work as a whole) ends by countering this accusation.*]

Concluding Remarks

In our history of the progress of geology, in the first volume, we stated that the opinion originally promulgated by Hutton, 'that the strata called *primitive* were mere altered sedimentary rocks,'[10] was vehemently opposed for a time, the main objection to the theory being its supposed tendency to promote a belief in the past eternity of our planet. Previously the absence of animal and vegetable remains in the so-called primitive strata, had been appealed to, as proving that there had been a period when the planet was uninhabited by living beings, and when, as was also inferred, it was uninhabitable, and, therefore, probably in a nascent state.

The opposite doctrine, that the oldest visible strata might be the monuments of an antecedent period, when the animate world was already in existence, was declared to be equivalent to the assumption, that there never was a beginning to the present order of things. The unfairness of this charge was clearly pointed out by Playfair, who observed, 'that it was one thing to declare that we had not yet discovered the traces of a beginning, and another to deny that the earth ever had a beginning.'[11]

We regret, however, to find that the bearing of our arguments in the first volume has been misunderstood in a similar manner, for we have been charged with endeavouring to establish the proposition, that 'the existing causes of change have operated with absolute uniformity from all eternity.'[12]

It is the more necessary to notice this misrepresentation of our views, as it has proceeded from a friendly critic whose theoretical opinions coincide in general with our own, but who has, in this instance, strangely misconceived the scope of our argument. With equal justice might an astronomer be accused of asserting, that the works of creation extend throughout *infinite* space, because he refuses to take for granted that the remotest stars now seen in the heavens are on the utmost verge of the material universe. Every improvement of the telescope has brought

436

thousands of new worlds into view, and it would, therefore, be rash and unphilosophical to imagine that we already survey the whole extent of the vast scheme, or that it will ever be brought within the sphere of human observation.

But no argument can be drawn from such premises in favour of the infinity of the space that has been filled with worlds; and if the material universe has any limits, it then follows that it must occupy a minute and infinitessimal point in infinite space. So, if in tracing back the earth's history, we arrive at the monuments of events which may have happened millions of ages before our times, and if we still find no decided evidence of a commencement, yet the arguments from analogy in support of the probability of a beginning remain unshaken; and if the past duration of the earth be finite, then the aggregate of geological epochs, however numerous, must constitute a mere moment of the past, a mere infinitessimal portion of eternity.

It has been argued, that as the different states of the earth's surface, and the different species by which it has been inhabited, have had each their origin, and many of them their termination, so the entire series may have commenced at a certain period. It has also been urged, that as we admit the creation of man to have occurred at a comparatively modern epoch – as we concede the astonishing fact of the first introduction of a moral and intellectual being, so also we may conceive the first creation of the planet itself.

We are far from denying the weight of this reasoning from analogy; but although it may strengthen our conviction, that the present system of change has not gone on from eternity, it cannot warrant us in presuming that we shall be permitted to behold the signs of the earth's origin, or the evidences of the first introduction into it of organic beings.

In vain do we aspire to assign limits to the works of creation in *space*, whether we examine the starry heavens, or that world of minute animalcules which is revealed to us by the microscope. We are prepared, therefore, to find that in *time* also, the confines of the universe lie beyond the reach of mortal ken. But in whatever direction we pursue our researches, whether in time or space, we discover everywhere the clear proofs of a Creative Intelligence, and of His foresight, wisdom, and power.

As geologists, we learn that it is not only the present condition of the globe that has been suited to the accommodation of myriads of living creatures, but that many former states also have been equally adapted to the organization and habits of prior races of beings. The disposition of the seas, continents, and islands, and the climates have varied; so it appears that the species have been changed, and yet they have all been so modelled, on types analogous to those of existing plants and animals, as to indicate throughout a perfect harmony of design and unity of purpose. To assume that the evidence of the beginning or end of so vast a scheme lies within the reach of our philosophical inquiries, or even of our speculations, appears to us inconsistent with a just estimate of the relations which subsist between the finite powers of man and the attributes of an Infinite and Eternal Being.

*Glossary**

ALGÆ. An order or division of the cryptogamic class of plants. The whole of the sea-weeds are comprehended under this division, and the application of the term in this work is to marine plants.

ALLUVIUM. Earth, sand, gravel, stones, and other transported matter which has been washed away and thrown down by rivers, floods, or other causes, upon land not *permanently* submerged beneath the waters of lakes or seas.

AMMONITE. An extinct and very numerous genus of the order of molluscous animals, called Cephalopoda, allied to the modern genus Nautilus, which inhabited a chambered shell, curved like a coiled snake. Species of it are found in all geological periods of the secondary strata; but they have not yet been seen in the tertiary beds. They are named from their resemblance to the horns on the statues of Jupiter Ammon.

AMYGDALOID. One of the forms of the Trap-rocks, in which agates and simple minerals appear to be scattered like almonds in a cake.

ANALOGUE. A body that resembles or corresponds with another body. A recent shell of the same species as a fossil-shell, is the analogue of the latter.

ANOPLOTHERE, ANOPLOTHERIUM. A fossil extinct quadruped belonging to the order Pachydermata, resembling a pig. It has received its name because the animal must have been singularly wanting in means of defence, from the form of its teeth and the absence of claws, hoofs, and horns.

ANTAGONIST POWERS. Two powers in nature, the action of the one counteracting that of the other, by which a kind of equilibrium or balance is maintained, and the destructive effect prevented that would be produced by one operating without a check.

ARENACEOUS. Sandy.

ARGILLACEOUS. Clayey, composed of clay.

AUGITE. A simple mineral of a dark green, or black colour, which forms a constituent part of many varieties of volcanic rocks.

---

* Abridged and adapted from the glossary which Lyell appended to the first edition of the *Principles* to 'render his work much more accessible to general readers'.

BASALT. One of the most common varieties of the Trap-rocks. It is a dark green or black stone, composed of augite and felspar, very compact in texture, and of considerable hardness, often found in regular pillars of three or more sides, called basaltic columns. Very remarkable examples of this kind of rock are seen at the Giant's Causeway, in Ireland, and at Fingal's Cave, in the island of Staffa, one of the Hebrides. The term is used by Pliny, and is said to come from *basal*, an Æthiopian word signifying iron, not an improbable derivation, inasmuch as the rock often contains much iron, and as many of the figures of the Egyptian temples are formed of basalt.

'BASIN' of Paris, 'BASIN' of London. Deposits lying in a great hollow or trough surrounded by low hills or high land, sometimes used in geology almost synonymously with 'formation.'

BELEMNITE. An extinct genus of the order of molluscous animals called Cephalopoda, that inhabited a long, straight, and chambered conical shell.

BITUMEN. Mineral pitch, of which the tar-like substance which is often seen to ooze out of the Newcastle coal when on the fire, and which makes it cake, is a good example.

BOWLDERS. A provincial term for large rounded blocks of stone lying on the surface of the ground, or sometimes imbedded in loose soil, different in composition from the rocks in their vicinity, and which have been therefore transported from a distance.

BRECCIA. A rock composed of angular fragments connected together by lime or other mineral substance.

CALCAIRE GROSSIER. An extensive stratum, or rather series of strata, belonging to the Eocene tertiary period, originally found in, and specially belonging to, the Paris Basin.

CALCAREOUS ROCK. Limestone. *Etym.*, *Calx*, lime.

CALCEDONY. A siliceous simple mineral, uncrystallized. Agates are partly composed of calcedony.

CARBONATE OF LIME. Lime combines with great avidity with carbonic acid, a gaseous acid only obtained fluid when united with water, – and all combinations of it with other substances are called *Carbonates*. All limestones are carbonates of lime, and quick lime is obtained by driving off the carbonic acid by heat.

CARBONATED SPRINGS. Springs of water, containing carbonic acid gas. They are very common, especially in volcanic countries, and sometimes contain so much gas, that if a little sugar be thrown into the water it effervesces like soda-water.

CARBONIC ACID GAS. A natural gas which often issues from the ground, especially in volcanic countries. *Etym.*, *Carbo*, coal, because the gas is obtained by the slow burning of charcoal. [In modern chemical language, carbon dioxide – *Ed.*]

CARBONIFEROUS. A term usually applied, in a technical sense, to the lowest group of strata of the secondary rocks; but any bed containing coal may be said to be carboniferous.

CATACLYSM. A deluge.

CEPHALOPODA. A class of molluscous animals, having their organs of motion arranged round their head.

CETACEA. An order of vertebrated mammiferous animals inhabiting the sea. The whale, dolphin, and narwal, are examples.

CHALK. A white earthy limestone, the uppermost of the secondary series of strata.

CHERT. A siliceous mineral, approaching in character to flint, but less homogeneous and simple in texture.

CHLORITIC SAND. Sand coloured green by an admixture of the simple mineral chlorite.

COAL FORMATION. This term is generally understood to mean the same as the Coal Measures. There are, however, 'coal formations' in all the geological periods, wherever any of the varieties of coal form a principal constituent part of a group of strata.

CONGENERS. Species which belong to the same genus.

CONGLOMERATE. Rounded water-worn fragments of rock, or pebbles, cemented together by another mineral substance, which may be of a siliceous, calcareous, or argillaceous nature.

CONIFERÆ. An order of plants which, like the fir and pine, bear cones or tops in which the seeds are contained.

CORNBRASH. A rubbly stone extensively cultivated in Wiltshire for growth of corn. It is a provincial term adopted by Smith. Brash is derived from breçan, Saxon, to break.

COSMOGONY, COSMOLOGY. Words synonymous in meaning, applied to speculations respecting the first origin or mode of creation of the earth. *Etym.*, κοσμος, *kosmos*, the world, and γονη, *gonce*, generation, or λογος, *logos*, discourse.

CRAG. A provincial name in Norfolk and Suffolk for a deposit, usually of gravel, belonging to the Older Pliocene period.

CRETACEOUS. Belonging to chalk.

CROP OUT. A miner's or mineral surveyor's term, to express the rising up or exposure at the surface of a stratum or series of strata.

CRUSTACEA. Animals having a shelly coating or crust which they cast periodically. Crabs, shrimps, and lobsters are examples.

CRYPTOGAMIC. A name applied to a class of plants in which the fructification, or organs of reproduction are concealed.

CRYSTALLINE. The internal texture which regular crystals exhibit when broken, or a confused assemblage of ill-defined crystals. Loaf-sugar and statuary-marble have a *crystalline* texture. Sugar-candy and calcareous spar are crystallized.

CYCADEÆ. An order of plants, which are natives of warm climates, mostly tropical, although some are found at the Cape of Good Hope. They have a short stem, surmounted by a peculiar foliage, termed pinnated fronds by botanists, which spreads in a circle. The growth of these plants is by a cluster of fresh fronds shooting from the top of the stem, and pushing the former fronds outwards. These last decay down to their bases, which are broad, and remain covering the sides of the stem. The term is derived from κυκας, *cycas*, a name applied by the ancient Greek naturalist Theophrastus to a palm, said to grow in Ethiopia.

DELTA. When a great river before it enters the sea divides into separate streams, they often diverge and form two sides of a triangle, the sea being the base. The land included by the three lines, and which is invariably alluvial, is called a delta from its resemblance to the letter of the Greek alphabet, which goes by that name Δ. Geologists extend the boundaries of the delta, so as to include all the alluvial land outside the triangle, which has been formed by the river.

DENUDATION. The carrying away of a portion of the solid materials of the land, by which the inferior parts are laid bare.

DICOTYLEDONOUS. A grand division of the vegetable kingdom, founded on the plant having two *cotyledons* or seed-lobes.

DIKES. When a mass of the unstratified or igneous rocks, such as granite, trap, and lava appears as if injected into a great rent in the stratified rocks, cutting across the strata, it forms a dike; and as they are sometimes seen running along the ground, and projecting, like a wall, from the strata on both sides of them being worn away, they are called in the north of England and in Scotland *dikes*, the provincial name for wall. It is not easy to draw the line between dikes and veins. The former are generally of larger dimensions, and have their sides parallel for considerable distances; while veins have generally many ramifications, and these often thin away into slender threads.

DILUVIUM. Those accumulations of gravel and loose materials which, by some geologists, are said to have been produced by the action of a diluvian wave or deluge sweeping over the surface of the earth.

DIP. When a stratum does not lie horizontally, but is inclined, the point of the compass towards which it sinks is called the dip of the stratum, and the angle it makes with the horizon is called the angle of dip or inclination.

DOLOMITE. A crystalline limestone, containing magnesia as a constituent part. Named after the French geologist Dolomieu.

EARTH'S CRUST. Such superficial parts of our planet as are accessible to human observation.

EOCENE. See explanation of this word, pp. 397-8.

ESTUARIES. Inlets of the land, which are entered both by rivers and the tides of the sea. Thus we have the estuaries of the Thames, Severn, Tay, &c.

FALUNS. A provincial name for some tertiary strata abounding in shells in Touraine, which resemble in lithological characters the 'crag' of Norfolk and Suffolk.

FAULT, in the language of miners, is the sudden interruption of the continuity of strata in the same plane, accompanied by a crack or fissure varying in width from a mere line to several feet, which is generally filled with broken stone, clay, &c., and such a displacement that the separated portions of the once continuous strata occupy different levels.

FAUNA. The various kinds of animals peculiar to a country constitute its FAUNA, as the various kinds of plants constitute its FLORA.

FELSPAR. A simple mineral, which constitutes the chief material of many of the unstratified or igneous rocks. The white angular portions in granite are felspar. It is originally a German miners' term.

FELSPATHIC. Of or belonging to felspar.

FERRUGINOUS. Anything containing iron.

FLOETZ ROCKS. A German term applied to the secondary strata by the geologists of that country, because these rocks were supposed to occur most frequently in flat horizontal beds.

FLORA. The various kinds of trees and plants found in any country constitute the Flora of that country in the language of botanists.

FLUVIATILE. Belonging to a river.

FORMATION. A group, whether of alluvial deposits, sedimentary strata, or igneous rocks, referred to a common origin or period.

FOSSIL. All minerals used to be called fossils, but geologists now use the word only to express the remains of animals and plants found buried in the earth. *Etym.*, *fossilis*, anything that may be dug out of the earth.

GALENA, a metallic ore, a compound of lead and sulphur. It has often the appearance of highly polished lead.

GARNET. A simple mineral generally of a deep red colour, crystallized, most commonly met with in mica slate, but also in granite and other igneous rocks.

GAULT. A provincial name in the east of England for a series of beds of clay and marl, the geological position of which is between the upper and the lower greensand.

GEOLOGY, GEOGNOSY. Both mean the same thing, but with an unnecessary degree of refinement in terms, it has been proposed to call our description of the structure of the earth *geognosy*. (*Etym.* γεα, *gea*, earth, and γινωσκω, *ginosco*, to know,) and our theoretical speculations as to its formation *geology*. (*Etym.*, γεα, and λογος, *logos*, a discourse.

GLACIS. A term borrowed from the language of fortification, where it means an easy insensible slope or declivity, less steep than a *talus*, which see.

GNEISS. A stratified primary rock, composed of the same materials as granite, but having usually a larger proportion of mica, and a laminated texture.

GRANITE. An unstratified or igneous rock, generally found inferior to or associated with the oldest of the stratified rocks, and sometimes penetrating them in the form of dikes and veins. It is composed of three simple minerals, felspar, quartz, and mica, and derives its name from having a coarse *granular* structure; *granum*, Latin for grain. Westminster, Waterloo, and London bridges, and the paving-stones in the carriage-way of the London streets are good examples of the most common varieties of granite.

GRAUWACKE, a German name, generally adopted by geologists for the lowest members of the secondary strata, consisting of sandstone and slate, and which form the chief part of what are termed by some geologists the *transition* rocks. The rock is very often of a grey colour, hence the name, *grau* being German for grey, and *wacke* being a provincial miner's term.

GREENSAND. Beds of sand, sandstone, limestone, belonging to the Cretaceous Period. The name is given to these beds, because they often, but not always, contain an abundance of green earth or chlorite scattered through the substance of the sandstone, limestone, &c.

GREENSTONE, a variety of trap, composed of hornblende and felspar.

GRIT, a provincial name for a coarse-grained sandstone.

GYPSUM, a mineral composed of lime and sulphuric acid, hence called also *sulphate of lime*. Plaster and stucco are obtained by exposing gypsum to a strong heat. It is found so abundantly near Paris, that Paris plaster is a common term in this country for the white powder of which casts are made. The term is used by Pliny for a stone used for the same purposes by the ancients. The derivation of it is unknown.

HORNBLENDE, a simple mineral of a dark green or black colour, which enters largely into the composition of several varieties of the trap rocks.

HYDROPHYTES. Plants which grow in water.

HYPOGENE ROCKS. For an explanation of this term, see p. 457n2.

ICHTHYOSAURUS, a gigantic fossil marine reptile, intermediate between a crocodile and a fish.

INDUCTION, a consequence, conclusion or inference, drawn from propositions or principles first laid down, or from the observation and examination of phenomena.

INFUSORY ANIMALCULES. Minute living creatures generated in many *infusions*; and the term *infusoria* has been given to all such animalcules whether found in infusions or in stagnant water, vinegar, &c.

INVERTEBRATED ANIMALS. Animals which are not furnished with a back-bone. For a further explanation, see 'Vertebrated Animals.'

ISOTHERMAL. Such zones or divisions of the land, ocean, or atmosphere, which have an equal degree of mean annual warmth, are said to be isothermal, from ισος, *isos*, equal, and θερμη, *therme*, heat.

JURA LIMESTONE. The limestones belonging to the oolite group constitute the chief part of the mountains of the Jura, between France and Switzerland, and hence the geologists of the Continent have given the name to the group.

LACUSTRINE, belonging to a lake.

LAMINÆ. Latin for plates; used in geology, for the smaller layers of which a stratum is frequently composed.

LAMELLIFEROUS. A stone composed of thin plates or leaves like paper.

LANDSLIP. A portion of land that has slid down in consequence of disturbance by an earthquake, or from being undermined, by water washing away the lower beds which supported it.

LAPIDIFICATION – Lapidifying process. Conversion into stone.

LAPILLI. Small volcanic cinders.

LAVA. The stone which flows in a melted state from a volcano.

LIAS. A provincial name, adopted in scientific language, for a particular kind of limestone, which being characterized, together with its associated beds, by peculiar fossils, is formed in this work into a particular group of the secondary strata.

LIGNITE. Wood converted into a kind of coal.

LITHODOMI. Molluscous animals which bore into solid rocks, and lodge themselves in the holes they have formed.

LITHOLOGICAL. A term expressing the stony structure or character of a mineral mass. We speak of the lithological character of a stratum as distinguished from its zoological character.

LITHOPHAGI. Molluscous animals which bore into solid stones.

LITTORAL. Belonging to the sea-shore.

LOAM. A mixture of sand and clay.

LYCOPODIACEÆ. Plants of an inferior degree of organization to Coniferæ, some of which they very much resemble in foliage, but all recent species are infinitely smaller. Many of the fossil species are as gigantic as recent coniferæ. Their mode of reproduction is analogous to that of ferns. In English they are called club-mosses, generally found in mountainous heaths in the north of England.

MADREPORE. A genus of corals, but generally applied to all the corals distinguished by superficial star-shaped cavities. There are several fossil species.

MAGNESIAN LIMESTONE. An extensive series of beds lying in geological position, immediately above the coal-measures, so called because the limestone, the principal member of the series, contains much of the earth magnesia as a constituent part.

MAMMILLARY. A surface which is studded over with rounded projections.

MAMMIFEROUS. Animals which give suck to their young.

MAMMOTH. An extinct species of the elephant (*E. primigenius*), of which the fossil bones are frequently met with in various countries.

GLOSSARY

MARL. A mixture of clay and lime; usually soft, but sometimes hard, in which case it is called indurated marl.

MARSUPIAL ANIMALS. A tribe of quadrupeds having a sack or pouch under the belly, in which they carry their young. The kangaroo is a well-known example.

MASTODON. A genus of fossil extinct quadrupeds allied to the elephant. So called from the form of the hind teeth or grinders, which have their surface covered with conical mammillary crests.

MATRIX. If a simple mineral or shell, in place of being detached, be still fixed in a portion of rock, it is said to be in its matrix.

MECHANICAL ORIGIN, Rocks of. When rocks are composed of sand, pebbles, or fragments, to distinguish them from those of an uniform crystalline texture, which are of chemical origin.

MEDUSÆ. A genus of marine radiated animals, without shells; so called because their organs of motion spread out like the snaky hair of the fabulous Medusa.

MEGALOSAURUS. A fossil gigantic amphibious animal of the saurian or lizard and crocodile tribe.

MEGATHERIUM. A fossil extinct quadruped, resembling a gigantic sloth.

MELASTOMA. A genus of MELASTOMACEA, an order of plants of the evergreen tree, and shrubby exotic kinds.

METAMORPHIC ROCKS. For an explanation of this term, see p. 457n2.

MICA. A simple mineral, having a shining silvery surface, and capable of being split into very thin elastic leaves or scales. It is often called *talc* in common life, but mineralogists apply the term talc to a different mineral. The brilliant scales in granite are mica.

MICA-SLATE, MICA-SCHIST, MICACEOUS SCHISTUS. One of the lowest of the stratified rocks, belonging to the primary class, which is characterized by being composed of a large proportion of mica, united with quartz.

MIOCENE. See an explanation of this term, p. 396.

MOLLUSCÆ, Molluscous Animals. Animals, such as shell-fish, which, being devoid of bones, have soft bodies.

MONITOR. An animal of the saurian or lizard tribe, species of which are found in both the fossil and recent state.

MONOCOTYLEDONOUS. A grand division of the vegetable kingdom, founded on the plant having only one *cotyledon*, or seed-lobe.

MOUNTAIN LIMESTONE. A series of limestone strata, of which the geological position is immediately below the coal measures, and with which they also sometimes alternate.

MUSCHELKALK. A limestone which, in geological position, belongs to the red sandstone group. This formation has not yet been found in England, and the German name is adopted by English geologists. The word means shell-limestone: *muschel*, shell, and *kalkstein*, limestone.

NEW RED SANDSTONE. A series of sandy, argillaceous, and often calcareous strata, the predominant colour of which is brick-red, but containing portions which are of a greenish grey. These occur often in spots and stripes, so that the series has sometimes been called the variegated sandstone. The European formation so called lies in a geological position immediately above the coal-measures.

NODULE. A rounded irregular-shaped lump or mass.

NORMAL GROUPS. Groups of certain rocks taken as a rule or standard.

OBSIDIAN. A volcanic product, or species of lava, very like common green bottle-glass, which is almost black in large masses, but semi-transparent in thin fragments. Pumice-stone is obsidian in a frothy state; produced most probably by water that was contained in or had access to the melted stone, and converted into steam. There are very often portions in a mass of solid obsidian, which are partially converted into pumice.

OLD RED SANDSTONE. A stratified rock belonging to the Carboniferous group.

OLIVINE. An olive-coloured, semi-transparent, simple mineral, very often occurring in the forms of grains and of crystals in basalt and lava.

OOLITE, Oolitic. A limestone, forming a characteristic feature of a group of the secondary strata. It is so named, because it is composed of rounded particles, like the roe or eggs of a fish.

ORGANIC REMAINS. The remains of animals and plants; *organized* bodies, found in a fossil state.

ORTHOCERATA. An extinct genus of the order of Molluscous Animals, called Cephalopoda, that inhabited a long chambered, conical shell, like a straight horn.

OUTLIERS. When a portion of a stratum occurs at some distance, detached from the general mass of the formation to which it belongs, some practical mineral surveyors call it an *outlier*, and the term is adopted in geological language.

OXYGEN. One of the constituent parts of the air of the atmosphere; that part which supports life. For a further explanation of the word, consult elementary works on chemistry.

OXIDE. The combination of a metal with oxygen; rust is oxide of iron.

PACHYDERMATA. An order of quadrupeds, including the elephant, rhinoceros, horse, pig, &c., distinguished by having thick skins.

PALÆOTHERIUM, PALEOTHERE. A fossil extinct quadruped, belonging to the order pachydermata, resembling a pig or tapir, but of great size.

PELAGIAN, PELAGIC. Belonging to the *deep* sea.

PEPERINO. An Italian name for a particular kind of volcanic rock, formed, like tuff, by the cementing together of volcanic sand, cinders, or scoriæ, &c.

PHANEROGAMIC PLANTS. A name given by Linnæus to those plants in which the reproductive organs are apparent.

PHYSICS. The department of science, which treats of the properties of natural bodies, laws of motion, &c., sometimes called Natural philosophy and mechanical philosophy.

PHYTOLOGY, PHYTOLOGICAL. The department of science which relates to plants – synonymous with botany and botanical.

PHYTOPHAGOUS. Plant eating.

PIT COAL. Ordinary coal; called so because it is obtained by sinking pits in the ground.

PITCH STONE. A rock of an uniform texture, belonging to the unstratified and volcanic classes, which has an unctuous appearance, like indurated pitch.

PLASTIC CLAY. One of the beds of the Eocene tertiary period. It is so called because it is used for making pottery.

PLESIOSAURUS. A fossil extinct amphibious animal, resembling the saurian, or lizard and crocodile tribe.

PLIOCENE. See explanation of this term, pp. 395–6.

POLYPARIA. CORALS. A numerous class of invertebrated animals, belonging to the great division called Radiata.

PORPHYRY. An unstratified or igneous rock. The term is as old as Pliny, and was applied to a red rock with small angular white bodies diffused through it, which are crystallized felspar, brought from Egypt. The term is hence applied to every species of unstratified rock, in which detached crystals of felspar are diffused through a base of other mineral composition.

PORTLAND LIMESTONE, PORTLAND BEDS. A series of limestone strata, belonging to the upper part of the Oolite group, found chiefly in England, in the Island of Portland on the coast of Dorsetshire. The great supply of the building stone used in London is from these quarries.

PRODUCTÆ. An extinct genus of fossil bivalve shells, occurring only in the older of the secondary rocks. It is closely allied to the living genus Terebratula.

PUMICE. – A light spongy lava, of a white colour, produced by gases, or watery vapour getting access to the particular kind of glassy lava called obsidian, when in a state of fusion – it may be called the froth of melted volcanic glass. The word comes from the Latin name of the stone, *pumex*.

PURBECK LIMESTONE, PURBECK BEDS. Limestone strata belonging to the Wealden group.

PYRITES (Iron). A compound of sulphur and iron, found usually in yellow shining crystals like brass, and in almost every rock stratified and unstratified. The shining metallic bodies, so often seen in common roofing slate, are a familiar example of the mineral.

QUADRUMANA. The order of mammiferous animals to which apes belong. *Etym.*, *quadrus*, a derivation of the Latin word for the number four, and *manus*, hand, – the four feet of those animals being in some degree usable as hands.

QUA-QUA-VERSAL DIP. The dip of beds to all points of the compass around a centre, as in the case of beds of lava round the crater of a volcano.

QUARTZ. A German provincial term, universally adopted in scientific language, for a simple mineral composed of pure silex, or earth of flints; rock-crystal is an example.

RED MARL. A term often applied to the New Red Sandstone, which is the principal member of the Red Sandstone group.

RETICULATE. A structure of cross lines, like a net, is said to be reticulated, from *rete*, a net.

SANDSTONE. Any stone which is composed of an agglutination of grains of sand, which may be either calcareous or siliceous.

SAURIAN. Any animal belonging to the lizard tribe.

SCHIST. Synonymous with slate.

SCHISTOSE ROCKS. Synonymous with *slaty* rocks.

SCORIÆ. Volcanic cinders. The word is Latin for cinders.

SEAMS. Thin layers which separate two strata of greater magnitude.

SECONDARY STRATA. An extensive series of the stratified rocks which compose the crust of the globe, with certain characters in common, which distinguish them from another series below them, called *primary*, and from a third series above them called *tertiary*.

SECULAR REFRIGERATION. The periodical cooling and consolidation of the globe, from a supposed original state of fluidity from heat.

SEDIMENTARY ROCKS, are those which have been formed by their materials having been thrown down from a state of suspension or solution in water.

SELENITE. Crystallized gypsum, or sulphate of lime – a simple mineral.

SERPENTINE. A rock usually containing much magnesian earth, for the most part unstratified, but sometimes appearing to be an altered or metamorphic stratified rock. Its name is derived from frequently presenting contrasts of colour, like the skin of some serpents.

SHALE. A provincial term, adopted in geological science, to express an indurated slaty clay.

SHELL MARL. A deposit of clay, peat, and other substances mixed with shells, which collects at the bottom of lakes.

SHINGLE. The loose and completely water-worn gravel on the sea-shore.

SILEX. The name of one of the pure earths, being the Latin word for *flint*, which is wholly composed of that earth. French geologists have applied it as a generic name for all minerals composed entirely of that earth, of which there are many of different external forms.

SILICA. One of the pure earths. *Etym.*, *silex*, flint, because found in that mineral.

SILICATE. A chemical compound of silica and another substance, such as silicate of iron. Consult elementary works on chemistry.

SILICEOUS. Of or belonging to the earth of flint. A siliceous rock is one mainly composed of silex.

SILICIFIED. Any substance that is petrified or mineralized by *siliceous* earth.

SILT. The more comminuted sand, clay, and earth, which is transported by running water. It is often accumulated by currents in banks. Thus we speak of the mouth of a river being silted up when its entrance into the sea is impeded by such accumulation of loose materials.

SIMPLE MINERAL. Individual mineral substances, as distinguished from rocks, which last are usually an aggregation of simple minerals. They are not simple in regard to their nature, for when subjected to chemical analysis, they are found to consist of a variety of different substances. Pyrites is a simple mineral in the sense we use the term, but it is a chemical compound of sulphur and iron.

SOLFATARA. A volcanic vent from which sulphur, sulphureous, watery, and acid vapours and gases are emitted.

SPORULES. The reproductory corpuscula (minute bodies) of cryptogamic plains.

STRATIFIED. Rocks arranged in the form of *strata*, which see.

STRATIFICATION. An arrangement of rocks in *strata*, which see.

STRATUM, STRATA. When several rocks lie like the leaves of a book, one upon another, each individual forms a *stratum*; – strata is the plural of the word. *Etym.*, *stratum*, part of a Latin verb signifying to strew or lay out.

STRIKE. The direction or line of bearing of strata, which is always at right angles to their prevailing dip.

SUBAPENNINES. Low hills which skirt or lie at the foot of the great chain of the Apennines in Italy. The term Subapennine is applied geologically to a series of strata of the Older Pliocene period.

SYENITE. A kind of granite, so called because it was brought from Syene in Egypt.

TALUS. When fragments are broken off by the action of the weather from the face of a steep rock, as they accumulate at its foot, they form a sloping heap, called a talus. The term is borrowed from the language of fortification, where *talus* means the outside of a wall of which the thickness is diminished by degrees, as it rises in height, to make it the firmer.

TERTIARY STRATA. A series of sedimentary rocks, with characters which distinguish them from two other great series of strata, – the secondary and primary, which lie *beneath* them.

TESTACEA. Molluscous animals, having a shelly covering. *Etym.*, *testa*, a shell, such as snails, whelks, oysters, &c.

THIN OUT. When a stratum, in the course of its prolongation in any direction, becomes gradually less in thickness, the two surfaces approach nearer and nearer; and when at last they meet, the stratum is said to thin out, or disappear.

TRACHYTE. A variety of lava essentially composed of glassy felspar, and frequently having detached crystals of felspar in the base or body of the stone, giving it the structure of porphyry. It sometimes contains hornblende and augite; and when these last predominate, the trachyte passes into the varieties of trap called greenstone, basalt, dolorite, &c.

TRAP and TRAPPEAN ROCKS. Volcanic rocks composed of felspar, augite, and hornblende. The various proportions and state of aggregation of these simple minerals, and differences in external forms, give rise to varieties, which have received distinct appellations, such as basalt, amygdaloid, dolorite, greenstone, and others. The term is derived from *trappa*, a Swedish word for stair, because in Sweden the rocks of this class often occur in large tabular masses, rising one above another, like the steps of a staircase.

TUFF, or TUFO. An Italian name for a variety of volcanic rock, of an earthy texture, seldom very compact, and composed of an agglutination of fragments or scoriæ and loose matter ejected from a volcano.

TUFACEOUS. A rock with the texture of tuff or tufo, which see.

VEINS, Mineral. Cracks in rocks filled up by substances different from the rock, which may either be earthy or metallic. Veins are sometimes many yards wide; and they ramify or branch off into innumerable smaller parts, often as slender as threads, like the veins in an animal, and hence their name.

VERTEBRATED ANIMALS. A great division of the animal kingdom, including all those which are furnished with a back-bone, as the mammalia, birds, reptiles, and fishes. The separate joints of the back-bone are called *vertebræ*, from the Latin verb *verto*, to turn.

VESICLE. A small circular inclosed space, like a little bladder.

VOLCANIC BOMBS. Volcanos throw out sometimes detached masses of melted lava, which, as they fall, assume rounded forms (like bomb-shells), and are often elongated into a pear shape.

VOLCANIC FOCI. The subterranean centres of action in volcanos, where the heat is supposed to be in the highest degree of energy.

ZOOPHYTES. Corals, sponges, and other aquatic animals allied to them, so called because, while they are the habitation of animals, they are fixed to the ground, and have the forms of plants. *Etym.*, ξωον, *zoon*, animal, and φυτον, *phyton*, plant.

Notes

1. J. Playfair, *Works* (1822), vol. 1, p. 415.

2. Probably derived from Playfair, *Works*, vol. 1, p. 131.

3. G. Cuvier, 'Eloge historique de Nicolas Desmarets', in *Recueil des éloges historiques lus dans les séances publiques de l'Institut Royal de France* (1819), vol. 2, p. 365.

4. Playfair, 'Biographical Account of James Hutton, M.D.', in *Works*, vol. 4, p. 50. Here and elsewhere Lyell cites Hutton, but in all cases his quotations derive from Playfair; see D. Dean, *James Hutton and the History of Geology* (1992), esp. pp. 238–41.

5. Not an actual quotation, but derives from Playfair, *Works*, vol. 4, pp. 51–3.

6. Dante, *Divina commedia*, *Inferno*, canto 3, lines 7–8; part of the inscription at the gates of Hell. Translation from Italian by Henry Francis Cary (1817), which Lyell added to later editions of the *Principles*.

7. Lyell's notorious misquotation of the most celebrated statement in the history of the earth sciences: 'The result, therefore, of our present enquiry is, that we find no vestige of a beginning, – no prospect of an end.' James Hutton, 'Theory of the Earth; or an Investigation of the Laws Observable in the Composition, Dissolution and Restoration of the Land upon the Globe', *Transactions of the Royal Society of Edinburgh* 1 (1788), p. 304; as usual, Lyell is using Playfair: see *Works*, vol. 1, p. 131.

8. Ibid. vol. 4, pp. 55–6.

9. J. Williams, *The Natural History of the Mineral Kingdom* (1789), vol. 1, pp. lvii, lix.

10. R. Kirwan, *Geological Essays* (1799), p. 3. Lyell does not mention that Kirwan begins from two fundamental 'laws of reasoning': 'the first is, that no effect shall be attributed to a cause whose *known* powers are inadequate to its production. The second is, that no cause should be adduced whose existence is not proved either by actual experience or approved testimony' (p. 2).

11. J. A. De Luc, *An Elementary Treatise on Geology* (1809), pp. 4, 5.

12. The entire passage refers to Virgil's panegyric on the country life in *Georgics*, bk 2, lines 490–93; translation from Latin original by John Dryden:

> 'Happy the Man, who, studying Nature's Laws,
> Thro' known Effects, can trace the secret Cause:
> His Mind possessing, in a quiet state,
> Fearless of Fortune, and resigned to Fate.'

13. Playfair, *Works*, vol. 1, p. 132.

14. Ibid. p. 350; Lyell's own footnote reference to Hutton's *Theory of the Earth* (1795) is similarly lifted from Playfair's *Illustrations*.

15. J. F. D'Aubuisson de Voisins, *Traité de géognosie* (1819), vol. 2, p. 253.

16. This is a reference, although Lyell does not say so, to Georges Cuvier's famous *Essay on the Theory of the Earth*, trans. R. Jameson (5th ed., 1827), pp. 23–4. The passage is worth quoting at length for the light it sheds on debates about the earth at the time of publication of the *Principles*: 'it has long been considered possible to explain the more ancient revolutions on its surface by means of these still existing causes; in the same manner as it is found easy to explain past events in political history, by an acquaintance with the passions and intrigues of the present day. But we shall presently see, that unfortunately the case is different in physical history: the thread of operations is here broken; the march of Nature is changed; and none of the agents which she now employs, would have been sufficient for the production of her ancient works.'

17. B. G. Niebuhr, *The History of Rome*, trans. J. C. Hare and C. Thirlwall (1828), vol. 1, p. 5.

18. Shakespeare, *Richard III*, act 1, sc. 1, lines 21–2.

19. Alexander Pope, *The Rape of the Lock*, canto 4, lines 13, 17, 14.

20. Lyell's note refers the reader to John Fleming, 'On the Value of the Evidence from the Animal Kingdom, Tending to Prove that the Arctic Region Formerly Enjoyed a Milder Climate than at Present', *Edinburgh New Philosophical Journal* 6 (1829), pp. 277–88, and 'Additional Remarks on the Climate of the Arctic Regions, in Answer to Mr. Conybeare', *Edinburgh New Philosophical Journal* 8 (1830), pp. 65–74.

21. C. Malte-Brun, *Universal Geography* (1822), vol. 1, p. 406.

22. H. Davy, *Consolations in Travel, or the Last Days of a Philosopher* (1830); quotation assembled from passages on pp. 149, 144–5, 146–7, 149.

23. Shakespeare, *Richard III*, act 1, sc. 4, lines 24–6, 33.

24. Virgil, *Eclogues*, bk 4, lines 34–6; translation from Latin original by Dryden. Lyell's note also refers the reader to D. Stewart, *Elements of the Philosophy of the Human Mind* (1814), vol. 2, sect. 4, and J. Prichard, *An Analysis of the Egyptian Mythology* (1819), p. 177.

25. W. Paley, *Natural Theology* (1802), pp. 483–4.

26. D. Stewart, *Elements of the Philosophy of the Human Mind* (1814), vol. 2, p. 230.

27. Virgil, *Georgics*, bk 4, lines 221–2; translation from Latin original by Dryden.

28. J. Farquharson, 'Some Notices of the Great Storm and Flood which Occurred in the Counties of Aberdeen, Banff, and Nairn . . .', n.s. 1, *Quarterly Journal of Science*, p. 331.

29. Pindar, *Pythian Odes*, 1 ('For Hieron of Aetna, Winner in the Chariot Race 470 BC'), line 20.

30. Ovid, *Metamorphoses*, bk 15, lines 340–41; translation from Latin original by Dryden.

31. Milton, *Paradise Lost*, bk 1, line 711.

32. An implicit attack on W. D. Conybeare, 'On the Valley of the Thames', *Proceedings of the Geological Society of London* 1 (1829), pp. 145−9.

33. W. Hamilton, *Campi Phlegraei* (1776), pp. 70−71.

34. Ibid. p. 75.

35. Lord Byron, *Childe Harold's Pilgrimage*, canto 4 (1818), lines 1634−8.

36. J. Michell, 'Conjectures Concerning the Cause, and Observations upon the Phænomena of Earthquakes', *Philosophical Transactions of the Royal Society of London* 51 (1760), pp. 568−9.

37. Ibid. p. 592.

38. Pliny, *Historia mundi*, bk 2, ch. 3 (107); translated from Latin original by J. Bostock and H. T. Riley.

39. Michell, 'Conjectures', pp. 600−601.

VOLUME II

1. J. Playfair, *Works*, vol. 1, pp. 458, 459. Lyell dedicates the second volume to his friend, the naturalist and Tory magistrate William John Broderip, and includes a brief 'Preface' noting that it had proved 'impossible to compress into two volumes, according to his original plan, the wide range of subjects which must be discussed, in order fully to explain his views respecting the causes of geological phenomena.'

2. J.-B. Lamarck, *Philosophie zoologique* (2 vols, 1809), vol. 1, p. 54.

3. Ibid. pp. 226−7.

4. Ibid. p. 229.

5. Ibid. passim, e.g. p. 234.

6. The preceding quotations are largely adapted from Lamarck, loc. cit. pp. 237, 255, 256−7.

7. Virgil, *Georgics*, bk 1, lines 199−200; translation from Latin original by Dryden. The *Telliamed* (1748), mentioned earlier in this paragraph, was by Benoît de Maillet, not by Julien Offroy de la Mettrie.

8. Lamarck, *Philosophie zoologique*, vol. 1, p. 357.

9. J. E. Smith, *Introduction to Physiological and Systematical Botany* (1807), p. 361.

10. I. F. Blumenbach, *A Manual of the Elements of Natural History* (1825), p. 15.

11. J. C. Prichard, *Researches into the Physical History of Mankind* (1819), vol. 1, p. 96; Prichard cites (as Lyell notes) the Russian naturalist and physician Johann Anton Güldenstädt.

12. B. Lacépède, J. B. Lamarck and G. Cuvier, 'Rapport des professeurs du Muséum, sur les collections d'histoire naturelle rapportées d'Egypte, par E. Geoffrey', *Annales du Muséum National d'Histoire Naturelle* 1 (1802), pp. 235−6.

13. W. Herbert, 'On the Production of Hybrid Vegetables, with the Result of Many Experiments Made in the Investigation of the Subject', *Transactions of the Horticultural Society of London* 4 (1822), p. 19.

14. F. Cuvier, 'Essay on the Domestication of Mammiferous Animals', *Edinburgh New Philosophical Journal* 4 (1828), p. 295.

15. J. Richardson, *Fauna Boreali-Americana* (1827), vol. 1, p. 273.

16. See Alexander Pope, *Essay on Man*, epistle 1, lines 221−4:

> 'How Instinct varies in the grov'ling swine,
> Compar'd, half-reas'ning elephant, with thine:
> 'Twixt that, and Reason, what a nice barrier;
> For ever sep'rate, yet for ever near!'

17. J. Hunter, 'Observations Tending to Shew that the Wolf, Jackal, and Dog, are All of the Same Species', *Philosophical Transactions of the Royal Society of London* 77 (1787), p. 253.

18. Virgil, *Georgics*, bk 3, lines 273−5; translation from Latin original by Dryden.

19. A.-P. De Candolle, *Essai élémentaire de geographie botanique (extrait du 18.e volume du Dictionnaire des sciences naturelles)*, [1820], pp. 60−61.

20. Shakespeare, *The Tempest*, act 4, sc. 1, line 249.

21. In Lyell's text the phrase in parentheses is a Latin quotation from Virgil, *Aeneid*, bk 5, line 320.

22. A. Humboldt, *Personal Narrative of Travels to the Equinoctial Regions of the New Continent* (1814−29), vol. 5, p. 180.

23. De Candolle, *Essai*, p. 44.

24. Elias Magnus Fries, quoted in J. Lindley, *An Introduction to the Natural System of Botany* (1830), p. 335.

25. P. Keith, *A System of Physiological Botany* (1816), vol. 2, p. 405.

26. Byron, *Childe Harold's Pilgrimage*, canto 3 (1816), lines 17−18.

27. J. Cook, *A Voyage Towards the South Pole, and Round the World* (1777), vol. 2. p. 69.

28. De Candolle, *Essai*, p. 50.

29. Ibid. pp. 50−51.

30. C. Linnaeus, *Oratio de telluris habitabilis incremento* (1746), p. 22. Translation of Latin original adapted from F. J. Brand, *Select Dissertations from the Amoenitates Academicæ* (1781), vol. 1, pp. 77−8.

31. G. Brocchi, *Conchiologia fossile subapennina* (1814), vol. 1, p. 229, referring to Noel Joseph von Necker, *Phytozoologie philosophique* (1790), p. 21.

32. Brocchi, *Conchiologia*, vol. 1, p. 229.

33. Ibid. p. 230.

34. Ibid. p. 229.

35. De Candolle, *Essai*, p. 26.

36. H. C. D. Wilcke, 'Politia naturæ', in C. Linnæus, *Amoenitates academicæ* (1764), vol. 6, pp. 25−6.

37. Ibid. pp. 26−7.

38. Milton, *Paradise Lost*, bk 1, lines 775−9.

39. Shakespeare, *Henry V*, act 5, sc. 2, line 45.

40. G. Graves, '*Otis Tarda*. Great Bustard', in his *British Ornithology* (1821), vol. 3.

41. T. Bewick, *A History of British Birds* (1821), vol. 1, p. 316.

42. A reference to Robert Burns, 'To a Mouse on Turning her up in her Nest with the Plough, November 1785', lines 7–8:

> 'I'm truly sorry man's dominion
> Has broken Nature's social union.'

43. Shakespeare, *As You Like It*, act 2, sc. 1, lines 61–3.

44. G. L. L. Comte de Buffon, *Histoire naturelle générale et particulière* (1756), vol. 6, p. 62. Buffon is discussing the effects of humans in relation to the degeneration of wild animals, not general issues of extinction, about which he is never very explicit. I am grateful to E. C. Spary and N. Jardine for help in finding this passage.

45. De Candolle, *Essai*, pp. 47–8.

46. Playfair, *Works*, vol. 1, p. 118.

47. A. Sedgwick, 'Address to the Geological Society', *Proceedings of the Geological Society of London* 1 (1831), pp. 303, 304.

48. Horace, *Odes*, bk 4, ode 2, lines 5–7; translation from Latin original by Richard Bentley.

49. A reference to Sedgwick, 'Address', pp. 303–4.

50. Ibid. pp. 307, 314.

51. Ibid. pp. 303, 304.

52. C. Maclaren, 'America', *Encyclopedia Britannica* (7th ed. 1832), vol. 2, pp. 605–54, pp. 610–11 (including, as Lyell mentions in a note, a map on p. 611). Maclaren's article appeared for the first time in this edition, which was in progress while the *Principles* was being published.

53. Shakespeare, *Henry V*, act 1, sc. 2, lines 163–5.

54. See this volume, p. 170; the reference is to G. Cuvier, *Essay on the Theory of the Earth* (5th ed. 1827), p. 120.

55. E. King, 'Account of a Petrifaction found on the Coast of East Lothian', *Philosophical Transactions of the Royal Society of London* 69 (1779), at pp. 40, 41.

56. Ibid. p. 37.

57. J. Davy, 'Observations on the Changes which have Taken Place in Some Ancient Alloys of Copper', *Philosophical Transactions of the Royal Society of London* pt ii (1826), p. 58.

58. A reference to Thucydides, *The Peloponnesian War*, bk 4: 40, although not a direct quotation.

59. A. Burnes, *Travels into Bokhara* (3 vols, 1834), vol. 3, p. 314. Lyell refers to a 'published account' deposited (as he says in a note) in the library of the Royal Asiatic Society. However, the work was not formally published by John Murray until 1834, so Lyell was evidently consulting a printed version not available for sale. Burnes's account contained sensitive material which had to be carefully vetted by the military authorities.

60. G. Berkeley, *Alciphron: Or, the Minute Philosopher. In Seven Dialogues* (1732), vol. 2, pp. 84–5.

61. H. Davy, *Consolations in Travel* (1831), p. 276.

62. F. Bacon, 'Of Vicissitude of Things', in *The Essays or Counsels, Civill and Morall*, ed. M. Kiernan (1985, first published 1625), p. 172.

63. O. von Kotzebue, *A Voyage of Discovery, into the South Sea and Beering's Straits* (1821), vol. 3, pp. 332–3.

64. F. W. Beechey, *Narrative of a Voyage to the Pacific and Beering's Strait* (1831), pt i, p. 188.

65. Ibid. p. 189.

66. J. MacCulloch, *A System of Geology* (1831), vol. 1, p. 219.

VOLUME III

1. The third volume is dedicated to the geologist Roderick Impey Murchison, who had travelled with Lyell in France and Italy. A twelve-page 'Preface' – omitted here – chronicles the events leading up to publication, and implicitly makes a case for Lyell's originality, especially in relation to his use of percentage methods for determining the age of strata.

2. Later in the *Principles* (vol. 3, pp. 374–5) Lyell writes that the term 'primary' should be replaced by 'hypogene', 'implying that granite and gneiss are "*nether-formed*" rocks, or rocks which have not assumed their present form and structure at the surface'. He then formally proposes the term "metamorphic" in the following way: 'We divide the hypogene rocks, then, into the unstratified, or plutonic, and the *altered* stratified. For these last the term "metamorphic" (from μεια, *trans*, and μορφη, *form*) may be used. The last-mentioned name need not, however, be often resorted to, because we may speak of hypogene *strata*, hypogene *limestone*, hypogene *schist*, and this appellation will suffice to distinguish the formations so designated from the plutonic rocks.'

3. Not included in this edition.

4. The quotation is from Horace, *Epode* 2, lines 11–12; translation of Latin original by Richard Bentley.

5. Walter Scott, *The Lady of the Lake; A Poem* (1810), canto 1, no. 14, lines 15–16.

6. Milton, *Paradise Lost*, bk 1, lines 180–83.

7. Ibid. lines 607–10.

8. An implicit reference to Sedgwick, 'Address', p. 314, and its support of the theories of Elie de Beaumont.

9. Ibid. p. 314.

10. This is a general statement, not a quotation from Hutton's or Playfair's writings.

11. Paraphrased from Playfair, *Works*, vol. 1, p. 131.

12. [George Poulett Scrope], review of Lyell's *Principles of Geology, Quarterly Review*

43 (1830), p. 463. Some other reviewers (e.g. *New Monthly Magazine* 40 (1834), p. 377) assumed that a 'super-orthodox' scriptural literalist must have written this anonymous essay, but Scrope was a close friend who had received detailed instructions from Lyell on what to say in his review.

Bibliography of Reviews

This is a list of all the known reviews of the *Principles* published in Britain between 1830 and 1835. Outside of famous notices by Scrope, Sedgwick and Whewell, the reception of Lyell's work in the periodical press has been little studied; nearly half of the present notices have never been cited before. It is hoped that this bibliography, although incomplete, will encourage interest in this important topic, not only in Britain but also on the Continent and in the United States, where Lyell's book was also widely noticed.

There were, of course, important reviews of later editions of the *Principles*, notably Richard Owen's discussion in the *Quarterly Review* 89 (Oct. 1851), pp. 412–51. These are not listed here, nor are books such as Henry De la Beche's *Researches in Theoretical Geology* (1834), John Herschel's *Preliminary Discourse on the Study of Natural Philosophy* (1831) or William Whewell's *History of the Inductive Sciences* (1837), which contain important commentaries on Lyell's work.

The volume number of the *Principles* under review is specified, as is the edition if other than the first. Most reviews in this period were anonymous; if known, the author is included in square brackets.

Athenaeum (31 July 1830, 25 Sept. 1830, 9 Oct. 1830), pp. 473, 595–7, 628–9 [Gillespie, probably Thomas Gillespie (1777–1844)]: I
Athenaeum (7 Jan. 1832), p. 5: II
Athenaeum (29 July 1833), pp. 409–11: III
Athenaeum (6 Dec. 1834), pp. 881–2: 3rd ed.
British Association for the Advancement of Science, Reports, 1832, pp. 365–414, at pp. 406–7 ('Report on the Progress, Actual State, and Ulterior Prospects of Geological Science'; William Daniel Conybeare): I
British Critic, Quarterly Theological Review, and Ecclesiastical Record 9 (Jan. 1831), pp. 180–206 [William Whewell]: I

British Critic, Quarterly Theological Review, and Ecclesiastical Record 15 (Apr. 1834), pp. 334–63: I–III

Eclectic Review n.s. 6 (July 1831), pp. 75–81

Edinburgh New Philosophical Journal 9 (Jan. 1831), pp. 209–29 ('On the Diluvial Theory, and on the Origin of the Valleys of Auvergne'; Charles Daubeny): I

Edinburgh New Philosophical Journal 9 (July 1830), p. 399 [?Robert Jameson]: I

Edinburgh New Philosophical Journal 12 (Jan. 1832), pp. 395–6 [?Robert Jameson]: II

Edinburgh New Philosophical Journal 15 (Oct. 1833), p. 399 [?Robert Jameson]: III

Gentleman's Magazine 100, pt ii (Sept. 1830), pp. 243–6: I

Gentleman's Magazine 102, pt i (Jan. 1832), pp. 43–7: II

Gentleman's Magazine n.s. 1 (Jan. 1834), pp. 77–8: III

Literary Gazette (24 July 1831, 7 Aug. 1830, 14 Aug. 1830), pp. 481, 505–6, 524–6: I

Literary Gazette (7 Jan. 1832, 28 Jan. 1832), pp. 8, 49: II

Literary Gazette (4 Oct. 1834), p. 671: 3rd ed.

Metropolitan Magazine 10 (May 1834), 'Literature', pp. 4–5: III

Monthly Magazine n.s. 10 (Dec. 1830), pp. 700–701: I

Monthly Review (Sept. 1830), pp. 38–49: I

Monthly Review (Mar. 1832), pp. 352–71: II

New Monthly Magazine 30 (1 Nov. 1830), p. 458: I

New Monthly Magazine 36 (1 June 1832), pp. 241–2: II

New Monthly Magazine 37 (1 Feb. 1833), pp. 366–8: I, 2nd ed.

New Monthly Magazine 40 (1 Mar. 1834), pp. 377–8: I–III

Philosophical Magazine n.s. 8 (Sept. 1830), pp. 215–19; n.s. 8 (Nov. 1830), pp. 359–62; n.s. 8 (Dec. 1830), pp. 401–6; n.s. 9 (Jan. 1831), pp. 19–23; n.s. 9 (Feb. 1832), pp. 111–17; n.s. 9 (Mar. 1831), pp. 188–97; n.s. 9 (Apr. 1831), pp. 258–70 ('An Examination of those Phaenomena of Geology, which seem to bear most directly on Theoretical Speculations'; William Daniel Conybeare): I; reply by Lyell in 9 (Jan. 1831), pp. 1–3

Presbyterian Review and Religious Journal 2 (July 1832), pp. 329–45: I, II

Proceedings of the Geological Society of London 1 (1831), pp. 281–316, at pp. 302–16 ('Address to the Geological Society'; Adam Sedgwick): I

Proceedings of the Geological Society of London 1 (1832), pp. 362–86, at pp. 373–6 ('Address to the Geological Society'; Roderick Murchison): II

Proceedings of the Geological Society of London 1 (1833), pp. 438–64, at pp. 443–4 ('Address to the Geological Society'; Roderick Murchison): III

Proceedings of the Geological Society of London 2 (1834), pp. 42–70, at pp. 64–70 ('Address Delivered at the Anniversary Meeting of the Geological Society'; George Bellas Greenough): I–III

Proceedings of the Geological Society of London 2 (1835), pp. 145–75, at pp. 169–70 ('Address Delivered at the Anniversary Meeting of the Geological Society'; George Bellas Greenough): 3rd ed.

Quarterly Review 43 (Oct. 1830), pp. 411–69 [George Poulett Scrope]: I

Quarterly Review 47 (Mar. 1832), pp. 103–32 [William Whewell]: II

Quarterly Review 53 (Apr. 1835), pp. 406–48 [George Poulett Scrope]: 3rd ed.

Scotsman (25 Sept. 1830), p. 1 [George Maclaren]: I

Spectator (14 Jan. 1832), p. 39: II

Index

Giraldus Cambrensis (Giraldus de Barri), 271
Glaciers, 46, 55, 83, 110
Glen Tilt, 15
God, see Creator
Gould, Stephen Jay, ix
Granite, origin of, 11, 15–16, 360, 381, 435, 444
Grant, Robert Edmond, xxx
Graves, George, 272
Gray, Asa, xxxvi
Great year, xviii–xix, 61–7, 287–9; see also Animals, Distribution of; Climate
Greenough, George Bellas, 461
Grimaldi, Francesco Antonio, 134, 137, 148
Guettard, Jean-Etienne, 12
Gulf stream, 54, 62, 370

Habitations of plants, 236; see also Plants, distribution of
Hall, Basil, xxviii, 114
Hall, James, 15, 173
Hamilton, Sir William, 135, 137, 141, 142, 144, 145, 159
Heat, central, see Central heat
Henderson, Ebenezer, 166, 267
Henslow, John Stevens, 209
Herbert, William, 209, 225–6
Herschel, John, xxii, xxv, xxxii, 459
Hipparchus, 81
History, relation to geology, xviii, xix, xxxix–xl, 5–8, 25, 29–30, 93, 433–4
Hitchcock, Edward, xxiv
Hoff, Karl Ernst von, xx
Hoffman, Friedrich, 408n
Holbach, Paul Henry Thiry, Baron d', x
Hooke, Robert, 16, 60, 132, 160
Hooker, Joseph, xxxvi

Hooker, William Jackson, 268
Hooykaas, Reijer, xix
Hope, Thomas, xxxi
Horace, xxviii, 304, 405n, 456n48, 457n4
Horner, Mary, see Lyell, Mary
Horsburgh, Captain James, 336, 343
Humans, see Man
Humboldt, Alexander von, 48, 52, 55, 62, 119, 121, 219, 235, 236, 273, 284, 294, 375
Hunter, John, 221–2, 229
Hutton, James, xxi, xxvi, 8, 9, 14–18, 19, 21, 23, 435, 436, 452n4; on catastrophes, 38, 173
Huttonians, xxxv, 18, 36–38, 307–9
Huxley, Thomas Henry, xxxvi, xxxvii
Hybrids, xxxii, 190, 197, 220–33

Icebergs, 46, 55, 57, 62–3, 116, 243
Iceland, xxxvi, 12, 29, 119, 130, 165–8, 267, 268, 275, 422–3
Ichthyosaurus, xviii, 67
Igneous causes, 117–79; and aqueous causes, 147–8, 173–9, 299–312, 314–17, 376, 414–15; passim; see also Earthquakes; Volcanoes
Iguandon, xviii, xix, 67
Induction, xxii, 15, 24, 96, 100, 102, 423, 444
Insects, 227–8, 259–63, 270
Instinct, xxii, 98, 210–19
Ippolito, Count, 134
Irving, Edward, xiii
Isothermal lines, 52–7, 444

Jameson, Robert, xv, xxx, xxxiii, 460
Java, 131, 414
Jorio, Andrea de, vii
Josselyn, John, 247

Keith, Patrick, 242

453n23, 455n20, 456n39, 456n43,
  456n53
Shipwrecks, 319–27
Sicily, causes of uplift of, 421–5;
  earthquake effects at, 137; fossil
  shells of, 43, 61, 404; geological
  structure of, 37, 369, 403–4; origin
  of Newer Pliocene of, 420–29
Sidonius Apollinaris, 431
Sloane, Hans, 242
Smith, William, 11, 22
Society for the Diffusion of Useful
  Knowledge, xiv
Soil, formation of, 300–306
Solander, Daniel Carl, 12
Solfatara, 149, 158, 160, 170, 450
Southcott, Joanna, xiii
Southwell, Charles, x
Spallanzani, Lazzaro, 242
Species, and altitude, 289–90;
  Brocchi on, 254–6; creation of
  new, xxxii, 233, 252–4, 294–8;
  definition of, 183–5, 196–202, 221,
  228, 232–3; effect of circumstances
  on, 186–91, 200–203, 210–19,
  278–91, 383–4, 427–9; extinction
  of, 99, 254–6, 286, 289–91, 307,
  376; Geoffroy Saint Hilaire on,
  xxx, xxxiii, 184; geographical
  distribution of, 234–50; and
  islands, 237–8, 243, 283; Hunter
  on, 221–2; interaction between
  different, 256–77, 289–91;
  Lamarck on, xxx, 184–97, 290;
  Linnaeus on, 196; total number of,
  295–6
Spectator, xv, xxxiii
Speculation in science, xxxiv, xxxviii–
  xxxix, 23–34, 36, 39, 351–6, 416,
  438
Spence, William, 227
Spencer, Herbert, xxxvi

Spix, Johann Baptist von, 270
Spontaneous generation, xxxii,
  198–9, 241
Springs, mineral, 40, 116, 300, 338,
  346–7; thermal, 164–8, 325
Stations of plants, 236, 256
Stewart, Dugald, xvi, xx, xxii, 98
Stonesfield mammals, xviii, 50, 89
Strabo, 161
Strata, correlation of, xi–xii, 137,
  380–94, 400–401; gaps in record
  of, 37, 372–9, 398–9; percentage
  method for dating, 400–401
Subsidence, *see* Elevation
Sumner, John Bird, xxiv, xxx

Targioni Tozzetti, Giovanni, 346
Taylor, John, 9
Temple of Serapis, Puzzuoli, vii,
  xxviii, xxxvi, 2, 152–61, 331, 422
Tennyson, Alfred Lord, x, xxxviii–
  xxxix, xxxix
Tertiary strata, classification of,
  394–402; conditions of formation
  of, 50, 76–9, 89–91, 371–6;
  history of discovery, 363–9;
  percentage method for classifying,
  400–401; synoptical table of, 402
Thames, xx, 149
Thucydides, xxviii, 328, 456n58
Tides, 116, 154–5, 341–2
Tiedemann, Fredrich, xxxiii, 231, 232
Time, perception of, xxv–xxvii,
  26–36 *passim*, 97, 307, 377, 378,
  415; scale of, xxix, xxxix, 17–18,
  290, 415, 431, 437
Toledo, Pietro Giacomo di, 159–60
Toriozzi, Cardinal, 21
Toulmin, George Hoggart, xxvii
Transition strata, 361–2
Transmutation, xxx–xxxv, 183–97,
  290–91; *see also* Lamarck; Species

READ MORE IN PENGUIN

In every corner of the world, on every subject under the sun, Penguin represents quality and variety – the very best in publishing today.

For complete information about books available from Penguin – including Puffins, Penguin Classics and Arkana – and how to order them, write to us at the appropriate address below. Please note that for copyright reasons the selection of books varies from country to country.

In the United Kingdom: Please write to *Dept. EP, Penguin Books Ltd, Bath Road, Harmondsworth, West Drayton, Middlesex UB7 0DA*

In the United States: Please write to *Consumer Sales, Penguin Putnam Inc., P.O. Box 12289 Dept. B, Newark, New Jersey 07101-5289.* VISA and MasterCard holders call 1-800-788-6262 to order Penguin titles

In Canada: Please write to *Penguin Books Canada Ltd, 10 Alcorn Avenue, Suite 300, Toronto, Ontario M4V 3B2*

In Australia: Please write to *Penguin Books Australia Ltd, P.O. Box 257, Ringwood, Victoria 3134*

In New Zealand: Please write to *Penguin Books (NZ) Ltd, Private Bag 102902, North Shore Mail Centre, Auckland 10*

In India: Please write to *Penguin Books India Pvt Ltd, 11 Community Centre, Panchsheel Park, New Delhi 110017*

In the Netherlands: Please write to *Penguin Books Netherlands bv, Postbus 3507, NL-1001 AH Amsterdam*

In Germany: Please write to *Penguin Books Deutschland GmbH, Metzlerstrasse 26, 60594 Frankfurt am Main*

In Spain: Please write to *Penguin Books S. A., Bravo Murillo 19, 1° B, 28015 Madrid*

In Italy: Please write to *Penguin Italia s.r.l., Via Benedetto Croce 2, 20094 Corsico, Milano*

In France: Please write to *Penguin France, Le Carré Wilson, 62 rue Benjamin Baillaud, 31500 Toulouse*

In Japan: Please write to *Penguin Books Japan Ltd, Kaneko Building, 2-3-25 Koraku, Bunkyo-Ku, Tokyo 112*

In South Africa: Please write to *Penguin Books South Africa (Pty) Ltd, Private Bag X14, Parkview, 2122 Johannesburg*

A CHOICE OF CLASSICS

Matthew Arnold	**Selected Prose**
Jane Austen	**Emma**
	Lady Susan/The Watsons/Sanditon
	Mansfield Park
	Northanger Abbey
	Persuasion
	Pride and Prejudice
	Sense and Sensibility
William Barnes	**Selected Poems**
Mary Braddon	**Lady Audley's Secret**
Anne Brontë	**Agnes Grey**
	The Tenant of Wildfell Hall
Charlotte Brontë	**Jane Eyre**
	Juvenilia: 1829–35
	The Professor
	Shirley
	Villette
Emily Brontë	**Complete Poems**
	Wuthering Heights
Samuel Butler	**Erewhon**
	The Way of All Flesh
Lord Byron	**Don Juan**
	Selected Poems
Lewis Carroll	**Alice's Adventures in Wonderland**
	The Hunting of the Snark
Thomas Carlyle	**Selected Writings**
Arthur Hugh Clough	**Selected Poems**
Wilkie Collins	**Armadale**
	The Law and the Lady
	The Moonstone
	No Name
	The Woman in White
Charles Darwin	**The Origin of Species**
	Voyage of the Beagle
Benjamin Disraeli	**Coningsby**
	Sybil

READ MORE IN PENGUIN

A CHOICE OF CLASSICS

Edward Gibbon	**The Decline and Fall of the Roman Empire** (in three volumes)
	Memoirs of My Life
George Gissing	**New Grub Street**
	The Odd Women
William Godwin	**Caleb Williams**
	Concerning Political Justice
Thomas Hardy	**Desperate Remedies**
	The Distracted Preacher and Other Tales
	Far from the Madding Crowd
	Jude the Obscure
	The Hand of Ethelberta
	A Laodicean
	The Mayor of Casterbridge
	A Pair of Blue Eyes
	The Return of the Native
	Selected Poems
	Tess of the d'Urbervilles
	The Trumpet-Major
	Two on a Tower
	Under the Greenwood Tree
	The Well-Beloved
	The Woodlanders
George Lyell	**Principles of Geology**
Lord Macaulay	**The History of England**
Henry Mayhew	**London Labour and the London Poor**
George Meredith	**The Egoist**
	The Ordeal of Richard Feverel
John Stuart Mill	**The Autobiography**
	On Liberty
	Principles of Political Economy
William Morris	**News from Nowhere and Other Writings**
John Henry Newman	**Apologia Pro Vita Sua**
Margaret Oliphant	**Miss Marjoribanks**
Robert Owen	**A New View of Society and Other Writings**
Walter Pater	**Marius the Epicurean**
John Ruskin	**Unto This Last and Other Writings**

READ MORE IN PENGUIN

A CHOICE OF CLASSICS

Walter Scott
The Antiquary
Heart of Mid-Lothian
Ivanhoe
Kenilworth
The Tale of Old Mortality
Rob Roy
Waverley

Robert Louis Stevenson
Kidnapped
Dr Jekyll and Mr Hyde and Other Stories
In the South Seas
The Master of Ballantrae
Selected Poems
Weir of Hermiston

William Makepeace
Thackeray
The History of Henry Esmond
The History of Pendennis
The Newcomes
Vanity Fair

Anthony Trollope
Barchester Towers
Can You Forgive Her?
Doctor Thorne
The Eustace Diamonds
Framley Parsonage
He Knew He Was Right
The Last Chronicle of Barset
Phineas Finn
The Prime Minister
The Small House at Allington
The Warden
The Way We Live Now

Oscar Wilde
Complete Short Fiction

Mary Wollstonecraft
A Vindication of the Rights of Woman
Mary and **Maria** (includes Mary Shelley's **Matilda**)

Dorothy and William
Wordsworth
Home at Grasmere